T0211952

Communications in Computer and Information Science 522

More information about this series at http://www.springer.com/series/7899

Piotr Gaj · Andrzej Kwiecień
Piotr Stera (Eds.)

Computer Networks

22nd International Conference, CN 2015
Brunów, Poland, June 16–19, 2015
Proceedings

 Springer

Editors
Piotr Gaj
Silesian University of Technology
Gliwice
Poland

Piotr Stera
Silesian University of Technology
Gliwice
Poland

Andrzej Kwiecień
Silesian University of Technology
Gliwice
Poland

ISSN 1865-0929 ISSN 1865-0937 (electronic)
Communications in Computer and Information Science
ISBN 978-3-319-19418-9 ISBN 978-3-319-19419-6 (eBook)
DOI 10.1007/978-3-319-19419-6

Library of Congress Control Number: 2015939670

Springer Cham Heidelberg New York Dordrecht London

Printed on acid-free paper

Springer International Publishing AG Switzerland is part of Springer Science+Business Media
(www.springer.com)

Preface

The regular and intense development of IT technology has had an impact on the rapid growth of the computer networks domain. Computer networks are the part of computer science whose development affects not only the other existing branches of technical science but also contributes in developing completely new areas.

Computer networks as well as the entire field of computer science are the subject of constant changes caused by the general development of technologies and a need of innovativeness in their applications. It produces a very creative and interdisciplinary interaction between computer science technologies and other technical activities, and directly leads to perfect solutions. New methods, together with tools for designing and modelling computer networks, are regularly extended. Above all, the essential issue is that the scope of computer network applications is increased thanks to the results of new research and new application proposals appearing regularly. Such solutions were not even taken into consideration in the past decades. While recent applications stimulate the progress of scientific research, the vast use of new solutions raise numerous problems, both practical and theoretical.

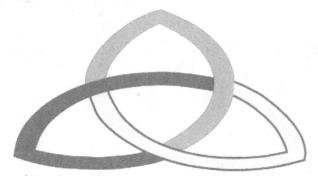

22nd International Science Conference *Computer Networks*

This book collates the research work made by scientists from numerous notable research centers. The included chapters refer to a wide spectrum of important issues regarding the computer networks and communication domain. It is a collection of papers presented at the 22nd edition of the International Science Conference: Computer Networks. The conference was held in Brunów Palace, located in Brunów, a small village near Lwówek Śląski, (Poland) during June 16–19, 2015. The conference, organized continuously since 1994 by the Institute of Informatics of Silesian University of Technology together with the Institute of Theoretical and Applied Informatics in Gliwice, is the oldest event of this kind in Poland. The current edition was the 22nd event in a row. The international status of the conference was attained seven years ago, thus in 2015 it was the eighth international edition. Just like its predecessors, the conference took place under the auspices of the Poland section of IEEE (technical co-sponsor). What is more, the conference partner is also iNEER (International Network for Engineering Education and Research).

The presented papers were accepted after careful reviews made by three independent reviewers in a double-blind way. The acceptance level was below 55 %. The chapters are organized thematically into several areas within the following tracks:

– Computer Networks
– Teleinformatics and Communications
– New Technologies
– Queueing Theory
– Innovative Applications

Each group includes very stimulating studies that may interest a wide readership.

In conclusion, on behalf of the Program Committee, we would like to express our acknowledgments to all authors for sharing their research results as well for their assistance in developing this monographic issue, which is in our judgment a reliable reference in the computer networks domain. We also want to thank the members of the Technical Program Committee for their participation in the reviewing process.

April 2015   Piotr Gaj
                                                                          Andrzej Kwiecień

Organization

CN2015 was organized by the Institute of Informatics from the Faculty of Automatic Control, Electronics and Computer Science, Silesian University of Technology (SUT) and supported by the Committee of Informatics of the Polish Academy of Sciences (PAN), Section of Computer Network and Distributed Systems in technical cooperation with the IEEE and consulting support of iNEER organization.

Executive Committee

All members of the Executive Committee are from the Silesian University of Technology, Poland.

Honorary Member:	Halina Węgrzyn
Organizing Chair:	Piotr Gaj
Technical Volume Editor:	Piotr Stera
Technical Support:	Aleksander Cisek
Technical Support:	Jacek Stój
Office:	Małgorzata Gładysz
Web Support:	Piotr Kuźniacki

Coordinators

PAN Coordinator:	Tadeusz Czachórski
IEEE PS Coordinator:	Jacek Izydorczyk
iNEER Coordinator:	Win Aung

Program Committee

Program Chair

Andrzej Kwiecień	Silesian University of Technology, Poland

Honorary Members

Win Aung	iNEER, USA
Klaus Bender	TU München, Germany
Adam Czornik	Silesian University of Technology, Poland
Andrzej Karbownik	Silesian University of Technology, Poland
Bogdan M. Wilamowski	Auburn University, USA

Technical Program Committee

Omer H. Abdelrahman	Imperial College London, UK
Anoosh Abdy	Realm Information Technologies, USA
Iosif Androulidakis	University of Ioannina, Greece

Tülin Atmaca	Institut National de Télécommunication, France
Rajiv Bagai	Wichita State University, USA
Zbigniew Banaszak	Warsaw University of Technology, Poland
Robert Bestak	Czech Technical University in Prague, Czech Republic
Leszek Borzemski	Wrocław University of Technology, Poland
Markus Bregulla	University of Applied Sciences Ingolstadt, Germany
Ray-Guang Cheng	National University of Science and Technology, Taiwan
Andrzej Chydziński	Silesian University of Technology, Poland
Tadeusz Czachórski	Silesian University of Technology, Poland
Andrzej Duda	INP Grenoble, France
Alexander N. Dudin	Belarusian State University, Belarus
Peppino Fazio	University of Calabria, Italy
Max Felser	Bern University of Applied Sciences, Switzerland
Holger Flatt	Fraunhofer IOSB-INA, Germany
Jean-Michel Fourneau	Versailles University, France
Rosario G. Garroppo	University of Pisa, Italy
Natalia Gaviria	Universidad de Antioquia, Colombia
Erol Gelenbe	Imperial College, UK
Roman Gielerak	University of Zielona Góra, Poland
Mariusz Głąbowski	Poznan University of Technology, Poland
Adam Grzech	Wrocław University of Technology, Poland
Edward Hrynkiewicz	Silesian University of Technology, Poland
Zbigniew Huzar	Wrocław University of Technology, Poland
Jacek Izydorczyk	Silesian University of Technology, Poland
Jürgen Jasperneite	Ostwestfalen-Lippe University of Applied Sciences, Germany
Jerzy Klamka	IITiS Polish Academy of Sciences, Gliwice, Poland
Demetres D. Kouvatsos	University of Bradford, UK
Stanisław Kozielski	Silesian University of Technology, Poland
Henryk Krawczyk	Gdańsk University of Technology, Poland
Wolfgang Mahnke	ABB, Germany
Francesco Malandrino	Politecnico di Torino, Italy
Aleksander Malinowski	Bradley University, USA
Kevin M. McNeil	BAE Systems, USA
Vladimir Mityushev	Pedagogical University of Cracow, Poland
Diep N. Nguyen	Macquarie University, Australia
Sema F. Oktug	Istanbul Technical University, Turkey
Michele Pagano	University of Pisa, Italy
Nihal Pekergin	Université de Paris, France
Piotr Pikiewicz	College of Business in Dąbrowa Górnicza, Poland
Jacek Piskorowski	West Pomeranian University of Technology, Poland
Bolesław Pochopień	Silesian University of Technology, Poland
Oksana Pomorova	Khmelnitsky National University, Ukraine
Silvana Rodrigues	Integrated Device Technology, Canada
Vladimir Rykov	Russian State Oil and Gas University, Russia

Alexander Schill	Technische Universität Dresden, Germany
Akash Singh	IBM Corp, USA
Mirosław Skrzewski	Silesian University of Technology, Poland
Tomas Sochor	University of Ostrava, Czech Republic
Maciej Stasiak	Poznan University of Technology, Poland
Kerry-Lynn Thomson	Nelson Mandela Metropolitan University, South Africa
Oleg Tikhonenko	Częstochowa University of Technology, Poland
Arnaud Tisserand	IRISA, France
Homero Toral Cruz	University of Quintana Roo, Mexico
Leszek Trybus	Rzeszów University of Technology, Poland
Adriano Valenzano	National Research Council of Italy, Italy
Bane Vasic	University of Arizona, USA
Peter van de Ven	Eindhoven University of Technology, The Netherlands
Miroslaw Voznak	VSB-Technical University of Ostrava, Czech Republic
Krzysztof Walkowiak	Wrocław University of Technology, Poland
Sylwester Warecki	Intel, USA
Jan Werewka	AGH University of Science and Technology, Poland
Tadeusz Wieczorek	Silesian University of Technology, Poland
Józef Woźniak	Gdańsk University of Technology, Poland
Hao Yu	Auburn University, USA
Grzegorz Zaręba	University of Arizona, USA

Additional Reviewers

Omer H. Abdelrahman	Edward Hrynkiewicz	Akash Singh
Iosif Androulidakis	Zbigniew Huzar	Mirosław Skrzewski
Tülin Atmaca	Jacek Izydorczyk	Tomas Sochor
Rajiv Bagai	Jürgen Jasperneite	Maciej Stasiak
Zbigniew Banaszak	Jerzy Klamka	Kerry-Lynn Thomson
Robert Bestak	Stanisław Kozielski	Oleg Tikhonenko
Leszek Borzemski	Henryk Krawczyk	Arnaud Tisserand
Ray-Guang Cheng	Andrzej Kwiecień	Homero Toral Cruz
Andrzej Chydziński	Wolfgang Mahnke	Leszek Trybus
Tadeusz Czachórski	Francesco Malandrino	Adriano Valenzano
Andrzej Duda	Aleksander Malinowski	Peter van de Ven
Alexander N. Dudin	Vladimir Mityushev	Bane Vasic
Peppino Fazio	Diep N. Nguyen	Miroslaw Voznak
Max Felser	Sema F. Oktug	Krzysztof Walkowiak
Holger Flatt	Michele Pagano	Sylwester Warecki
Jean-Michel Fourneau	Nihal Pekergin	Jan Werewka
Rosario G. Garroppo	Piotr Pikiewicz	Tadeusz Wieczorek
Natalia Gaviria	Jacek Piskorowski	Józef Woźniak
Erol Gelenbe	Oksana Pomorova	Hao Yu
Roman Gielerak	Silvana Rodrigues	Grzegorz Zaręba
Mariusz Głąbowski	Vladimir Rykov	
Adam Grzech	Alexander Schill	

Sponsoring Institutions

Organizer: Institute of Informatics, Faculty of Automatic Control, Electronics and Computer Science, Silesian University of Technology
Coorganizer: Committee of Informatics of the Polish Academy of Sciences, Section of Computer Network and Distributed Systems
Technical cosponsor: IEEE Poland Section

Technical Partner

Conference partner: iNEER

Contents

Waterfall Traffic Identification: Optimizing Classification Cascades

Paweł Foremski[1]($\boxtimes$), Christian Callegari[2], and Michele Pagano[2]

[1] The Institute of Theoretical and Applied Informatics of the Polish Academy of Sciences, Bałtycka 5, 44-100 Gliwice, Poland
pjf@iitis.pl

[2] Department of Information Engineering, University of Pisa, Via Caruso 16, 56122 Pisa, Italy
{c.callegari,m.pagano}@iet.unipi.it

Abstract. The Internet transports data generated by programs which cause various phenomena in IP flows. By means of machine learning techniques, we can automatically discern between flows generated by different traffic sources and gain a more informed view of the Internet.

In this paper, we optimize Waterfall, a promising architecture for cascade traffic classification. We present a new heuristic approach to optimal design of cascade classifiers. On the example of Waterfall, we show how to determine the order of modules in a cascade so that the classification speed is maximized, while keeping the number of errors and unlabeled flows at minimum. We validate our method experimentally on 4 real traffic datasets, showing significant improvements over random cascades.

Keywords: Network management · Traffic classification · Machine learning · Cascade classification

1 Introduction

Internet traffic classification is a well-known problem in computer networks. Since introduction of Peer-to-Peer (P2P) networking and encrypted protocols we have seen a rapid growth of classification methods that apply statistical analysis and machine learning to various characteristics of IP traffic, e.g. [1–3]. Survey papers list many existing methods grouped in various categories [4,5], yet each year still brings new publications in this field. Some authors suggested connecting several methods in *multi-classifier* systems as a future trend in traffic classification [6,7]. For example, in [8], the authors showed that *classifier fusion* can increase the overall classification accuracy. In [9], we proposed to apply the alternative of *classifier selection* instead, showing that *cascade classification* can successfully be applied to traffic classification. This paper builds on top of that.

In principle, cascade traffic classification works by connecting many classifiers in a single system that evaluates feature vectors in a sequential manner.

© Springer International Publishing Switzerland 2015
P. Gaj et al. (Eds.): CN 2015, CCIS 522, pp. 1–10, 2015.
DOI: 10.1007/978-3-319-19419-6_1

Our research showed that by using just 3 simple modules working in a cascade, it is possible to classify over 50 % of IP flows using the first packet of a network connection. We also showed that by adding more modules one can reduce the total amount of CPU time required for system operation. However, the problem still largely unsolved is how to choose from a possible large pool of modules, and how to order them properly so that the classification performance is maximized. In this paper, we propose a solution to this problem.

The contribution of our paper is as follows:

1. We propose a new solution to the cascade optimization problem, tailored to traffic classification (Sect. 3).
2. We give a quick method for estimating performance of a Waterfall system (Sect. 3).
3. We experimentally validate our proposal on 4 real traffic datasets, demonstrating that our algorithm works and can bring significant improvements to system performance (Sect. 4).

The rest of the paper is organized as follows. In Sect. 2, we give background on the Waterfall architecture and on existing methods for building optimal cascade classifiers. In Sect. 3, we describe our contribution, which is validated experimentally in Sect. 4. Section 5 concludes the paper.

2 Background

We introduced cascade traffic classification in [9]. Our Waterfall architecture integrates many different classifiers in a single "chain" of modules. The system sequentially evaluates module selection criteria and decides which modules to use for a given classification problem x. If a particular module is selected and provides a label for x, the algorithm finishes. Otherwise, the process advances to the next module. If there are no more modules, the flow is labeled as "Unknown". The algorithm is illustrated in Fig. 1. We refer the reader to [9] for more details.

Cascade classification is a *multi-classifier* machine learning system, which follows the *classifier selection* approach [10]. Although presented in 1998 by Alpaydin and Kaynak [11], so far few authors considered the problem of optimal cascade configuration that would match the Waterfall architecture. In a 2006 paper [12], Chellapilla et al. propose a cascade optimization algorithm that only updates the rejection thresholds of the constituent classifiers. The authors apply an optimized depth first search to find the cascade that satisfies given constraints on time and accuracy. However, comparing with our work, the system does not optimize the module order. In another paper published in 2008 [13], Sherif proposes a greedy approach for building cascades: start with a generic solution and sequentially prepend a module that reduces CPU time. Comparing with our work, the approach does not evaluate all possible cascade configurations and thus can lead to suboptimal results.

In this paper, we propose a new solution to the cascade classification problem, which is better suited for traffic classification than [12,13]. However, we

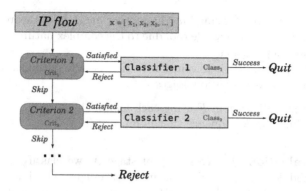

Fig. 1. The Waterfall architecture. A flow enters the system and is sequentially examined by the modules. In case of no successful classification, it is rejected.

assume no confidence levels on the classification outputs, thus we do not consider rejection thresholds as input values to the optimization problem. One can consider the same classifier parametrized with various thresholds as a set of separate modules available to build the cascade from.

3   Optimal Classification

Let us consider the problem of optimal cascade structure: we have n modules in set E that we want to use for cascade classification of IP flows in set F in an optimal way. In other words, we need to find a sequence of modules X that minimizes a cost function C:

$$E = \{1, \ldots, n\}, \tag{1}$$

$$X = (x_1, \ldots, x_m) \quad m \le n, \ x_i \in E, \ \forall_{i \ne j} \, x_i \ne x_j, \tag{2}$$

$$C(X) = f(t_X) + g(e_X) + h(u_X). \tag{3}$$

The terms t_X, e_X, and u_X respectively represent the total amount of CPU time used, the number of errors made, and the number of flows left unlabeled while classifying F with X. The terms f, g, and h are arbitrary real-valued functions. Because $m \le n$, some modules may be skipped in the optimal solution. Note that u_X does not depend on the order of modules, because unrecognized flows always traverse till the end of the cascade.

3.1   Proposed Solution

To find the optimal cascade, we propose to quickly check all possible X. We propose an approximate method, because for an accurate method one would need to run the full classification process for each X, i.e. experimentally evaluate all permutations of all combinations in E. This would take S experiments, where

$$S = \sum_{i=1}^{n} \frac{n!}{(n-i)!}, \tag{4}$$

which is impractical even for small n. On another hand, fully theoretical models of the cost function seem infeasible too, due to the complex nature of the cascade and module inter-dependencies.

Thus, we propose a heuristic solution to the cascade optimization problem. The algorithm has two evaluation stages:

(A) Static: classify all flows in F using each module in E, and
(B) Dynamic: find the X sequence that minimizes $C(X)$.

A. Static Evaluation. In every step of stage A, we classify all flows in F using single module x, $x \in E$. We measure the average CPU time used for flow selection and classification: $t_s^{(x)}$ and $t_c^{(x)}$. We store each output flow in one of the three outcome sets, depending on the result: $F_S^{(x)}$, $F_O^{(x)}$, or $F_E^{(x)}$. These sets hold respectively the flows that were skipped, properly classified, and improperly classified. Let us also introduce $F_R^{(x)}$:

$$F_R^{(x)} = F \setminus \left(F_S^{(x)} \cup F_O^{(x)} \cup F_E^{(x)} \right), \tag{5}$$

the set of rejected flows. See Fig. 2 for an illustration of the module measurement procedure. As the result of every step, the performance of module x on F is fully characterized by a tuple of $P^{(x)}$:

$$P^{(x)} = \left(t_s^{(x)}, t_c^{(x)}, F_S^{(x)}, F_O^{(x)}, F_E^{(x)} \right). \tag{6}$$

Finally, after n steps of stage A, we obtain n tuples: the input to stage B.

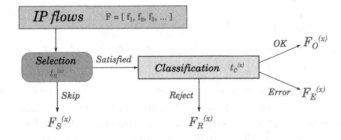

Fig. 2. Measuring performance of module $x \in E$

B. Dynamic Evaluation. Having all of the required experimental data, we can quickly estimate $C(X)$ for arbitrary X. Because f, g, and h are used only for adjusting the cost function, we focus on their arguments: t_X, e_X, and u_X.

Let $X = (x_1, \ldots, x_i, \ldots, x_m)$ represent certain order and choice of modules, and G_i represent the set of flows entering the module number i ($G_1 = F$). We estimate the cost factors using the following procedure:

$$t_X \approx \sum_{i=1}^{m} |G_i| \cdot t_s^{(x_i)} + \left| G_i \setminus F_S^{(x_i)} \right| \cdot t_c^{(x_i)}, \tag{7}$$

$$e_X = \sum_{i=1}^{m} \left| G_i \cap F_E^{(x_i)} \right|, \tag{8}$$

$$u_X = |G_{m+1}|, \tag{9}$$

where

$$G_{i+1} = G_i \setminus \left(F_O^{(x_i)} \cup F_E^{(x_i)} \right) \quad i \le m. \tag{10}$$

The difference operation in Eq. (10) is crucial, because we need to remove the flows that were classified in the previous step. In stage A, our algorithm evaluates static performance of every module, but in stage B we need to simulate cascade operation. The difference operator in Eq. (10) connects the static cost factors (t_X, e_X, u_X) with the dynamic effects of cascade classification.

Module performance depends on its position in the cascade because preceding modules alter the distribution of traffic classes in the flows conveyed onward. For example, a module designed for P2P traffic running before a port-based classifier can improve its accuracy, by removing the flows that run on non-standard ports or abuse the traditional port assignments.

3.2 Discussion

In our optimization algorithm we simplified the original problem to n experiments and several operations on flow sets. We can speed up the search for the best X because the algorithm is additive:

$$C(X + x_i) = C(X) + C(x_i). \tag{11}$$

Thus, we can apply the branch and bound algorithm [14].

Note that the results depend on F: the optimal cascade depends on the protocols represented in the traffic dataset, and on the ground-truth labels. The presented method cannot provide the ultimate solution that would be optimal for every network, but it can optimize a specific cascade system working in a specific network. In other words, it can reduce the amount of required CPU power, the number of errors, and the number of unlabeled flows, given a set of modules and a set of flows. We evaluate this issue in Sect. 4 (Table 2).

We assume that the flows are independent of each other, i.e. labeling a particular flow does not require information on any other flow. In case such information is needed, e.g. DNS domain names for the dnsclass module, it should be extracted before the classification process starts. Thus, traffic analysis and flow classification must be separated to uphold this assumption. We successfully implemented such systems for our DNS-Class [15] and Mutrics [9] classifiers.

In the next section, we experimentally validate our method and show that it perfectly predicts e_X and u_X, and approximates t_X properly (see Fig. 3). The simulated cost accurately follows the real cost, hence we argue that our proposal is valid and can be used in practice. In the next section, we analyze the trade-offs between speed, accuracy, and ratio of labeled flows (Fig. 4), but the final choice of the cost function should depend on the purpose of the classification system.

4 Experimental Validation

In this section, we use real traffic datasets to demonstrate that our method is effective and gives valid results. We ran 4 experiments:

1. Comparing simulated t_X, e_X, and u_X to real values, which proves validity of Eqs. (7)–(9);
2. Analyzing the effect of f, g, and h on the results, which proves that parameters influence the optimization process properly;
3. Optimizing the cascade on one dataset and testing it on another dataset, which verifies robustness in time and space;
4. Comparing optimized cascades to random configurations, which demonstrates that our work is meaningful.

We used 4 real traffic datasets, as presented in Table 1. Datasets *Asnet1* and *Asnet2* were collected at the same Polish ISP company serving <500 users, with an 8 month time gap. Dataset *IITiS1* was collected at an academic network serving <50 users, at the same time as *Asnet1*. Dataset *Unibs1* was also collected at an academic network (University of Brescia[1]), but a few years earlier and without packet payloads. We established ground-truth using Deep Packet Inspection (DPI) and trained the modules using 60 % of flows chosen randomly – as described in our original work [9]. The remaining flows were used for evaluating our proposal. We used the following Waterfall modules: `dstip`, `dnsclass` [15], `npkts`, `port`, and `portsize`. We handled *Unibs1* differently, because the dataset has no packet payloads and has all IP addresses anonimized: we used the `stats` module instead of `dnsclass`.

Experiment 1. In the first experiment, we compare simulated cost factors with the reality. We randomly selected 100 000 flows from each dataset and ran the static evaluation on them. Next, we generated 100 random cascades, and for each cascade we ran real classification and the dynamic evaluation stage of our optimization algorithm. As a result, we obtained pairs of real and estimated values of t_X, e_X, and u_X for same X values. The results for t_X are presented in Fig. 3. For e_X and u_X we did not observe a single error, i.e. our method perfectly predicted the real values. For CPU time estimations, we see a high

Table 1. Datasets used for experimental validation

Dataset	Start	Duration	Src. IP	Dst. IP	Packets	Bytes	Avg. Util	Avg. Flows (/5 min.)	Payload
Asnet1	2012-05-26 17:40	216 h	1,828 K	1,530 K	2,525 M	1,633 G	18.0 Mbps	7.7 K	92 B
Asnet2	2013-01-24 16:26	168 h	2,503 K	2,846 K	2,766 M	1,812 G	25.7 Mbps	12.0 K	84 B
IITiS1	2012-05-26 11:19	220 h	32 K	46 K	150 M	95 G	1.0 Mbps	753.7	180 B
Unibs1	2009-09-30 11:45	58 h	27	1 K	33 M	26 G	0.9 Mbps	111.7	0 B

---

[1] Downloaded from http://www.ing.unibs.it/ntw/tools/traces/.

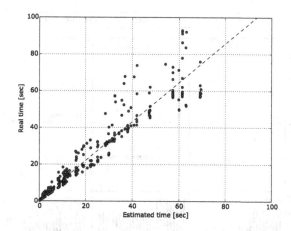

Fig. 3. Experiment 1. Estimated classification time vs real classification time. Dashed line shows least-squares approximation, the correlation coefficient is 0.95.

correlation of 0.95, with little under-estimation of the real value. For all datasets, the estimation error was below 20 % for majority of evaluated cascades (with respect to the real value). The error was above 50 % only for 5 % of evaluated cascades. We conclude that in general our method properly estimates the cost factors and we can use it to simulate different cascade configurations.

Experiment 2. In our second experiment, we want to show the effect of tuning the cost function for different goals: minimizing the computation time, minimizing errors, and labeling as many flows as possible. We chose the following cost function:

$$C(X) = f(t_X) + g(e_X) + h(u_X) = t_X^a + e_X^b + u_X^c. \tag{12}$$

Next, we separately varied the a, b, c exponents in range of 0–10, and observed the performance of the optimal cascade found by applying such cost function. We ran the experiment for *Asnet1*, *Asnet2*, and *IITiS1*. In Fig. 4, we present the results: dependence of CPU time, number of errors, number of unlabeled flows, and module count on $f(t_X)$, $g(e_X)$, and $h(u_X)$. As expected, higher f exponent leads to faster classification, with fewer number of modules in the cascade (more unclassified flows) and usually less errors. Optimizing for accuracy leads to reduction of errors and CPU time, at the cost of higher number of flows left without a label. Note that we observed more errors than in the case of time optimization – probably because the number of errors was low, thus the g exponent had less impact on such values. Finally, if we choose to classify as much traffic as possible, the system will use all available modules, at the cost of increasing the CPU time. We conclude that our proposal works, i.e. by varying the parameters we optimize the cascade for different goals.

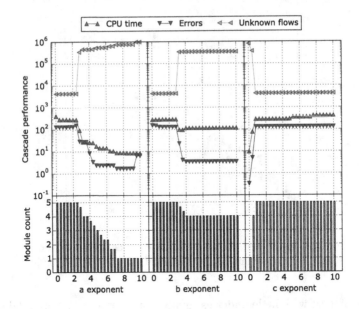

Fig. 4. Experiment 2. Optimizing the cascade for different goals: best classification time (a exponent), minimal number of errors (b exponent), and the lowest number of unlabeled flows (c exponent): the plot shows the averages for 3 datasets.

Table 2. Experiment 3. Optimization stability: increase in the cost $C(X)$, depending on the reference dataset used for determining the optimal cascade.

Reference		Test dataset		
		Asnet1	Asnet2	IITiS1
Asnet1	portsize, dstip, dnsclass, npkts, port	-	0.46%	6.54%
Asnet2	dstip, dnsclass, portsize, npkts, port	0.49%	-	8.01%
IITiS1	dnsclass, port, dstip, portsize, npkts	0.04%	0.03%	-

Experiment 3. In the third experiment, we verify if the result of optimization is stable in time and space, i.e. if the optimal cascade stays optimal with time and changes of the network. We ran optimization for 3 datasets (all flows in *Asnet1*, *Asnet2*, and *IITiS1*), obtaining different cascade configuration for each dataset. Next, we evaluated these configurations on the other datasets, i.e. *Asnet1* on *Asnet2* and *IITiS1*, etc. We measured the increase in the value of the cost function $C(X)$ and compared it with the original value. Table 2 presents the results. We see that our proposal yielded results that are stable in time: the cascades found for *Asnet1* and *Asnet2*, which are 8 months apart, are very similar and can be exchanged with little decrease in performance. However, the cascades found for *Asnet1* and *Asnet2* gave 7 % and 8 % decrease in performance compared with *IITiS1*. We conclude that optimization results are quite specific to the network, but also stable in time, for the evaluated datasets.

Table 3. Experiment 4. Average improvements compared to random cascade selection. We evaluated 100 random cascades on 100 000 random flows for each of 4 datasets.

Dataset		CPU time	Errors	Unknown flows
Asnet1	Random:	23.7	0	2,750
	Optimal:	7.6	0	26
	Improvement:	68%	0%	99%
Asnet2	Random:	34.5	26.4	10,350
	Optimal:	28.4	15.0	363
	Improvement:	18%	43%	96%
IITiS1	Random:	28.5	0	9,203
	Optimal:	12.9	0	1,327
	Improvement:	55%	0%	86%
Unibs1	Random:	6.7	25.8	356
	Optimal:	1.6	20.0	267
	Improvement:	77%	23%	25%
Average improvement:		55%	17%	77%

Experiment 4. In our last experiment, we compare our proposal with random choice of the modules, i.e. a situation in which we have a possibly large number of "black box" modules to build the cascade from. For example, we could have a large number of npkts modules trained on different flow samples, and with different parameters. We used the data collected in Experiment 1 (100 000 flows and 100 random cascades for each of 4 datasets) and calculated the average t_X, e_X, and u_X values. Next, we run our optimization algorithm on the same 100 000 flows for each dataset and measured the improvements with respect to the average performance of random cascades. We used the cost function given in Eq. (12), for $a = 0.95$, $b = 1.75$, and $c = 1.20$. In Table 3, we present obtained results: in every case, our algorithm optimized the classification system to work better, significantly reducing the amount of CPU time required for operation. Thus, we conclude that our work is meaningful and can help a network administrator to configure a cascade classification system properly.

5 Conclusions

In this paper, we presented a new method for optimizing cascade classifiers, on the example of the Waterfall traffic classification architecture. The method evaluates the constituent classifiers and quickly simulates cascade operation in every possible configuration. By searching for the cascade that minimizes a custom cost function, the method finds the best configuration for given parameters, which corresponds to minimizing required CPU time, number of errors, and number of unclassified IP flows. We experimentally validated our proposal on 4 real traffic datasets, demonstrating method validity, effectiveness, stability, and improvements with respect to random choices.

Not only does our proposal apply to traffic classification, but it can be also applied in the field of machine learning (for multi-classifier systems). However, our approach does not consider rejection thresholds of the classifiers, which is a certain limitation for application in other fields. We release an open source implementation of our proposal as an extension to the *Mutrics* classifier[2].

References

1. Karagiannis, T., Papagiannaki, K., Faloutsos, M.: Blinc: multilevel traffic classification in the dark. In: ACM SIGCOMM Computer Communication Review, vol. 35, pp. 229–240. ACM (2005)
2. Finamore, A., Mellia, M., Meo, M., Rossi, D.: KISS: stochastic packet inspection classifier for UDP traffic. IEEE/ACM Trans. Netw. **18**(5), 1505–1515 (2010)
3. Bermolen, P., Mellia, M., Meo, M., Rossi, D., Valenti, S.: Abacus: accurate behavioral classification of P2P-TV traffic. Comp. Netw. **55**(6), 1394–1411 (2011)
4. Nguyen, T.T., Armitage, G.: A survey of techniques for internet traffic classification using machine learning. Commun. Surv. Tutor. IEEE **10**(4), 56–76 (2008)
5. Callado, A., Kamienski, C., Szabó, G., Gero, B., Kelner, J., Fernandes, S., Sadok, D.: A survey on internet traffic identification. Commun. Surv. Tutor. IEEE **11**(3), 37–52 (2009)
6. Dainotti, A., Pescape, A., Claffy, K.C.: Issues and future directions in traffic classification. Netw. IEEE **26**(1), 35–40 (2012)
7. Foremski, P.: On different ways to classify Internet traffic: a short review of selected publications. Theor. Appl. Inf. **25**(2), 119–136 (2013)
8. Dainotti, A., Pescapé, A., Sansone, C.: Early classification of network traffic through multi-classification. In: Domingo-Pascual, J., Shavitt, Y., Uhlig, S. (eds.) TMA 2011. LNCS, vol. 6613, pp. 122–135. Springer, Heidelberg (2011)
9. Foremski, P., Callegari, C., Pagano, M.: Waterfall: rapid identification of IP flows using cascade classification. In: Kwiecień, A., Gaj, P., Stera, P. (eds.) CN 2014. CCIS, vol. 431, pp. 14–23. Springer, Heidelberg (2014)
10. Kuncheva, L.I.: Combining Pattern Classifiers: Methods and Algorithms. Wiley (2004)
11. Alpaydin, E., Kaynak, C.: Cascading classifiers. Kybernetika **34**(4), 369–374 (1998)
12. Chellapilla, K., Shilman, M., Simard, P.: Optimally combining a cascade of classifiers. Proceed. SPIE **6067**, 207–214 (2006)
13. Abdelazeem, S.: A greedy approach for building classification cascades. In: Seventh International Conference on Machine Learning and Applications, ICMLA 2008, pp. 115–120. IEEE (2008)
14. Land, A.H., Doig, A.G.: An automatic method of solving discrete programming problems. Econometrica **28**(3), 497–520 (1960)
15. Foremski, P., Callegari, C., Pagano, M.: DNS-class: immediate classification of IP flows using DNS. Int. J. Netw. Manag. **24**(4), 272–288 (2014)

---

[2] See https://github.com/iitis/mutrics/tree/bks.

Estimating the Intensity of Long-Range Dependence in Real and Synthetic Traffic Traces

Joanna Domańska[1], Adam Domański[2],
and Tadeusz Czachórski[1(✉)]

[1] Institute of Theoretical and Applied Informatics,
Polish Academy of Sciences, Baltycka 5, 44-100 Gliwice, Poland
{joanna,tadek}@iitis.gliwice.pl
[2] Institute of Informatics, Silesian Technical University,
Akademicka 16, 44-100 Gliwice, Poland
adamd@polsl.pl

Abstract. This paper examines various techniques for estimating the intensity of Long-Range Dependence (LRD). Trial data sets with LRD are generated using Fractional Gaussian noise and Markov modulated Poisson process. The real data set collected in IITiS PAN is also used.

Keywords: Self-similarity · Long-range dependence · Hurst parameter · Fractional gaussian noise · Markov modulated poisson process

1 Introduction

Many empirical and theoretical studies have shown the self-similar and LRD characteristics of the network traffic [1–4]. These features have a great impact on a network performance [3,5,6]. They enlarge mean queue lengths at buffers and increase packet loss probability, reducing this way the quality of services provided by a network [7]. That is why it is necessary to take into account this feature when you want to create a realistic model of traffic sources [8,9].

Various modelling techniques are currently used for generating the LRD traffic. The majority of them is based on non-Markovian approach [10–13]. The advantage of these models is that they give a good description of the traffic with the use of few parameters. Their drawbacks consist in the fact that they do not allow the use of traditional and well known queueing models and modeling techniques for computer networks performance analysis. There are also Markov based models to generate a LRD traffic over a finite number of time scales [8,14–17]. This approach makes possible the adaptation of traditional Markovian queueing models to evaluate network performance.

In this work we investigate long-range dependence in the synthetic data generated from Fractional Gaussian noise traffic source and Markovian traffic source. We also use for our analysis the realistic packet traffic trace collected in IITiS PAN.

© Springer International Publishing Switzerland 2015
P. Gaj et al. (Eds.): CN 2015, CCIS 522, pp. 11–22, 2015.
DOI: 10.1007/978-3-319-19419-6_2

Section 2 briefly describes the distinction between the terms: Self-similarity and Long-Range Dependence. Section 3 demontrates Fractional Gaussian noise as an example of exactly self-similar traffic source. Section 4 shortly describes the Markovian traffic source which is sufficient to generate long-range dependent traffic over several time scales. Section 5 presents the intensity of LRD estimation for real and synthetic traffic traces. Some conclusions are presented in Sect. 6.

2 The Distinction Between Self-Similarity and Long-Range Dependence

Over the last two decades, long-range dependency (LRD), self-similarity and heavy-tailed distributions have dominated Internet traffic analysis. Extensive measurements have revealed these phenomena in network traffic.

In the literature the terms long-range dependence and self-similarity are often used without distinction, although they are not equivalent concepts [18].

A continuous time process $Y(t)$ is exactly self-similar with the Hurst parameter H if it satisfies the following condition [19]:

$$Y(t) \stackrel{d}{=} a^{-H} Y(at)$$

for $t \geq 0$, $a \geq 0$ and $0 < H < 1$. The above equality is in the sense of finite dimensional distributions and the Hurst parameter expresses the degree of the self-similarity [20]. The process $Y(t)$ may be nonstationary [21].

In the case of network traffic one usually has to deal with time series rather than a continuous process. In that context the above definition can be summarized as follows. Let $X(t)$ be a stationary sequence representing increment process (e.g. in bytes/second). The corresponding aggregated sequence having level of aggregation m:

$$X^{(m)}(k) = \frac{1}{m} \sum_{i=1}^{m} X((k-1)m + i), \quad k = 1, 2, \ldots$$

is obtained by averaging $X(t)$ over nonoverlapping blocks of length m. The following condition is satisfied for a self-similar process:

$$X \stackrel{d}{=} m^{1-H} X^{(m)}$$

for all integers m. A stationary sequence X is second-order self-similar if for all m, $m^{1-H} X^{(m)}$ has the same variance and auto-correlation as X. A stationary sequence X is asymptotically second-order self-similar if $m^{1-H} X^{(m)}$ has the same variance and auto-correlation as X as $m \to \infty$.

Asymptotically second-order self-similar processes are also called long-range dependent processes and this is a property exhibited by network traffic [19]. Long-range dependence of data means that the behavior of a time-dependent process shows statistically significant correlations across large time scales and

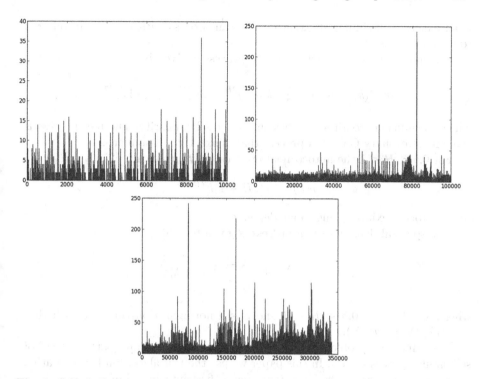

Fig. 1. Self-similar processes exhibit similar fluctuations over different time scales (IITiS trace)

self-similarity describes the phenomenon in which the behavior of a process is preserved irrespective of scaling in space or time [2]. Figure 1 shows as an example the fluctuation over differents time scales for IITiS trace described in Sect. 5.

The autocorrelation function od LRD process decays very slowly and if the Hurst parameter is between 0.5 and 1, the autocorrelation has an asymptotically behavior. If the Hurst parameter is smaller or equal to 0.5, the process exhibits SRD (Short-Range Dependence). Second-order stationary process whose auto-correlation function decays hyperbolically is asymptotically second-order self-similar [22]. For this reason the term self-similarity is often used interchangeably with the long-range dependence [19].

3   Fractional Gaussian Noise

Fractional Gaussian noise (fGn) has been proposed as a model [23] for the long-range dependence postulated to occur in a variety of hydrological and geophysical time series. Nowadays, fGn is one of the most commonly used self-similar processes in network performance evaluation [18]. Let $B_h(t)$ be a fractional Brownian motion process. Then the sequence of increments:

$$X(t) = B_h(t) - B_h(t-1)$$

is an exactly self-similar stationary Gaussian process with zero mean, referred to as fGn process.

The autocorrelation function of fGn process is given by [2]:

$$\rho^{(m)}(k) = \rho(k) = \frac{1}{2}\left[(k+1)^{2H} - 2k^{2H} + (k-1)^{2H}\right],$$

which is sufficient condition for second-order selfsimilarity. The fGn process is the only stationary Gaussian process that is exactly self-similar [24].

For $0.5 < H < 1$ the autocorrelation decays hyperbolically [25]:

$$\rho(k) \sim H(2H-1)k^{2H-2}$$

so the process exhibits long-range dependence.

The spectral density of fGn process is given by [18]:

$$f(\lambda) = c|e^{J\lambda} - 1|^2 \sum_{i=-\infty}^{\infty} |2\pi i + \lambda|^{2H-1},$$

where $\lambda \in [-\pi, \pi]$, $0.5 < H < 1$ and c is a normalization constant such that $\int_{-\pi}^{\pi} f(\lambda)d\lambda = Var(X)$.

The important problem is the systhetic generation of sample paths (traces) of self-similar processes [18]. In this paper we use the fast algorithm for generating approximate sample paths for a fGn process, first introduced in [26].

4   Markovian Model of LRD Traffic

Many researchers discussed the suitability of Markovian models to describe IP network traffic that exhibits self-similarity and long-range dependence [8,14,15,17,21]. They concluded that matching LRD is only required within the time scales of interest to the system under research [21,27,28].

In consequence, more traditional traffic models such as Markov Modulated Poisson Process may still be used for modeling LRD traffic. Markov Modulated Poisson Process (MMPP) is a widely used tool for the teletraffic models analysis. To generate LRD traffic we use in this paper the MMPP model proposed in [29], and precisely described in [17]. This model consists of d two-state MMPP models. The i-th MMPP ($1 \le i \le d$) can be parameterized by two square matrices:

$$\mathbf{D_0^i} = \begin{bmatrix} -(c_{1i} + \lambda_{1i}) & c_{1i} \\ c_{2i} & -(c_{2i} + \lambda_{2i}) \end{bmatrix}$$

$$\mathbf{D_1^i} = \begin{bmatrix} \lambda_{1i} & 0 \\ 0 & \lambda_{2i} \end{bmatrix}.$$

The element c_{1i} is the transition rate from state 1 to 2 of the i-th MMPP and c_{2i} is the rate out of state 2 to 1. λ_{1i} and λ_{2i} are the traffic rate when the i-th

MMPP is in state 1 and 2 respectively. The sum of $\mathbf{D_0}^i$ and $\mathbf{D_1}^i$ is an irreducible infinitesimal generator $\mathbf{Q}^i$ with the stationary probability vector:

$$\overrightarrow{\pi}_i = \left(\frac{c_{2i}}{c_{1i} + c_{2i}}, \frac{c_{1i}}{c_{1i} + c_{2i}} \right).$$

The superposition of these two-state MMPPs is a new MMPP with 2^d states and its parameter matrices, $\mathbf{D_0}$ and $\mathbf{D_1}$, can be computed using the Kronecker sum of those of the d two-state MMPPs [30]:

$$(\mathbf{D_0}, \mathbf{D_1}) = \left(\oplus_{i=1}^d \mathbf{D_0}^i, \oplus_{i=1}^d \mathbf{D_1}^i \right).$$

Let N_t^i be a number of arrivals from the i-th MMPP in time slot $(0, t]$. The variance time for this MMPP can be expressed as:

$$Var\{N_t^i\} = (\lambda_i^* + 2k_{1i})t - \frac{2k_{1i}}{k_{2i}} \left(1 - e^{-k_{2i}t} \right)$$

where:

$$\lambda_i^* = \frac{c_{2i}\lambda_{1i} + c_{1i}\lambda_{2i}}{c_{1i} + c_{2i}}$$

$$k_{1i} = (\lambda_{1i} - \lambda_{2i})^2 \frac{c_{1i}c_{2i}}{(c_{1i} + c_{2i})^3}$$

$$k_{2i} = c_{1i} + c_{2i}.$$

The second-order properties are determined by this three entities: λ_i^*, k_{1i} and k_{2i}. The covariance function of the number of arrivals in two time slots of size Δt is expressed by [29]:

$$
\begin{aligned}
\gamma_i(k) &= \frac{(\lambda_{1i} - \lambda_{2i})^2 c_{1i}c_{2i} e^{-((c_{1i}c_{2i})(k-1)\Delta t)}}{(c_{1i} + c_{2i})^4} \cdot \left(1 - 2e^{-((c_{1i}+c_{2i})\Delta t)} + e^{-((c_{1i}+c_{2i})2\Delta t)} \right) \\
&= \frac{k_{1i}}{k_{2i}} e^{-(k_{2i}(k-1)\Delta t)} \cdot \left(1 - 2e^{(-k_{2i}\Delta t)} + e^{(-k_{2i}2\Delta t)} \right) \\
&\approx \frac{(\Delta t)^2 (\lambda_{1i} - \lambda_{2i})^2 c_{1i}c_{2i} e^{-((c_{1i}+c_{2i})(k-1)\Delta t)}}{(c_{1i} + c_{2i})^2} = (\Delta t)^2 k_{1i}k_{2i}e^{-(k_{2i}(k-1)\Delta t)}.
\end{aligned}
$$

Article [29] illustrates that a superposition of four described above two-state MMPP models suffices to replicate second-order self-similar behaviour over several time scales.

5 Long-Range Dependence Estimation

The Hurst parameter characterizes a process in terms of the degree of self-similarity and LRD [31]. The degree of self-similarity and LRD increases with increasing of H [32]. A Hurst value smaller or equal to 0.5 means the lack of self-similarity or the presence of SRD [33]. A Hurst parameter greater than 0.5 means the existance of LRD [31].

In our works three types of LRD traffic traces were considered. Two of them: fGn traffic source and MMPP traffic source have been described in Sects. 3 and 4. We also used a trace of real Internet traffic collected from the network of IITiS institute serving < 50 academic users. This data set has been collected during the whole May 2012 on the Internet gateway of our Institute [4, 34]. The datasets contains different subsets of network protocols. Table 1 gives a summary description of the IITiS traffic data analyzed in the paper.

Table 1. The IITiS traffic classification

	9-31.05	10.05	12.05	25.05	27.05	31.05
Protocol	Part of the traffic [%]					
SSLv3	0.04	<0.01	0.01	0.09	0.04	0.02
SSHv2	3.69	<0.01	<0.01	0.36	0.07	17.66
HTTP	17.86	8.73	10.94	21.00	18.25	10.01
POP	0.01	<0.01	<0.01	<0.01	<0.01	<0.01
TLSv1	2.45	1.98	1.23	2.54	9.38	1.05
IGMP	0.07	0.23	0.04	0.11	0.08	0.05
GRE	0.46	0.15	<0.01	0.23	2.09	0.01
ICMP	0.13	0.28	0.09	0.22	0.15	0.09
PPP	5.68	7.04	<0.01	2.10	30.81	0.02
FTP-DATA	0.05	<0.01	0.66	<0.01	<0.01	<0.01
FTP	<0.01	0.11	0.01	<0.01	<0.01	<0.01
TPKT	0.29	<0.01	0.12	<0.01	<0.01	<0.01
SSH	2.87	<0.01	<0.01	<0.01	<0.01	18.63
T.125	0.05	0.31	0.03	<0.01	<0.01	<0.01
SNMP	0.03	<0.01	<0.01	0.04	0.09	<0.01
TCP (unrec. app.)	62.82	73.77	85.07	68.60	35.80	48.86
Jabber/XML	0.05	0.16	0.03	0.07	0.05	0.03
LLMNR	0.02	0.02	0.01	0.03	0.03	0.01
SSDP	0.06	0.04	0.02	0.12	0.07	0.07
SMTP	0.54	2.09	0.16	1.23	0.63	0.38
DNS	1.83	2.82	1.28	2.67	1.69	1.33
SSLv2	0.14	0.18	0.01	0.04	0.03	<0.01
MDNS	0.03	<0.01	0.02	0.06	0.03	0.04
NTP	0.04	0.05	0.02	0.11	0.09	0.03
PIMv2	0.03	0.10	0.02	0.04	0.03	0.02
SSL	0.09	0.07	0.06	0.09	0.08	0.05
DB-LSP-DISC	0.05	<0.01	<0.01	0.07	0.05	0.03
UDP (unrec. app.)	0.51	0.26	0.08	0.16	0.38	1.58
X.224	0.04	0.28	0.03	<0.01	<0.01	<0.01
IMAP	<0.01	<0.01	0.01	<0.01	<0.01	<0.01
PPTP	<0.01	<0.01	<0.01	<0.01	0.02	<0.01
IPv6	<0.01	<0.01	<0.01	<0.01	0.03	<0.01

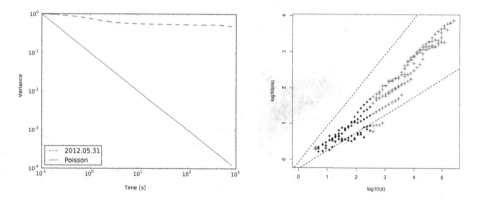

Fig. 2. Variance-time (left) and R-S (right) plots (IITiS trace)

There are many methods of estimating this parameter. In order to estimate the true values of the Hurst parameter we used in our works five estimators:

- aggregate variance method,
- R-S Plot,
- method based on periodogram,
- local Whittle's estimator,
- Wavelet based method.

We used the statistical package R mentioned in [35].

The aggregate variance method was described in [35–38]. This technique is time-domain based. It considers $Var(X^{(m)})$, where the aggregated process X^m is a time series derrived from X by aggregating it over blocks of size m. The aggregated sequence is then plotted versus m after taking logarithm. The estimated value of Hurst parameter is obtained by fitting a simple least squares line through the resulting points in the plane. The asymptotic slope between -1 and 0 suggests LRD and estimated Hurst parameter is given by $H = 1 - \text{slope}/2$. Figure 2 shows as an example the variance-time plot for IITiS trace.

The *R-S Plot* method [35,37,39] is one of the oldest Hurst parameter estimator. Let $R(n)$ be the range of the data aggregated over blocks of length n and $S^2(n)$ be the sample variance of data aggregated at the same scale. For a stochastic process X the rescaled range of X over a time interval n is defined as the ratio R/S:

$$\frac{R}{S}(n) = S^{-1}(n) \left[\max_{0 \leq t \leq n} \left(X(t) - t\overline{X}(n) \right) - \min_{0 \leq t \leq n} \left(X(t) - t\overline{X}(n) \right) \right]$$

where $\overline{X}(n)$ is the sample mean over the time interval n and $S(n)$ is standard deviation. For LRD processes, the ratio has the following characteristic for large n:

$$\frac{R}{S} \sim \left(\frac{n}{2} \right)^H.$$

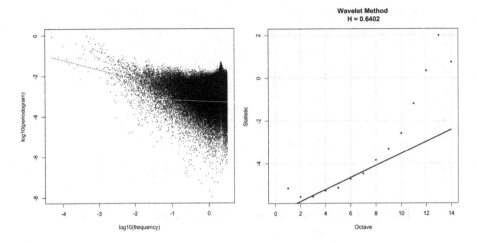

Fig. 3. Periodogram (left) and Wavelet (right) based analysis (IITiS trace)

A log-log plot of $\frac{R}{S}(n)$ versus n should have a constant slope H as n becomes large. Figure 2 shows as an example the R-S plot for IITiS trace.

The method using *Periodogram* [35, 37, 40] is frequency domain method, the periodogram is defined by:

$$I_X(\omega) = \frac{1}{2\pi n} \left| \sum_{j=1}^{n} X_j e^{ij\omega} \right|^2 .$$

A log-log plot $I_X(\omega_{n,k})$ versus $\omega_{n,k} = \frac{2\pi k}{n}$ should have a slope of $1 - 2H$ around $\omega = 0$. Figure 3 shows as an example the periodogram for IITiS trace.

Local Whittle's estimator assumes only a functional form for the spectral density at frequencies near zero. To estimate Hurst parameter one should minimize the function:

$$Q(H) = \sum_j \left[\log f_j(\omega_j) + \frac{\log I_X(\omega_j)}{f_j(\omega_j)} \right]$$

where $I_X(\omega)$ is the periodogram and $f(\omega) = c\omega^{2H-1}$.

Wavelet analysis is successfully used to measure the Hurst parameter [35]. Wavelets can be thought of as akin to Fourier series, but using waveforms other than sine waves. Figure 3 shows as an example the wavelet analysis for IITiS trace. The estimator fits a straight line to a frequency spectrum derived using wavelets.

Table 2 gives the the obtained Hurst parameters for one day IITiS traces:

- trace 1 – 6 002 874 samples,
- trace 2 – 13 874 610 samples,
- trace 3 – 36 135 490 samples.

In the case of Wavelets based method a 95 % confidence interval should be interpreted only as a confidence interval on the fitted line. Our previous work [4] did not confirm the relationship between the degree of LRD and the number of transmitted packet of a given type. One can see a little agreement between the estimators.

Table 2. Hurst parameter estimates for IITiS data traces

	Trace 1	Trace 2	Trace 3
Estimator	Hurst parameter		
R/S method	0.74	0.655	0.763
Aggregate variance method	0.912	0.817	0.933
Periodogram method	0.781	0.715	0.84
Whittle method	0.714	0.599	0.761
Wavelet-based method	0.681 +/− 0.013	0.61 +/− 0.027	0.71 +/− 0.017

Table 3 shows the obtained Hurst parameters for MMPP data traces:

- MMPP 1 – input parameters: $d = 5$, $n = 6$, $\lambda^* = 3.5$, $H = 0.75$ and $\rho = 0.6$,
- MMPP 2 – input parameters: $d = 5$, $n = 6$, $\lambda^* = 3.5$, $H = 0.6$ and $\rho = 0.6$,
- MMPP 3 – input parameters: $d = 4$, $n = 5$, $\lambda^* = 9.985$, $H = 0.79$ and $\rho = 0.0213$.

The fGn model and the MMPP 3 model were used to simulate the LRD data using a theoretical Hurst parameter and the same theoretical mean as the IITiS data. Different Hurst parameters have been chosen to represent a low and a high level of long-range dependence in data. The models are run to produce 200 000 packets. The MMPP 1 and MMPP 2 models have the example parameters described above.

Table 3. Hurst parameter estimates for MMPP data traces

	MMPP 1	MMPP 2	MMPP 3
Estimator	Hurst parameter		
R/S method	0.757	0.659	0.798
Aggregate variance method	0.665	0.586	0.715
Periodogram method	0.83	0.549	0.831
Whittle method	0.678	0.57	0.728
Wavelet-based method	0.851 +/− 0.036	0.601 +/− 0.011	0.841 +/− 0.036

As can be seen the MMPP model is more inconsistent than the fGn model (see Table 4).

Table 4. Hurst parameter estimates for fGn data traces

	fGn 1	fGn 2	fGn 3
Estimator	Hurst parameter		
R/S method	0.605	0.701	0.939
Aggregate variance method	0.507	0.642	0.88
Periodogram method	0.521	0.661	0.991
Whittle method	0.688	0.75	0.882
Wavelet-based method	0.574 +/− 0.028	0.698 +/− 0.017	0.937 +/− 0.009

6 Conclusions

The article confirms that although the Hurst parameter is well defined mathematically, it is problematic to measure it properly [35]. There are several methods to estimate the Hurst parameter but they often produce conflicting results [2]. Many researchers conclude that wavelet technique generally is faring well in comparative studies [31,35].

Our results demonstrate the best agreement among the estimators results in the case of fGn model. In the case of real traffic the differences estimators predictions are more visible. This confirms the opinion that the problems with real-life data are worse than those faced when measuring and characterizing synthetic data [35]. Real data can have periodicity, trends and quantisation effect and obtained results are more dependent on the sampling rate than in the case of synthetic traffic traces.

References

1. Crovella, M., Bestavros, A.: Self-similarity in world wide web traffic: evidence and possible causes. IEEE/ACM Trans. Netw. **5**, 835–846 (1997)
2. Karagiannis, T., Molle, M., Faloutsos, M.: Long-range dependence: ten years of internet traffic modeling. IEEE Internet Comput. **8**(5), 57–64 (2004)
3. Domański, A., Domańska, J., Czachórski, T.: The impact of self-similarity on traffic shaping in wireless LAN. In: Balandin, S., Moltchanov, D., Koucheryavy, Y. (eds.) NEW2AN 2008. LNCS, vol. 5174, pp. 156–168. Springer, Heidelberg (2008)
4. Domańska, J., Domański, A., Czachórski, T.: A few investigation of long-range dependence in network traffic. In: Czachórski, T., Gelenbe, E., Lent, R. (eds.) Information Science and Systems 2014, pp. 137–144. Springer International Publishing, Switzerland (2014)
5. Domańska, J., Domański, A.: The influence of traffic self-similarity on QoS mechanism. In: International Symposium on Applications and the Internet, SAINT. Trento, Italy (2005)
6. Domańska, J., Augustyn, D.R., Domański, A.: The choice of optimal 3-rd order polynomial packet dropping function for NLRED in the presence of self-similar traffic. Bull. Pol. Acad. Sci. Tech. Sci. **60**(4), 779–786 (2012)

7. Stallings, W.: High-Speed Networks: TCP/IP and ATM Design Principles. Prentice-Hall, New Jersey (1998)
8. Muscariello, L., Mellia, M., Meo, M., Ajmone Marsan, M., Lo Cigno, R.: Markov models of internet traffic and a new hierarchical MMPP model. Comput. Commun. **28**(16), 1835–1851 (2005)
9. Foremski, P., Gorawski, M., Grochla, K.: Source model of TCP traffic in LTE networks. In: Czachórski, T., Gelenbe, E., Lent, R. (eds.) Information Science and Systems 2014, pp. 125–135. Springer International Publishing, Switzerland (2014)
10. Erramilli, A., Singh, R.P., Pruthi, P.: An application of deterministic chaotic maps to model packet traffic. Queueing Syst. **20**(1–2), 171–206 (1995)
11. Gallardo, J.R., Makrakis, D., Orozco-Barbosa, L.: Use of α-stable self-similar stochastic processes for modeling traffic in broadband networks. Perform. Eval. **40**(1–3), 71–98 (2000)
12. Harmantzis, F.C., Hatzinakos, D.: Heavy network traffic modeling and simulation using stable FARIMA processes. In: 19th International Teletraffic Congress, pp. 300–303. Beijing, China (2005)
13. Laskin, N., Lambadatis, I., Harmantzis, F.C., Devetsikiotis, M.: Fractional Levy motion and its application to network traffic modeling. In: Computer Networks, vol. 40, no. 3, pp. 363–375. Elsevier Science Publishers B.V. (2002)
14. Robert, S., Boudec, J.Y.L.: New models for pseudo self-similar traffic. Perform. Eval. **30**(1–2), 57–68 (1997)
15. Clegg, R.G.: Markov-modulated on/off processes for long-range dependent internet traffic. In: Computing Research Repository (2006). CoRR. arXiv:cs/0610135
16. Domańska, J., Domański, A., Czachórski, T.: Internet traffic source based on hidden markov model. In: Balandin, S., Koucheryavy, Y., Hu, H. (eds.) NEW2AN 2011 and ruSMART 2011. LNCS, vol. 6869, pp. 395–404. Springer, Heidelberg (2011)
17. Domańska, J., Domański, A., Czachórski, T.: Modeling packet traffic with the use of superpositions of two-state MMPPs. In: Kwiecień, A., Gaj, P., Stera, P. (eds.) CN 2014. CCIS, vol. 431, pp. 24–36. Springer, Heidelberg (2014)
18. Lopez-Ardao, J.C., Lopez-Garcia, C., Suarez-Gonzalez, A., Fernandez-Veiga, M., Rodriguez-Rubio, R.: On the use of self-similar processes in network simulation. ACM Trans. Model. Comput. Simul. **10**(2), 125–151 (2000)
19. Gong, W.-B., Liu, Y., Misra, V., Towsley, D.: Self-similarity and long range dependence on the internet: a second look at the evidence, origins and implications. Comput. Netw. **48**(3), 377–399 (2005)
20. Bhattacharjee, A., Nandi, S.: Statistical analysis of network traffic inter-arrival. In: 12th International Conference on Advanced Communication Technology, pp. 1052–1057. USA (2010)
21. Nogueira, A., Salvador, P., Valadas, R., Pacheco, A.: Markovian modelling of internet traffic. In: Kouvatsos, D.D. (ed.) Next Generation Internet: Performance Evaluation and Applications. LNCS, vol. 5233, pp. 98–124. Springer, Heidelberg (2011)
22. Tsybakov, B., Georganas, N.D.: On self-similar traffic in ATM queues: definitions, overflow probability bounds, and cell delay distribution. IEEE/ACM Trans. Netw. **5**(3), 397–409 (1997)
23. Mandelbrot, B.B., Ness, J.V.: Fractional brownian motions, fractional noises and applications. SIAM Rev. **10**(4), 422–437 (1968)
24. Samorodnitsky, G., Taqqu, M.S.: Stable Non-Gaussian Random Processes: Stochastic Models with Infinite Variance. Chapman and Hall (1994)
25. Cox, D.R.: Long-range dependance: a review. In: Statistics: An Appraisal (1984)

26. Paxson, V.: Fast, approximate synthesis of fractional Gaussian noise for generating self-similar network traffic. ACM SIGCOMM Comput. Commun. Rev. **27**(5), 5–18 (1997)

27. Grossglauser, M., Bolot, J.C.: On the relevance of long-range dependence in network traffic. IEEE/ACM Trans. Netw. **7**(5), 629–640 (1999)

28. Nogueira, A., Valadas, R.: Analyzing the relevant time scales in a network of queues. In: SPIE Proceedings, vol. 4523 (2001)

29. Andersen, A.T., Nielsen, B.F.: A markovian approach for modeling packet traffic with long-range dependence. IEEE J. Sel. Areas Commun. **16**(5), 719–732 (1998)

30. Fischer, W., Meier-Hellstern, K.: The Markov-modulated poisson process (MMPP) cookbook. Perform. Eval. **18**(2), 149–171 (1993)

31. Stolojescu, C., Isar, A.A.: Comparison of some Hurst parameter estimators. In: 13th International Conference on Optimization of Electrical and Electronic Equipment, pp. 1152–1157. Brasov, Romania (2012)

32. Rutka, G.: Neural network models for internet traffic prediction. Electron. Electr. Eng. **4**(68) (2006)

33. Abry, P., Veitch, D.: Wavelet analysis of long-range-dependent traffic. IEEE Trans. Inf. Theory **44**(1), 2–15 (1998)

34. Foremski, P., Callegari, C., Pagano, M.: Waterfall: rapid identification of IP flows using cascade classification. In: Kwiecień, A., Gaj, P., Stera, P. (eds.) CN 2014. CCIS, vol. 431, pp. 14–23. Springer, Heidelberg (2014)

35. Clegg, R.G.: A practical guide to measuring the hurst parameter. Int. J. Simul. **7**(2), 3–14 (2006)

36. Beran, J.: Statistics for Long-Memory Processes. Chapman and Hall (1994)

37. Taqqu, M.S., Teverovsky, V.: On estimating the intensity of long-range dependence in finite anf infinite variance time series. In: A Practical Guide To Heavy Tails: Statistical Techniques and Applications, pp. 177–217. Birkhauser Boston Inc., Boston (1998)

38. Park, C., Hernandez-Campos, F., Long, L., Marron, J., Park, J., Pipiras, V., Smith, F., Smith, R., Trovero, M., Zhu, Z.: Long range dependence analysis of internet traffic. J. Appl. Stat. **38**(7), 1407–1433 (2011)

39. Mandelbrot, B.B., Wallis, J.: Computer experiments with fractional gaussian noises. Water Resour. Res. **5**(1), 228–241 (1969)

40. Geweke, J., Porter-Hudak, S.: The estimation and application of long memory time series models. J. Time Ser. Anal. **4**(4), 221–238 (1983)

Data Suppression Algorithms for Surveillance Applications of Wireless Sensor and Actor Networks

Bartłomiej Płaczek[✉] and Marcin Bernas

Institute of Computer Science, University of Silesia,
Będzińska 39, 41-200 Sosnowiec, Poland
{placzek.bartlomiej,marcin.bernas}@gmail.com

Abstract. This paper introduces algorithms for surveillance applications of wireless sensor and actor networks (WSANs) that reduce communication cost by suppressing unnecessary data transfers. The objective of the considered WSAN system is to capture and eliminate distributed targets in the shortest possible time. Computational experiments were performed to evaluate effectiveness of the proposed algorithms. The experimental results show that a considerable reduction of the communication costs together with a performance improvement of the WSAN system can be obtained by using the communication algorithms that are based on spatiotemporal and decision aware suppression methods.

Keywords: Wireless sensor and actor networks · Data suppression · Target tracking · Surveillance applications

1 Introduction

Wireless sensor and actor networks (WSANs) are composed of sensor nodes and actors that are coordinated via wireless communications to perform distributed sensing and acting tasks. In WSANs, sensor nodes collect information about the physical world, while actors use the collected information to take decisions and perform appropriate actions upon the environment. The sensor nodes are usually small devices with limited energy resources, computation capabilities and short wireless communication range. In contrast, the actors are equipped with better processing capability, stronger transmission powers and longer battery life. The number of actors in WSAN is significantly lower than the number of sensor nodes [1, 2].

The WSANs technology has enabled new surveillance applications, where sensor nodes detect targets of interest over a large area. The information collected by sensor nodes allows mobile actors to achieve surveillance goals such as target tracking and capture. Several examples of the WSAN-based surveillance applications can be found in the related literature, including land mine destruction [3], chasing of intruders [4], and forest fires extinguishing [5].

© Springer International Publishing Switzerland 2015
P. Gaj et al. (Eds.): CN 2015, CCIS 522, pp. 23–32, 2015.
DOI: 10.1007/978-3-319-19419-6_3

The surveillance applications of WSANs require real-time data delivery to provide effective actions. A fast response of actors to sensor inputs is necessary. Moreover, the collected information must be up to date at the time of acting. On the other hand, the sensor readings have to be transmitted to the mobile actors through multi-hop communication links, which results in transmission delays, failures and random arrival times of packets. The energy consumption, transmission delay, and probability of transmission failure can be reduced by decreasing the amount of transmitted data [6–8]. Thus, minimization of data transmission is an important research issue for the development of the WSAN-based surveillance applications [2]. It should be noted that other methods can be used in parallel to alleviate the above issues, e.g., optimisation of digital circuits design for network nodes [9].

This paper introduces an approach to reduce the data transmission in WSAN by means of suppression methods that were originally intended for wireless sensor networks (WSNs). The basic idea behind data suppression methods is to send data to actors only when sensor readings are different from what both the sensor nodes and the actors expect. In the suppression schemes, a sensor node reports only those data readings that represent a deviation from the expected behaviour. Thus, the actor is able to recognize relevant events in the monitored environment and take appropriate actions.

The data suppression methods available in the literature were designed for monitoring applications of WSNs. In such applications, a sink node needs to collect information describing a given set of parameters with a defined precision or recognize predetermined events. These state-of-the-art suppression methods are based on an assumption that a large subset of sensor readings does not need to be reported to the sink as these readings can be inferred from the other transferred data [10–13]. In order to infer suppressed data, the sink uses a predictive model of the monitored phenomena. The same model is used by sensor nodes to decide if particular data readings have to be transmitted. A sensor node suppresses transmission of a data reading only when it can be inferred within a given error bound.

Temporal suppression techniques exploit correlations between current and historical data readings of a single sensor node. The simplest scheme uses a naïve model, which assumes that current sensor reading is the same as the last reported reading [14]. When using this method, a sensor node transmits its current reading to sink only if difference between the current reading and previously reported reading is above a predetermined threshold.

Parameters monitored by WSNs usually exhibit correlations in both time and space [15]. Thus, several more sophisticated spatiotemporal suppression methods were proposed that combine the basic temporal suppression with detection of spatially correlated data from nearby nodes [12, 13, 16]. According to the spatiotemporal approach, sensor nodes are clustered based on spatial correlations. Sensor readings within each cluster are collected at a designated node (cluster head), which then uses a spatiotemporal model to decide if the readings have to be transmitted to sink.

In previous work of the first author [17] a decision-aware data suppression approach was proposed, which eliminates transfers of sensor readings that are not useful for making control decisions. This approach was motivated by an observation that for various control tasks large amounts of sensor readings often do not have to be transferred to the sink node as control decisions made with and without these data are the same. The decision-aware suppression was used for optimizing transmission of target coordinates from sensor nodes to a mobile sink which has to track and catch a moving target. According to that approach only selected data are transmitted that can be potentially useful for reducing the time in which the target will be reached by the sink.

According to the authors' knowledge, there is a lack of data suppression methods in the literature dedicated for the surveillance applications of WSANs. In this paper the available data suppression methods are adapted to meet the requirements of the WSANs. Effectiveness of these methods is evaluated by using a model of WSAN, where mobile actors have to capture randomly distributed targets in the shortest possible time.

The paper is organized as follows. Details of the WSAN model are discussed in Sect. 2. Section 3 introduces algorithms that are used by actors to navigate toward targets as well as algorithms of sensor – actor communication that are based on the data suppression concept. Results of simulation experiments are presented in Sect. 4. Finally, conclusions are given in Sect. 5.

2   Network Model

In this study a model of WSAN is considered, which includes 16 actors and 40 000 sensor nodes. The monitored area is modelled as a grid of 200×200 square segments. Discrete coordinates (x, y) are used to describe positions of segments, sensor nodes, actors, and targets ($x = 0, 1, \ldots, 199$, $y = 0, 1, \ldots, 199$). The sensor nodes are placed in centres of the segments. Each sensor node detects presence of a target in a single segment. Communication range of a sensor node covers the segment where this node is located as well as the eight neighbouring segments. Radius of the actor's communication range equals 37 segments. In most cases, the sensor nodes have to use multi-hop transmission for reporting their readings to actors. Due to the long communication range, each actor can transmit data to a large number of nodes (up to 4293) directly in one hop.

The task of sensor nodes is to detect stationary targets in the monitored area and report their positions to actors. On the basis of the received information, each actor selects the nearest target and moves toward it. This process is executed in discrete time steps. Maximum speed of actor equals two segments per time step. At each time step three new targets are created at random positions. A target is eliminated if an actor reaches the segment in which the target was detected. The targets may correspond to fires, intruders, landmines, enemy units, etc.

Default (initial) positions of actors were determined to ensure that the communication ranges of the 16 actors cover the entire monitored area (Fig. 1).

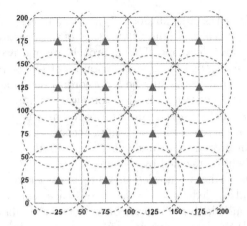

Fig. 1. Default positions (triangles) and communication ranges (circles) of actors

An actor, which has not received information from sensor nodes about current target locations moves toward its default position. Such situation occurs when there is no target within the actor's range or the information about detected target is suppressed by sensor node.

For the above WSAN model, data communication cost is evaluated by using two metrics: number of data transfers (packets sent from sensor nodes to actors), and total hop count. The hop count is calculated assuming that the shortest path is used for each data transfer. Performance of the targets elimination by actors is assessed on the basis of average time to capture, i.e., the time from the moment when a target is created to the moment when it is captured by an actor and eliminated.

3 Actor Navigation and Data Communication Algorithms

During target chasing, the mobile actors decide their movement directions based on the navigation algorithm, which is presented in Fig. 2. Each actor holds a target map TM to collect the information delivered by particular sensor nodes. An element of target map $TM(x,y)_i$ equals 1 if the i-th actor has received information that there is target detected in segment (x,y). In opposite situation, the target map element equals 0. The following symbols are used in the pseudo-code of the navigation algorithm: $(x_C, y_C)_i$ denotes a segment in which the i-th actor is currently located, $(x_N, y_N)_i$ is the nearest target according to the information collected by i-th actor in its target map, $(x_S, y_S)_i$ is the selected destination segment, and $(x_D, y_D)_i$ denotes the default actor position. It should be remembered that discrete coordinates (x,y) are used to identify the segments. Current position of actor, the destination segment as well as the targets map are broadcasted by the actor to all sensor nodes in its communication range.

An actor moves toward its default position unless a target is registered in its target map. Each actor takes a decision regarding the segment $(x_C^+, y_C^+)_i$ into

```
1   at each time step do
2       TM(xC, yC)i := 0
3       broadcast (xC, yC)i, (xS, yS)i, and TMi
4       collect data from sensor nodes and update TMi
5       if at least one target is registered in TMi then
6           find the nearest target (xN, yN)i in TMi
7           (xS, yS)i := (xN, yN)i
8       else
9           (xS, yS)i := (xD, yD)i
10      move toward (xS, yS)i
```

Fig. 2. Pseudo-code of navigation algorithm executed by i-th actor

```
1 at each time step do
2      collect (xC, yC)i, (xS, yS)i, and TMi from actors and update AL(x,y)
3      if target is detected and was not reported to any actor then
4          find actor i* in AL(x,y) which satisfies actor selection condition
5          if suppression condition is not satisfied then
6              report target position to actor i*
7      if target was reported to actor i* and is no longer detected then
8          if actor i* is available in AL(x,y) and TM(x,y)i* := 1 then
9              report elimination of target to actor i*
```

Fig. 3. Pseudo-code of data communication algorithm executed by sensor node (x, y)

which it will move during the next time step. The actor's decision is taken by solving the following optimization problem:

$$\text{minimzed}((x_C^+, y_C^+)_i, (x_S, y_S)_i)$$
$$\text{subject to} d((x_C^+, y_C^+)_i, (x_C, y_C)_i) \leq v_{\max} \tag{1}$$

where $d(\cdot)$ denotes the Euclidean distance between segments and $v_{\max} = 2$ (segments per time step) is the maximum speed of actor.

The sensor nodes report their readings to the actors by using the algorithm presented in Fig. 3. Each sensor node holds a list of actors $AL_{(x,y)}$ which includes IDs (i) of actors that have communicated their status data to the sensor node during current time step. It means that the sensor placed in segment (x, y) knows the actual positions, destinations and target maps of the actors listed in $AL_{(x,y)}$.

In order to minimize the data communication cost it was assumed that at each time step a sensor node may transmit data to one selected actor (i^*). Two simple actor selection conditions are considered in this study. According to the first actor selection condition, a sensor node reports its reading to the nearest known actor. The rationale behind this condition is that the nearest actor is expected to capture the target in shortest time and the additional benefit is that the transmission requires minimum hop count. The second actor selection condition aims at balancing the actors' workload. When using this condition, a sensor node selects that actor for which the number of targets registered in the target map is minimal.

The data communication algorithm presented in Fig. 3 utilizes the temporal data suppression method to minimize the amount of transmitted data. The temporal suppression is introduced by the if-then statements in lines 3 and 7 of the pseudo-code (Fig. 3). This basic data suppression method is extended to spatiotemporal and decision-aware suppression by using additional condition (5-th line of the pseudo-code).

According to the spatiotemporal data suppression method (STS), a sensor node in segment (x, y) will suppress reporting the position of detected target to the selected actor i^* if the information which is currently available for the actor indicates that there is at least one target within distance d_{STS} from segment (x, y). The distance threshold d_{STS} is calculated according to the following formula:

$$d_{STS} = \alpha \cdot d((x_C, y_C)_{i^*}, (x, y)), \qquad (2)$$

where α is a parameter of the algorithm.

The above approach is based on the heuristic rule that the actor does not need the precise information about target location to chase the target effectively when the distance to target is large. The closer to target, the higher precision of the localization has to be obtained [18].

In case of the decision aware suppression method, the transmission of target coordinates (x, y) from the sensor node to actor i^* is suppressed if it can be expected that this information will not influence the actor's decision. It means that the sensor node suppresses data transmission if, according to the information available for the node, the actor will move into the same segment $(x_C^+, y_C^+)_i$ regardless of whether the information about the new target in segment (x, y) is transmitted or not. For instance, the suppression is performed when there is a target located in the actor's destination segment $(x_S, y_S)_{i^*}$ and the new detected target (x, y) is more distant to the actor than segment $(x_S, y_S)_{i^*}$.

Similarly like for the spatiotemporal suppression, the distance between sensor and actor is taken into account when using the decision aware suppression method. The suppression is performed only if this distance is above a predetermined threshold value d_{DAS} (in segments).

4   Experiments

Computational experiments were performed to compare the data communication cost and target chasing performance for six algorithms that use different combinations of the actor selection and data suppression approaches (Table 1). The comparison was made by taking into account three criteria: time to capture, hop count and number of data transfers. The WSAN model presented in Sect. 2 was used for a simulation-based evaluation of the above metrics. During simulations the targets were created at random time steps and locations. The simulation is finished when the number of eliminated targets reaches 2700. This section discusses the experimental results that were obtained from 20 simulation runs for each algorithm and parameter setting.

Table 1. Compared algorithms

Algorithm	Actor selection condition	Data suppression method
TS-1	Nearest actor	Temporal
STS-1		Spatiotemporal
DAS-1		Decision aware
TS-2	Actor with minimum number of reported targets	Temporal
STS-2		Spatiotemporal
DAS-2		Decision aware

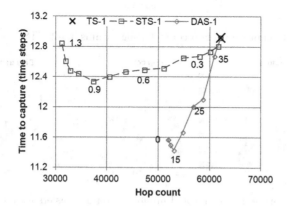

Fig. 4. Time to capture vs. hop count for algorithms TS-1, STS-1, and DAS-1

Figures 4 and 5 depict the dependencies between average values of hop count and time to catch for the compared algorithms. Labels of data points in the charts show values of the algorithm parameters. The results of STS-1 and STS-2 algorithms are presented for α ranging between 0 and 1.4. In case of DAS-1 and DAS-2 the distance threshold d_{DAS} was changed from 0 to 40 in steps of 5 segments. In general, lower communication cost (hop count) and higher performance (shorter time to capture) were achieved by the algorithms from the first group (TS-1, STS-1, and DAS-1), in which the target is always reported to the nearest actor. It can be observed in Figs. 4 and 5 that the spatiotemporal and decision aware suppression methods improve the results of the temporal suppression.

It should be noted that for $\alpha = 0$ the STS algorithm corresponds to TS. Similarly, DAS corresponds to TS for $d_{DAS} > 37$ as the radius of actor's communication range equals 37. The shortest time to capture was achieved by using the DAS-1 algorithm with $d_{DAS} = 15$ segments. The lowest hop counts were obtained for the STS algorithms with high α values.

Figures 6 and 7 present detailed results for selected settings that allow the compared algorithms to achieve the maximum performance, i.e., minimum average

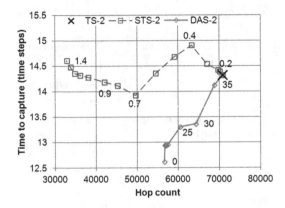

Fig. 5. Time to capture vs. hop count for algorithms TS-2, STS-2, and DAS-2

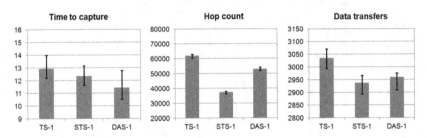

Fig. 6. Time to capture, hop count and number of data transfers for algorithms TS-1, STS-1 ($\alpha = 0.9$), and DAS-1 ($d_{DAS} = 15$)

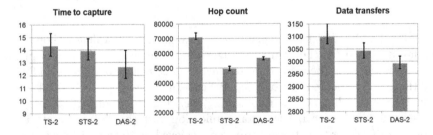

Fig. 7. Time to capture, hop count and number of data transfers for algorithms TS-2, STS-2 ($\alpha = 0.7$), and DAS-2 ($d_{DAS} = 0$)

time to capture. The error bars show the range between minimum and maximum of the metrics obtained from the 20 simulation runs. Average values are depicted as columns.

According to the presented results, it can be concluded that the spatiotemporal and decision aware suppression methods reduce the number of data transfers and hop counts in comparison with the temporal suppression. Moreover, these approaches decrease the average time in which actors eliminate the targets. The effect of decreased time to capture is especially visible for the algorithms that are based on the decision aware suppression. The reason underlying these results arises from the fact that when using the STS and DAS algorithms the sensor nodes do not report the detected targets if it is not necessary for effective navigation of actors. The information about target is transmitted from a sensor node to an actor when the distance between them is shorter. Therefore, the probability that a new target will appear closer to the selected actor before it reaches the previously reported target is diminished and there is smaller chance that the assignment of targets to actors will be non-optimal.

5 Conclusion

Reduction of data transmission is an important issue for the development of WSAN-based surveillance applications that require real-time data delivery, energy conservation, and effective utilization of the bandwidth-limited wireless communication medium. In this paper an approach is introduced to reduce the data transmission in WSAN by means of suppression methods that were originally intended for wireless sensor networks. Communication algorithms based on temporal, spatiotemporal, and decision aware data suppression methods are proposed for a WSAN system in which mobile actors have to capture distributed targets in the shortest possible time.

Effectiveness of the proposed data communication algorithms was verified in computational experiments by using a WSAN model. The experimental results show that the spatiotemporal and decision aware suppression methods reduce the number of data transfers and hop counts in comparison with the temporal suppression, which ensures that the actors receive complete information about targets detected in their communication ranges. Further research will be conducted to test the proposed approach in more complex network scenarios. Moreover, an interesting topic for future works is to investigate the impact of transmission failures on performance of the presented algorithms.

References

1. Kamali, M., Laibinis, L., Petre, L., Sere, K.: Formal development of wireless sensor-actor networks. Sci. Comput. Program. **80**, 25–49 (2014)
2. Akyildiz, I.F., Kasimoglu, I.H.: Wireless sensor and actor networks: research challenges. Ad Hoc Netw. **2**(4), 351–367 (2004)
3. Khamis, A., ElGindy, A.: Minefield mapping using cooperative multirobot systems. J. Robot. **2012**, 1–7 (2012). article ID 698046
4. Vedantham, R., Zhuang, Z., Sivakumar, R.: Mutual exclusion in wireless sensor and actor networks. In: 3rd Annual IEEE Communications Society on Sensor and Ad Hoc Communications and Networks SECON 2006, vol. 1, pp. 346–355 (2006)
5. Kumar, M.S., Rajasekaran, S.: Detection and extinguishing forest fires using wireless sensor and actor networks. Int. J. Comput. Appl. **24**(1), 31–35 (2011)
6. Płaczek, B.: Selective data collection in vehicular networks for traffic control applications. Transp. Res. Part C Emer. Technol. **23**, 14–28 (2012)

7. Płaczek, B.: Uncertainty-dependent data collection in vehicular sensor networks. In: Kwiecień, A., Gaj, P., Stera, P. (eds.) CN 2012. CCIS, vol. 291, pp. 430–439. Springer, Heidelberg (2012)

8. Bernaś, M.: WSN power conservation using mobile sink for road traffic monitoring. In: Kwiecień, A., Gaj, P., Stera, P. (eds.) CN 2013. CCIS, vol. 370, pp. 476–484. Springer, Heidelberg (2013)

9. Porwik, P.: The spectral test of the boolean function linearity. Int. J. Appl. Math. Comput. Sci. 13(4), 567–576 (2003)

10. Alippi, C., Anastasi, G., Di Francesco, M., Roveri, M.: An adaptive sampling algorithm for effective energy management in wireless sensor networks with energy-hungry sensors. IEEE Trans. Instrum. Meas. 59(2), 335–344 (2010)

11. Zhang, Y., Lum, K., Yang, J.: Failure-aware cascaded suppression in wireless sensor networks. IEEE Trans. Knowl. Data Eng. 25(5), 1042–1055 (2013)

12. Zhou, X., Xue, G., Qian, C., Li, M.: Efficient data suppression for wireless sensor networks. In: 14th IEEE International Conference on Parallel and Distributed Systems ICPADS 2008, pp. 599–606 (2008)

13. Evans, W.C., Bahr, A., Martinoli, A.: Distributed spatiotemporal suppression for environmental data collection in real-world sensor networks. In: IEEE International Conference on Distributed Computing in Sensor Systems DCOSS, pp. 70–79 (2013)

14. Silberstein, A., Gelfand, A., Munagala, K., Puggioni, G., Yang, J.: Making sense of suppressions and failures in sensor data: a bayesian approach. In: Proceedings of the 33rd International Conference on Very Large Data Bases, pp. 842–853 (2007)

15. Puggioni, G., Gelfand, A.E.: Analyzing space-time sensor network data under suppression and failure in transmission. Stat. Comput. 20(4), 409–419 (2010)

16. Yigitel, M.A., Incel, O.D., Ersoy, C.: QoS-aware MAC protocols for wireless sensor networks: a survey. Comput. Netw. 55(8), 1982–2004 (2011)

17. Płaczek, B.: Communication-aware algorithms for target tracking in wireless sensor networks. In: Kwiecień, A., Gaj, P., Stera, P. (eds.) CN 2014. CCIS, vol. 431, pp. 69–78. Springer, Heidelberg (2014)

18. Płaczek, B., Bernaś, M.: Optimizing data collection for object tracking in wireless sensor networks. In: Kwiecień, A., Gaj, P., Stera, P. (eds.) CN 2013. CCIS, vol. 370, pp. 485–494. Springer, Heidelberg (2013)

Energy Aware Object Localization in Wireless Sensor Network Based on Wi-Fi Fingerprinting

Marcin Bernas[⊠] and Bartłomiej Płaczek

Institute of Computer Science, University of Silesia,
Bedzinska 39, 41-200 Sosnowiec, Poland
{marcin.bernas,placzek.bartlomiej}@gmail.com

Abstract. The usage of GPS systems for indoor localization is limited, therefore multiple indirect localization techniques were proposed over the years. One of them is a localization method based on Wi-Fi (802.11) access point (AP) signal strength (RSSI) measurement. In this method, a RSSI map is constructed via Localization Fingerprinting (LF), which allows localizing object on the basis of a pattern similarity. The drawback of LF method is the need to create the RSSI map that is used as a training dataset. Therefore, in this study a Wireless Sensor Network (WSN) is used for this task. The introduced in this paper energy aware localization method allows to acquire the actual RSSI map or broadcast a localization signal, if there is not sufficient information to perform the localization by using nearby APs. To localize objects in a given cell, various classifiers were used and their localization accuracy was analyzed. Simulations were performed to compare the introduced solution with a state-of-the-art approach. The experimental results show that the proposed energy aware method extends the lifetime of WSN and improves the localization accuracy.

Keywords: WSN · RSSI · Fingerprint · Localization · Indoor area

1 Introduction

The IEEE 802.11 wireless technology (Wi-Fi) became a standard technology that is providing Internet wireless services. The Wi-Fi signal strength, from multiple sources, can also be used to estimate the position of Wi-Fi enabled devices like computers, mobile phones or robots. The utilization of this localization technique can be found in many applications, such as advertisement in malls or medical information system in hospitals. The standard Wi-Fi enabled devices cannot measure direction of the signal, therefore only value of RSSI signal is used in this research to provide a universal solution. One of the methods that exploit the RSSI signal strength at a given position is the Localization Fingerprint (LF) method. The drawback of this approach is a need to collect the LF map. Moreover, due to changes in indoor infrastructure the LF map has to be updated over time to be precise. Therefore, in this paper the Wireless Sensor Network (WSN) is used to create the up-to-date LF map. To extend lifetime of

© Springer International Publishing Switzerland 2015
P. Gaj et al. (Eds.): CN 2015, CCIS 522, pp. 33–42, 2015.
DOI: 10.1007/978-3-319-19419-6_4

a WSN an energy efficient localization method was proposed. Simulations were performed using information collected by sensor nodes from a real-world indoor environment. Various classification methods were analyzed to find the classifier, which gives the best accuracy of the object localization.

The rest of this paper is organized as follows: Sect. 2 describes state-of-the-art solutions. Section 3 presents the WSN implementation, data collection algorithm and the classifiers used to find the location of a device. In Sect. 4 simulation results are presented and discussed. Finally, conclusions are given in Sect. 5.

2   Related Works

The localization techniques like GPs [1,2], acoustic [3,4] and light-based approaches [5] are widely used in open and relatively flat environments. However, this approach is less effective in non-line-of-sight (NLOS) environments like mountainous or indoor areas.

The disadvantage of light-based solution is strong dependency to contrast and background light intensity [4,6]. For indoor environments, where multiple obstacles can be found, the indirect localization becomes more and more popular. One of its implementations is the Wi-Fi (IEEE 802.11) based localization. This solution takes advantage over the fact of its common implementations in public places, like: airports, malls or campuses. The most common practice in Wi-Fi-enabled devices is to apply one of two types of the location-sensing techniques: propagation based method [2,7] and location fingerprinting (LF) [8,9]. Propagation-based techniques measure angle of arrival (AOA) and time difference of arrival (TDOA) of received signals. Using this data and a mathematical model it is possible to determine the location of a device. A drawback of the propagation-based methods is a need to take under consideration environmental obstacles that can cause the signal to blend, as well as exact position of each access point. The WSN Wi-Fi fingerprinting was introduced in [10,11], however the first work applies a simple kNN classifier for the localization, while the second one uses a support vector machine (SVM) as a main localization mechanism. The energy utilization of WSN is also a vital issue which was considered in previous research on energy consumption [12] and energy models [13]. Finally, most works perform analysis within a one level of a building [11,14]. In this paper six level building will be used as test area. The effectiveness of the proposed method are compared with other, most commonly used solutions – the k-nearest neighborhood (kNN) variants [15], perceptron neural network [16], support vector machine [11,17], random forest [18], Bayes Nets method [19] and fuzzy models [20].

3   Localization Using WSN Fingerprint

In the proposed solution the sink is gathering data from sensors $(s_{ij}, i = 1, \ldots m$ $j = 1, \ldots o)$ and it is executing the localization at the same time in response to an object request. The localization is based on LF map gathered from WSN nodes.

Fig. 1. The sensor node used in research

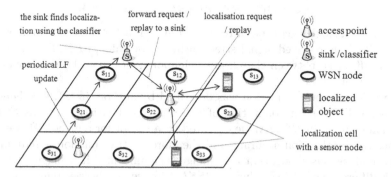

Fig. 2. The localization method overview

The nodes are enabled for Wi-Fi communication. Every node reads the available Wi-Fi networks with their RSSI value and sends these values to the sink. The standard omnidirectional antenna is used in most of the mobile devices; therefore the direction is not taken under consideration. A sensor node presented in Fig. 1 was used in the experiments. It is constructed from microcontroller, Wi-Fi module and 7200 J energy source. The sensors was used to collect RSSI values of APs in a researched building. The obtained data were used in simulation to verify the method proposed in this paper.

For the sake of simplicity, it was assumed that the localization is limited to areas identified by single wireless nodes (cells). The maximal cell size is limited by indoor obstacles and Wi-Fi standard. In this research it is limited to 5 m radius by distribution of sensors. For further simplification the sensors are equally distributed in a mesh. Thus it is convenient to use a matrix of sensor readings: $S = [s_{ij}]$. The matrix S is compared by the sink with the localization request of a Wi-Fi enabled device (object). The requests from object are managed by AP to reduce the energy consumption of WSN. The sink performs localization using classification algorithm to find a cell id and send it to a device. The solution is presented in Fig. 2.

3.1   WSN as Fingerprint Location Map

The proposed method allows to localize an object even if it is not in range of sufficient number of APs. Thus, the sensors were programmed to work in two modes.

Nodes, in first mode, are monitoring the changes in Wi-Fi signal strength of APs and perform actualization, if the change in $RSSI$ value is significant. In second mode, if the data are insufficient to localize a device (number of registered APs is too low), the nodes start to broadcast its position to simplify the localization process. To stop the fast depletion of sensors energy, the sensors are hibernated and waked up every dt seconds. Each sensor is storing LF local vector s_{ij}, where i and j are the sensor position in a grid. Vector s_{ij} is defined as follows:

$$s_{ij} = \{a_1, a_2, \ldots, a_p, \ldots, a_n\}, \tag{1}$$

$$a_p = <MAC, RSSI, dRSSI>, \quad p = 1, \ldots n, \tag{2}$$

where: a_p – is a single access point record, MAC – is a MAC address of an access point, $RSSI$ – is received signal strength indicator value, $dRSSI$ – the signal strength difference between the closest neighbor nodes, n – the number of AP in proximity of a node.

The model of data reduction was described by authors in [21,22]; therefore the paper will focus on adaptation of two observed heuristics. It was noticed that small changes in $RSSI$ value observed by a sensor for a given AP does not influence the localization accuracy and therefore its transmission is redundant. Further analysis has shown that the change is significant if it is bigger than the smallest difference of $RSSI$ values ($dRSSI_{\min}$) observed between analyzed sensor and its closest neighborhood. Intuitively, if a change is bigger than $dRSSI_{\min}$, the $RSSI$ value of this sensor can become bigger or smaller than readings from neighbor sensor. Every sensor, in proposed method, makes decision independently, so the observed change in analyzed node and its neighbor node in total should not be higher than $dRSSI_{\min}$. Therefore, the half of $dRSSI_{\min}$ is taken as a threshold value. Second observation is based on the number of AP needed for correct localization. The initial research has shown that localization using only one or two APs has to high localization error, so minimal number of monitored AP was set to $k = 3$. The described heuristics was used to formulate an update algorithm (Fig. 3) of the FL mapping for a given s_{ij} node.

The energy usage of the proposed algorithm was verified using network implemented in ns2 simulation tool and real data obtained from distributed sensors. The data was collected from 6 level building of The Institute of Computer Science, at The University of Silesia. The dimensions of building are approximately 80 m (width) by 18 m (height). The radio frequency channels of IEEE 802.11 are in b, g and a standard working in the 2.4 GHz/5 MHz band. The received signal strength during data acquisition was limited to value between -93 dBm and -27 dBm. The data from 42 places within a building was measured, uniformly by 7 on every floor level, therefore simulation was also limited to that number.

During simulation the nodes send data that are used to create the LF map of the building. The data contain real APs readings obtained from the researched building to take under consideration its special characteristics e.g. obstacles. To reduce the energy consumption, the schedule routine was implemented into simulation. The result of energy consumption was presented in Fig. 4. For comparison purposes two methods was implemented. The first method represents

1 Find available networks, read their $RSSI$ value (s_{ij}) and send it to a sink.
   Set $dRSSI$ of all a_p, $p = 1, \ldots n$ to $-10\,$dB.
2 Listen for data from neighbor nodes s_{ef}, where $|i - e| <= 1$ and $|j - f| <= 1$.
   Store their vectors (up to 8).
3 If there are access points with the same MAC then calculate $dRSSI = dRSII_{min}$
   of a s_{ij} node as a minimum of difference between $RSSI$ value of s_{ij} node
   and $RSSI$ values of stored neighbours' nodes (s_{ef}).
4 If cardinality of detected access point is lower than k, the sensor starts
   to broadcast its $ID(ij)$. (rule included in Algorithm 1')
5 Read the available network $MACs$ and their $RSSI$ values (s'_{ij}).
6 If new access point was found or value of the $RSSI$ changed more than $dRSSI/2$
   for at least one a_p in s_{ij} then send new values to a sink and set $s_{ij}=s'_{ij}$.
7 Hibernate for dt seconds.
8 Go to step 2

Fig. 3. Algorithm 1 for sensor data transmission

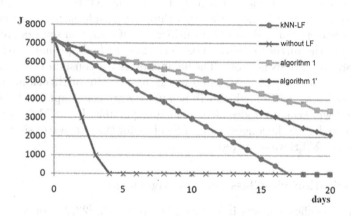

Fig. 4. The simulated energy utilization for $dt = 10\,$s

the simplest approach, called without-LF, where sensors become access points
(AP) and broadcast its position while not in hibernation mode. The second
method is based on WSN LF mapping and kNN classification, proposed in [10]
(called kNN-LF). These methods were compared against the two versions of the
proposed Algorithm 1, described in Fig. 3. The modification, where point 3 of
Algorithm 1 was included, was marked as Algorithm 1'. The hibernate/wake up
time dt was set to 10's. Energy resources of nodes was set to 7200 J to simulate
the working time of the used sensor nodes.

In case of the without-LF method, the WSN stayed alive for 4 days. The
simulation shows that using WSN as a fingerprinting mechanism can extend its
life-time significantly from 4 to 17 days. Moreover, the proposed algorithm is
more energy efficient because sensors are transmitting the LF map update only
if needed. The small difference in results of Algorithm 1 and Algorithm 1' is

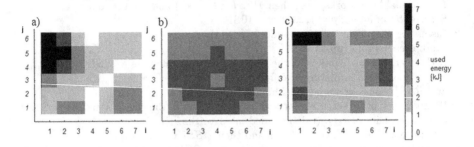

Fig. 5. The utilization of sensor s_{ij} energy using algorithms: LEACH-C (a), MTE (b), PEGASIS (c)

caused by large number of AP working within analyzed building – up to 100. In the worst case scenario for Algorithm 1', when no AP is available, the nodes will work similarly like without-LF algorithm and broadcast its positions.

The previous research showed that utilization of energy in sensors are uneven [23], therefore the further research was performed to find optimal routing algorithm for proposed WSN network. The balance of energy consumption for selected routing algorithms [24] was presented in Fig. 5. The aim of this simulation was to find a routing algorithm, which will balance energy consumption of the sensor network and will keep all nodes alive as long as possible. Using MTE algorithm, all nodes stayed alive the longest time. Therefore for the purpose of simulation on small area, the MTE was selected.

3.2   Localization via Classification Algorithm

The idea of classification was introduced by authors in [23,25] for VANET network and it is adopted for WSN. The classification process is used as localization mechanism. The localization is computed by finding the cell at which the device is positioned. The cell is connected with a sensor. Each cell has 5 m radius and is characterized by a data provided from $s_{ij}(i = 1,\ldots 7, j = 1,\ldots 6)$ sensors. The sensors are enumerated within a building, where index j defines the floor level and index i corresponds to a sensor locations counting from one side of a building. To find the matching cell for an incoming request (s_r) the sink is comparing both s_r and s_{ij} readings to find best match. In the proposed implementation each sensor has its class and is described by all APs $RSSI$ values. The AP is identified by its MAC address. If the AP is not registered by the sensor the 0 value is set. The learning dataset was presented in Table 1. The prepared dataset is used to train a classifier. Similarly, the request vector s_r is transformed to perform classification and is treated as a test vector. The result of classification is a localization narrowed to a cell. In the performed research the classifiers commonly used for localization and classification were selected. There are kNN [15], neural networks [16], Random Forest [18], Bayes Net [19], and support vector machine [11,17].

Table 1. The training set table representation

Sensor/MAC	00::6C:1D	00::1W:00	00::4D:02	00::C3:04	00::76:40	00::41:42	00::AB:44
s11	−70.0	−82.0	−84.0	−84.0	−84.0	−81.0	−79.0
s12	−60.0	−73.0	−64.0	−64.0	−64.0	−72.0	−73.0
s13	−70.0	−73.0	−61.0	−60.0	−60.0	−73.0	−73.0
s14	−80.0	0.0	−76.0	−77.0	−76.0	−63.0	−63.0
s15	0.0	0.0	−84.0	−83.0	−83.0	−48.0	−48.0

4 Simulation Results

The simulation was performed using data collected from the building of The Institute of Computer Science, at University of Silesia. The sensor nodes were distributed over the building to measure the APs $RSSI$ value. At the same time the mobile device was carried out between cells and the request of localization was performed by obtaining the test values of APs $RSSI$. The acquired data was used as an input to a simulation to verify effectiveness of the proposed method. The location returned from a sink was compared with the location of cell at which the request was made. The accuracy was measured for the exact localization (the same cell) and localization, where nearby cells were also accepted. The results obtained for the examined methods are presented in Table 2.

Table 2. The accuracy of two versions of Algorithm 1

Method	Algorithm 1		Algorithm 1'	
	Accuracy [%]	Accuracy [%] (with neighborhood)	Accuracy [%]	Accuracy [%] (with neighborhood)
kNN	75.8	93.1	78.3	96.1
Random Forest	92.3	98.2	93.8	99
Bayes Net	95.5	99.4	96.1	100
PNN	78.4	90.2	82.2	95.2
SVM	88.8	96.5	91.7	99.4

The best accuracy was obtained using Bayes Nets and Random Forest method. In the literature the kNN [10] and SVM [11] solutions were preferred, however as the initial research shows, using various classifiers can give better results and should be investigated further. The misclassification for given nodes was presented in Fig. 6. The marked diagonal presents the correctly classified objects. The values outside diagonal present the object misclassification that is usually done between floors or neighbor sensors. Finally the comparison of the proposed model was performed via simulation. The proposed model was based on the

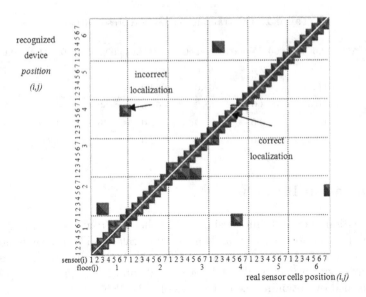

Fig. 6. Localization of the device to a correct cell for Bayes Nets classifier

Bayes Net classifier and energy aware model defined as Algorithm 1. To compare the results, two measures were defined: the life-time of WSN and localization accuracy. The results were presented in Table 3. The comparative method was based on [10], as the kNN-LF method, where the kNN classifier was used. The without-LF method uses sensor nodes to broadcast its position. Application of the WSN nodes that are broadcasting their positions increases the localization accuracy greatly (up to 100 %) at the cost of energy consumption. The hibernate procedure extends the operation time of a WSN up to three days using 7200 J of energy for each sensor. However, the utilization of WSN only for fingerprinting purposes extends the network lifespan up to 12 days. The time could be much longer, however for testing purposes the hibernation time was shortened to $dt = 10$ s. The proposed method for reduction of transmission number increases the WSN network lifetime up to 25 days, while using Bayesian Net algorithm the accuracy was increased by 14.9 % in comparison to kNN-LF method used in [10]. The obtained results are encouraging and the further research will be performed to find an optimal balance between accuracy and energy consumption.

Table 3. The comparison of localization method

Method	Without LF	kNN-LF	Algorithm 1'	Algorithm 1
Accuracy [%]	100	80.6	96.1	95.5
Uptime [days]	3	12	20	25

5   Conclusion

The paper presents results of a performance analysis of various classifiers used for localization of Wi-Fi enabled devices. The simulation based on ns2 was performed and accuracy of the localization was measured using real-world data obtained from sensors within a test site. To simplify the LF mapping, the energy aware model of sensor network was proposed. The obtained map via LF mapping with Bayesian Net classifier allowed to localize a device with accuracy of 96.1 %. It is worth noticing that majority of incorrect localization are made with neighbor cells. The proposed and implemented heuristic method allowed to extend the lifetime of a WSN network from 12 up to 25 days. Furthermore, the research showed that selection of classifier has a major impact on the localization accuracy. Further research will focus on implementing a fully operational WSN and extending the experiments to other locations. Furthermore, the research will be conducted to find an effective method to the operation mode of sensor network (fingerprinting or beacon mode).

References

1. Liu, H., Darabi, H., Banerjee, P., Liu, J.: Survey of wireless indoor positioning techniques and systems. IEEE Trans. Syst. Man, Cybern. Part C **37**(6), 1067–1080 (2007)
2. Chen, L., Li, B., Zhao, K., Rizos, C., Zheng, Z.: An indoor geolocation system for wireless LANs. In: Parallel Processing Workshops, pp. 29–34 (2003)
3. Stoleru, R., He, T., Stankovic, J., Luebke, D.: A high-accuracy, low-cost localization system for wireless sensor networks. In: Proceedings of the 3rd International Conference on Embedded Networked Sensor Systems, pp. 13–26. ACM Press, New York (2005)
4. Stoleru, R., Vicaire, P., Hey, T., Stankovic, J.: StarDust: a flexible architec-ture for passive localization in wireless sensor networks. In: Proceedings of the 4th International Conference on Embedded Networked Sensor Systems, pp. 57–70. ACM Press, New York (2006)
5. Chen, W., Hou, J., Sha, L.: Dynamic clustering for acoustic target tracking in wireless sensor networks. IEEE Trans. Mob. Comput. **3**(3), 258–271 (2004)
6. Wang, J., Zha, H., Cipolla, R.: Coarse-to-fine vision-based localization by indexing scale-invariant features. IEEE Trans. Syst. Man, Cybern. Part B **36**, 413–421 (2006)
7. Farjow, W., Chehri, A., Hussein, M., Fernando, X.: Support vector machines for indoor sensor localization. In: Wireless Communications and Networking Conference IEEE (2011). doi:10.1109/WCNC.2011.5779231
8. Kwon, J., Dundar, B., Varaiya, P.: Hybrid algorithm for indoor positioning using wireless LAN. In: Vehicular Technology Conference, pp. 4625–4629 (2004)
9. Prasithsangaree, P., Krishnamurthy, P., Chrysanthis, P.: On indoor position location with wireless LANs. In: Personal, Indoor and Mobile Radio Communications,vol. 2, pp. 720–724 (2002)
10. Tatatr, Y., Yildrim, G.: An alternative indoor localization technique based on fingerprint in wireless sensor networks. Int. J. Adv. Res. Comput. Commun. Eng. **2**(2), 1288–1294 (2013)

11. Figuera, C., Rojo-Álvarez, J., Wilby, M., Mora-Jiménez, M., Caamaño, A.: Advanced support vector machines for 802.11 indoor location. Sig. Process. **92**(9), 2126–2136 (2012)
12. Płaczek, B.: Uncertainty-dependent data collection in vehicular sensor networks. In: Kwiecień, A., Gaj, P., Stera, P. (eds.) CN 2012. CCIS, vol. 291, pp. 430–439. Springer, Heidelberg (2012)
13. Zhou, H., Luo, D., Gao, Y., Zuo, D.: Modeling of node energycon-sumption for wireless sensor networks. Wirel. Sens. Netw. **3**, 18–23 (2011).doi:10.4236/wsn.2011.31003
14. Chan, E., Baciu, G., Mak, S.: Using wi-fi signal strength to localize in wireless sensor networks. In: 2009 International Conference on Communications and Mobile Computing, vol. 1, pp. 538–542 (2009)
15. Bernas, M., Placzek, B., Porwik, P., Pamula, T.: Segmentation of vehicle detector data for improved k-nearest neighbours-based traffic flow prediction. In: IET Intelligent Transport Systems, pp. 1–11 (2014). doi:10.1049/iet-its.2013.0164
16. Berthold, M., Diamond, J.: Constructive training of probabilistic neural networks. Neuro-computing **19**(1–3), 167–183 (1998)
17. Wang, C., Chen, W., Sun, Y.: Sensor network localization using kernel spectral regression. In: Communication and Wireless Computing, (2009). doi:10.1002/wcm.820
18. Rreiman, L.: Random forests. Mach. Learn. **45**(1), 5–32 (2001)
19. Langley, J.: Estimating continuous distributions in Bayesian classifiers. In: Proceedings of the 11th Conference on Uncertainty in Artificial Intelligence, pp. 338–345 (1995)
20. Kudlacik, P., Porwik, P.: A new approach to signature recognition using the fuzzy method. Pattern Anal. Appl. **17**(3), 451–463 (2014)
21. Płaczek, B., Bernas, M.: Optimizing data collection for object tracking in wireless sensor networks. In: Kwiecień, A., Gaj, P., Stera, P. (eds.) CN 2013. CCIS, vol. 370, pp. 485–494. Springer, Heidelberg (2013)
22. Placzek, B., Bernas, M.: Uncertainty-based information extraction in wireless sensor networks for control applications. Ad Hoc Netw. **14C**, 106–117 (2014)
23. Bernas, M.: WSN power conservation using mobile sink for road traffic monitoring. In: Kwiecień, A., Gaj, P., Stera, P. (eds.) CN 2013. CCIS, vol. 370, pp. 476–484. Springer, Heidelberg (2013)
24. Francesco, D., Das, K., Anastasi, G.: Data collection in wireless sensor networks with mobile elements: a survey. ACM Trans. Sens. Netw. **8**(1), 7:1–7:31 (2012)
25. Bernas, M.: VANETs as a part of weather warning systems. In: Kwiecień, A., Gaj, P., Stera, P. (eds.) CN 2012. CCIS, vol. 291, pp. 459–466. Springer, Heidelberg (2012)

LTE or WiFi? Client-Side Internet Link Selection for Smartphones

Paweł Foremski[✉] and Krzysztof Grochla

The Institute of Theoretical and Applied Informatics of the Polish Academy
of Sciences, Bałtycka 5, 44-100 Gliwice, Poland
{pjf,kgrochla}@iitis.pl

Abstract. Current mobile phones and tablets are equipped with two
technologies for accessing the Internet: WiFi and Cellular. Deciding which
of these two interfaces provides faster data transfer is often non-trivial,
but most of the currently used devices use a simple priority scheme that
prefers WiFi to Cellular.

In this paper, we propose a novel system that automatically selects
the best link available in the current user location. The system periodically probes the bandwidth available on both links and makes statistical predictions, while avoiding excessive data and battery usage. We
experimentally validated our approach using a dedicated application for
Android.

Keywords: Cognitive networks · Available bandwidth estimation ·
Cellular networks · LTE · SON · Mobility management · Offloading

1 Introduction

Current mobile phones and tablets are equipped with multiple wireless interfaces. Apart from using the UMTS or LTE technology, they transmit data using
WiFi in areas where it is available. The choice of the wireless interface is typically static. The Android and iOS systems employ a priority-based selection
scheme, which chooses WiFi whenever possible, or the Cellular link instead.
However, this simple policy is inefficient in terms of energy use and achieved
transmission speed: it requires both interfaces to be enabled virtually all of the
time, which decreases battery lifetime, and there are many situations in which
WiFi is slower than Cellular, e.g. in heavily loaded public WiFi networks and in
locations connected over a low-speed ADSL line.

Some users manually disable the WiFi interface for most of their time, and
manually enable it only when in range of a known AP. Such solution increases
battery lifetime, but requires manual control and takes time. To automate this
process, a few tools were proposed – e.g. Sony location-based WiFi or Smart
WiFi Toggler [1] – but they do not evaluate the available Internet bandwidth.

In this paper, we propose a method to automatically select the optimal interface in terms of maximum transmission speed, with minimal energy usage and

P. Gaj et al. (Eds.): CN 2015, CCIS 522, pp. 43–53, 2015.
DOI: 10.1007/978-3-319-19419-6_5

data transfer. We introduce a lightweight tool for estimating available bandwidth of WiFi and Cellular links, and we present a novel algorithm to select the best interface in the current user location.

The rest of the paper is organized as follows: in Sect. 2 we present the background work and the motivation for the proposed solution, in Sect. 3 we describe the proposed methodology to select optimal interface for data transmission. Section 4 covers the experimental validation of the method using a smartphone application and analysis of the stability of the link selection. We finish the paper with a short conclusion in Sect. 5.

2 Background

The rapid growth in the number of wireless devices used for accessing the Internet substantially increased the traffic load on mobile networks. The global mobile traffic grew by 81 % in 2013 [2], which results in high load on the currently deployed wireless networks, and – as a consequence – to decreased quality of service in some locations during peak hours. These changes make it hard for the user to manually select the optimal Internet connection.

Most of the currently produced smartphones are equipped with two radio interfaces: 3G/LTE (Cellular) and IEEE 802.11 (WiFi) [3]. These two types use different frequencies and media access methods: while WiFi is based on random channel access, the Cellular networks channel access is managed by the base station. WiFi operates in ISM band, in which anyone can easily start transmission, while Cellular networks use licensed bands, in which transmission is controlled by network operator. In both technologies the throughput of the transmission is limited by the radio signal propagation conditions, radio bandwidth available, and the amount of devices sharing the same radio resources.

The two most widely used operating systems for smartphones, Android and iOS, by default use WiFi if both connections are available. According to [3], 2/3 of consumers prefer WiFi to Cellular. WiFi is free in most cases, while the Cellular data plan requires a monthly fee. WiFi is also often perceived as more efficient than Cellular. While the IEEE 802.11 standard offers very high transmission rates of up to 300 Mbps in local networks, the actual bandwidth is often limited by an ADSL link to which the WiFi access point is connected, e.g. between 2 and 25 Mbps. In crowded locations, where many users share the same backhaul connection or where many interfering APs are deployed, the WiFi performance is heavily degraded [4]. On the other hand, the average throughput offered by LTE networks varies between 9 Mbps [5] and 13 Mbps [6], which is higher than the throughput of a low-cost ADSL link. The performance of an LTE connection depends on the distance to the base station and on the number of users transmitting data through it, so it can significantly change in space and time. The measurement presented in [5] shows that it may change between 0.6 Mbps per 5 percentile to 24 Mbps for the 95 percentile. Thus, deciding whether WiFi or LTE offers a faster transmission is not an easy task for the user.

The problem of transferring the data traffic from Cellular to WiFi was heavily investigated in the literature, but most of these works evaluated it from the network operator perspective, which aims at offloading transmissions towards the unlicensed bands [7]. The 3GPP Release 10 defined data offloading as a key solution to cope with the constantly increasing load on packet data networks [8]. Offloading to WiFi is considered jointly with small cell deployments [9]. However, implementation of the infrastructure to manage the offloading from the operator perspective is costly and requires economical relations between the network operator and the owners of access points [10].

From the client perspective, users want to simply use the interface that offers the fastest data transfer in their current location. Users may manually enable or disable the WiFi interface, but this consumes time and introduces burden. There are applications that automate this process and disable the WiFi interface in locations where the user configured the phone to do so, e.g. Smart WiFi Toggler [1]. However, the selection of the best link in specific location should estimate the available bandwidth on both WiFi and LTE interfaces, and realize the selection quickly and in an energy efficient way. This can be achieved by measuring on the client side which of the two interfaces provides faster access to the Internet. The throughput estimation should minimize the amount of data transferred, to minimize the cost of its use and minimize the energy utilization. To the best of our knowledge, there is no such tool currently available.

3   Selecting the Best Link Automatically

In this section, we describe our method for selecting the best Internet link on mobile devices. The basic idea is to make periodic and lightweight measurements of the instantaneous download speeds in locations where the user has active WiFi connection. We summarize these measurements using statistics and select Cellular link if it performs better than WiFi in the current location.

3.1   Available Bandwidth Estimation

Available end-to-end bandwidth is an important metric of an Internet path, which has high impact on the quality of an Internet link in general. Numerous methods for measuring this metric were proposed in the literature, under the name of Available Bandwidth Estimation (ABE) [11–13]. In [14], the authors experimentally compared 9 ABE tools in the same networking environment, in terms of intrusiveness, response time, and accuracy in presence of different cross-traffic streams. However, only Spruce [12], pathChirp [11], and Assolo [13] generated less than 500 KB of traffic per measurement. We did not consider the other methods because mobile operators put monthly limits on Cellular data transfer. We chose Assolo as a state-of-the-art method, because it is an optimized version of pathChirp, and has lower intrusiveness than Spruce.

However, in [15] the authors showed that current ABE tools will not work in large-scale distributed systems. The authors reported a significant underestimation of the available bandwidth, with divergence of estimations vs. real values. In our paper, we experimentally confirm these results in Sect. 4: Assolo cannot reliably estimate the bandwidth of an artificially limited Internet link (see Fig. 2). Thus, we propose a new lightweight ABE tool: pik.

The basic idea behind pik is to send a short peak of UDP data and measure its duration at the receiver. We also estimate the Round-Trip Time (RTT) for better accuracy and to work-around various network buffers. The measurement process is as follows. The client, which is the receiver part wanting to know its available bandwidth, registers at the server and obtains a random password used for further authentication. Next, the client sends a PING request, to which the server replies with a 50 B response. This step is repeated by default 5 times with a 1-s timeout, and the average time between sending the request and receiving the response is treated as the link RTT. Finally, the client sends a START request, to which the server replies with a peak of data, by default 100 packets of 1 KB length. The server sends the data to the network as fast as possible, in a single loop without any pauses. The client assumes reception of the first packet at the time of the START request plus RTT, but the time of the last packet is measured. The duration of the peak at the receiver side is calculated, and finally the available bandwidth is the amount of data received divided by the peak duration. This final stage has a time limit of 3 s by default.

The pik tool works well for Internet links with artificial limits, i.e. bandwidth caps set by an ISP operator, but we need to repeat the measurement several times to gain reliable information on the link performance.

3.2   Link Selection

Basing on experimental evaluation, we propose the following condition to select the Cellular connection instead of WiFi:

$$0.75 \cdot \overline{M_c^{(L)}} > \overline{M_w^{(L)}}, \tag{1}$$

$$M_c^{(L)} = \{c_1^{(L)}, \ldots, c_n^{(L)}\} \quad n \geq 5, \tag{2}$$

$$M_w^{(L)} = \{w_1^{(L)}, \ldots, w_m^{(L)}\} \quad m \geq 5, \tag{3}$$

$$L = (\text{SSID}, \text{BSSID}), \tag{4}$$

where $M_c^{(L)}$ is a set of pik measurements for Cellular at location L, $M_w^{(L)}$ is the same for WiFi, and L is a tuple of SSID and BSSID for the associated WiFi AP.

The goal of Eq. 1 is to select Cellular if on average it performs much better than WiFi in the current location, e.g. if it is a few Mbps faster. We highlight that the goal of our work is not estimating the available bandwidth, but choosing the better performing link on mobile devices. The bandwidth available to the user depends on many factors, e.g. the number of contending hosts in a WLAN network or on the scheduling algorithm in LTE, which is dynamic. Thus, we propose to periodically repeat the measurements several times and make the decision using link statistics. Hence, we present a heuristic approach validated through experiments, instead of comparing the bandwidths directly.

We propose to update $M_c^{(L)}$ and $M_w^{(L)}$ periodically in the background, without disrupting normal operation of the device. A link selection system can schedule bandwidth measurements each few hours, provided that the screen is off and a configured WiFi network is available. New data should replace old measurements. However, there is a trade-off between frequent updates and the amount of transferred data, hence the update rate should be chosen wisely. Finally, we believe that updating $M_c^{(L)}$ and $M_w^{(L)}$ without active measurements, e.g. by observing the interface byte counters, is prone to errors. One could not assume that current transmissions are not band-limited at the sender side (e.g. video streams). Mobile devices allow for only one active connection, so it would be cumbersome to passively profile two links at the same time.

3.3 Practical Application

We implemented a practical link selection system for the Android platform as "BX Network" application[1]. The application has two operation modes: (1) *Active*, when the device screen is on and unlocked, and (2) *Sync*, when the screen is off. Basically, BX Network runs the link selection algorithm (Eq. 1) when entering the Active mode and uses the Android API to apply the results. When leaving this mode, all links are switched off. However, while in the Sync mode, the application periodically enables both links for a short period of time, letting the Android synchronization to run. In such cases, BX Network also collects pik measurements if possible and desired.

We implemented pik for Android using native API (NDK), for performance reasons. Measurements are governed by a scheduler that runs pik at most every 3 h in the same location. If there are less than 5 measurements in the last 7 days for the current location, the scheduler allows more frequent updates, to collect the data for Eq. 1 as fast as possible. On the other hand, if no configured WiFi network is available, no pik measurements are made. Results are stored in a `sqlite` database. Whenever the system needs to select the best link, it fetches location information from the Android WiFi API, and queries the database for pik results in the current user location, for the last 7 days.

The final effect of running BX Network on a smartphone is that it automatically switches to the best available Internet connection in a few seconds after unlocking the screen. The delay is due to the WiFi scanning procedure implemented in recent versions of Android.

4 Experimental Validation

We base our experimental validation on 3 data sources: (DS1) bandwidth tests to two distant hosts based in US (New York) and Poland (Poznan), repeated 100 times in 24 h, (DS2) bandwidth tests over an artificially limited link to the host based in US, repeated 480 times during 3 h, and (DS3) measurements collected during typical usage of BX Network on a single smartphone for 40 days. For DS1 and DS2 we used three idle and stationary Internet uplinks – LTE, WiFi, and Ethernet – of which the Ethernet link was the fastest. All data was collected during Nov 2014–Jan 2015. We conducted 4 experiments described below.

---

[1] See https://play.google.com/store/apps/details?id=com.bxlabs.network.

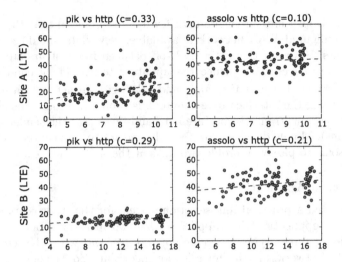

Fig. 1. HTTP download speed (horizontal axes, Mbps) versus speed estimated by pik and Assolo (vertical axes, Mbps). Measurements repeated 100 times for 24 h using LTE. Correlation coefficients shown in c, least-squares fit shown as dashed lines.

Experiment 1: Estimating Link Speed Using Pik and Assolo. In Fig. 1, we use DS1 to compare ABE tools against real bandwidth attained while downloading a 3 MB file from a web server, over an LTE link. We tested if a single run of a lightweight ABE method can estimate the real speed available to user while downloading a medium-sized web object. We repeated the experiment 100 times, running pik, Assolo, and an HTTP client (wget) immediately one after another. We did not limit the server link nor the LTE link in any way: we assume that the bandwidth changes were due to network congestion and load on the LTE base station. The results show that neither pik nor a state-of-the-art tool can estimate the HTTP download speed reliably. We see some correlation, but the results are generally random, especially for Assolo. Both methods over-estimate the HTTP speed by a factor of 1.5–5 (pik) or 4–12 (Assolo). However, pik results are generally more stable and closer to reality.

In Fig. 2, we show situation in which the link is artificially limited on the server side using Token Bucket Filter (TBF), with rates increasing from 1 till 16 Mbps (DS2). Again, we compare the HTTP download speeds with the values estimated by pik and Assolo. The results show almost perfect estimation using pik (correlation equal to 1.0) and mediocre results using Assolo (correlation equal to 0.45). The pik method slightly under-estimated the HTTP speed, while Assolo over-estimated the real values by an order of magnitude (for small TBF rates). Thus, in case the link is limited by the network operator, pik reliably estimates the available bandwidth, while Assolo does not. We conclude Experiment 1 that it is difficult to predict the HTTP download speed using small amount of data, but evaluated tools demonstrated some ability, of which pik was better and more reliable than Assolo. Thus, we can use pik for quick measurements of instantaneous link speeds.

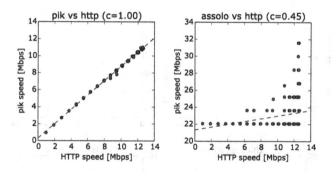

Fig. 2. HTTP speed vs estimates by pik and Assolo. Bandwidth limited using Token Bucket Filter (1–16 Mbps). Measurements repeated 30 times over an Ethernet link.

Experiment 2: Selecting the Best Link with Statistics. In Fig. 3, we search for the statistical method that would select the best link in DS1 reliably. We take random samples of all pik measurements for 3 Internet links and we apply statistical measures of location: arithmetic mean and quartiles (Q1, Q2, and Q3). The link with the highest value gets selected as the best link. Whenever the algorithm selects Ethernet, we treat it as a success. We repeated the experiment 100 times for various sample sizes, presenting the average for two distant hosts. The results show that in most cases the bigger is the sample, the better. We obtained the best results by using the value of the average and the first quartile. We recommend using the arithmetic mean for its simplicity and popularity (it is available in `sqlite`). The sample size should be at least 5.

In Fig. 4, we present results obtained from DS2: LTE and WiFi speeds measured using pik for increasing TBF rates. The upper plots show raw measurements, and the lower plot presents their statistics. Basing on the results presented in Fig. 3, we chose to apply the arithmetic mean of 15-element random samples. We see that the accuracy for LTE is generally better than for WiFi. The mean absolute error was 0.74 Mbps for LTE and 2.3 Mbps for WiFi, which is reflected in the fact that the estimated bandwidth for LTE much closely follows the TBF limit. Thus, we conclude that Eq. 1 can be used to choose LTE in favor of WiFi, but if it is much faster. By applying statistical analysis to raw pik measurements we obtain meaningful results.

Experiments 3 and 4: Stability and Costs. In Fig. 5, we evaluate stability of our link selection method. Using data in DS3, we make a link selection decision basing on a random sample of all measurements. Then, we simulate the impact of a new measurement on the decision, by adding one more measurement to the sample and evaluating the arithmetic mean once again. If the decision does not change, we treat the algorithm as stable. Figure 5 presents results obtained for two user locations, repeated 10 000 times for sample size ranging from 1 till 15. The results show the bigger is the sample size, the better. For given data source, the algorithm is stable in 99 % of cases for 5-element random sample, and 100 % for at least 10-element random sample.

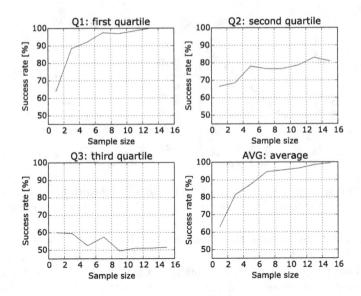

Fig. 3. Probability of choosing the best link for various statistics of pik measurements. Experiments repeated 100 times for each sample size.

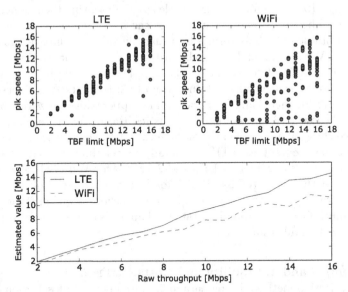

Fig. 4. Estimating TBF-limited link speed on LTE and WiFi. The upper plots show raw pik measurements, while the lower plot presents their statistics.

In our last experiment, we extracted real Cellular data usage from DS3, which should roughly illustrate the real monthly costs of using BX Network. The application was active for 40 days and used the scheduler described in Sect. 3 for measuring LTE and WiFi speeds on a typical smartphone. During the test period,

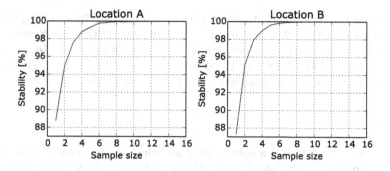

Fig. 5. Probability that a new pik measurement does not change the decision on the best link. Experiment repeated on real data 10 000 times for each sample size.

Table 1. Rough simulation of Cellular data costs

	Typical user		Active user	
	Work days	Holidays	Work days	Holidays
Amount per year	230	135	265	100
Time at home [h]	12	16	10	12
Time at work [h]	8	0	9	0
Other activities [h]	4	8	5	12
Used WLANs per day	3	1	5	2
Measurements per day	7	5	10	7
Data usage per day [KB]	714	510	1020	714
Average per month [KB]	19,170		28,080	

the application made 220 Cellular measurements. For each measurement, pik used 100 data packets of 1000 B and 10 ping packets of 50 B, which corresponds to 102 KB per measurement (including packet overhead). Thus, it used 561 KB a day, or 16 MB per month. In Table 1 we simulate monthly Cellular data costs for two scenarios: typical user and active user. We assume that the user has various network usage patterns for work days and weekends, and in different places [16]. Assuming the user accesses a WLAN at home, at work, and sometimes in travel, we propose various amounts of measurements per day. However, even for an active user, monthly Cellular data usage should stay below 30 MB.

In this work we don't target the energy usage minimization as the goal of the optimization. Measurements in [6] show that transfer of the same amount of data on LTE requires 5.4–12 times more energy than on WiFi, so if we calculate only the energy used for transfer it is always more efficient to turn on WiFi when it is available. However the total energy usage also depends on the time spent during waiting for the data. This time is linearly proportional to the network bandwidth available for both LTE and WiFi. The amount of energy consumed by proposed application, given our assumptions on Cellular data costs (Table 1), is proportional to transferring a few MB per day. Existing mobile applications for testing Internet speed, like SpeedTest [17], transfer an order of magnitude more data per single measurement, which requires much more energy. Our application also improves energy efficiency by switching off the wireless interfaces when they are not needed, reducing the energy usage in standby mode. In summary, the comparison of the total energy consumed with or without automatic link selection require more depth analysis, which we leave for further study.

5   Conclusions

In this paper, we presented a new method for selecting the optimal interface to access the Internet on mobile devices. We show that Available Bandwidth Estimation techniques available in the literature, e.g. Assolo, do not provide reliable results for modern mobile networks. To overcome this problem, we developed our own method to measure which of the two interfaces – WiFi or Cellular – performs better that executes quickly and uses small amounts of data.

The proposed method was implemented as a free application for Android and tested within a public LTE network. Our application also improves energy efficiency of data transfer on mobile devices by switching off the Internet links when they are not needed, at the cost of energy used to perform the measurements. We release an open source implementation of pik at https://github.com/iitis/pik.

We proposed a link selection method based on the available bandwidth, because it directly affects the web page and file download times perceived by the user. We leave the evaluation of other metrics that could be used for choosing between Cellular and WiFi, e.g. minimizing link latency, or maximizing the strength of the radio signal for further study.

Acknowledgment. This work was funded by the Polish National Centre for Research and Development, under research grant nr LIDER/10/194/L-3/11/: project "Optimization and load balancing in next generation wireless networks", http://projekty.iitis.pl/zosb.

References

1. Aguehian, S.: Smart WiFi Toggler. https://play.google.com/store/apps/details?id=com.sebouh00.smartwifitoggler&hl=en. Accessed: 29 January 2015
2. Cisco Visual Networking Index: Global mobile data traffic forecast update, 2014–2019, Cisco Systems Inc. (2014)
3. Witteeveen, D., Ward, S.: Global Mobile Consumer Survey 2013 (2013). http://goo.gl/P6xJvy. Accessed: 29 January 2015
4. Abinader, F.M., Almeida, E.P., Choudhury, S., Sousa, V.A., Cavalcante, A.M., Chaves, F.S., Tuomaala, E., Vieira, R.D., Doppler, K.: Performance evaluation of IEEE 802.11n WLAN in dense deployment scenarios. In: Vehicular Technology Conference (VTC Fall), pp. 1–5. IEEE (2014)
5. Huang, J., Qian, F., Guo, Y., Zhou, Y., Xu, Q., Mao, Z.M., Sen, S., Spatscheck, O.: An in-depth study of LTE: effect of network protocol and application behavior on performance. In: ACM SIGCOMM 2013 Proceedings, pp. 363–374. ACM (2013)

6. Huang, J., Qian, F., Gerber, A., Mao, Z.M., Sen, S., Spatscheck, O.: A close examination of performance and power characteristics of 4G LTE networks. In: ACM MobiSys 2012 Proceedings, pp. 225–238. ACM (2012)
7. Dimatteo, S., Hui, P., Han, B., Li, V.O.: Cellular traffic offloading through wifi networks. In: 2011 IEEE 8th International Conference on Mobile Adhoc and Sensor Systems (MASS), pp. 192–201. IEEE (2011)
8. Sankaran, C.: Data offloading techniques in 3gpp rel-10 networks: A tutorial. Commun. Mag. **50**(6), 46–53 (2012). IEEE
9. Bennis, M., Simsek, M., Czylwik, A., Saad, W., Valentin, S., Debbah, M.: When cellular meets wifi in wireless small cell networks. Commun. Mag. **51**(6), 44–50 (2013). IEEE
10. Gao, L., Iosifidis, G., Huang, J., Tassiulas, L.: Economics of mobile data offloading. In: 2013 IEEE Conference on Computer Communications Workshops (INFOCOM WKSHPS), pp. 351–356. IEEE (2013)
11. Ribeiro, V., Riedi, R., Baraniuk, R., Navratil, J., Cottrell, L.: pathchirp: Efficient available bandwidth estimation for network paths. In: Passive and Active Measurement Workshop, vol. 4 (2003)
12. Strauss, J., Katabi, D., Kaashoek, F.: A measurement study of available bandwidth estimation tools. In: Proceedings of the 3rd ACM SIGCOMM Conference on Internet Measurement, pp. 39–44. ACM (2003)
13. Goldoni, E., Rossi, G., Torelli, A.: Assolo, a new method for available bandwidth estimation. In: Fourth International Conference on Internet Monitoring and Protection, ICIMP 2009, pp. 130–136. IEEE (2009)
14. Goldoni, E., Schivi, M.: End-to-end available bandwidth estimation tools, an experimental comparison. In: Ricciato, F., Mellia, M., Biersack, E. (eds.) TMA 2010. LNCS, vol. 6003, pp. 171–182. Springer, Heidelberg (2010)
15. Croce, D., Mellia, M., Leonardi, E.: The quest for bandwidth estimation techniques for large-scale distributed systems. ACM SIGMETRICS Perform. Eval. Rev. **37**(3), 20–25 (2010)
16. Gorawski, M., Grochla, K.: Review of mobility models for performance evaluation of wireless networks. In: Gruca, A., Czachórski, T., Kozielski, S. (eds.) Man-Machine Interactions 3. AISC, vol. 242, pp. 573–584. Springer, Heidelberg (2014)
17. Ookla: Speedtest.net Mobile. http://www.speedtest.net/mobile/. Accessed: 16 February 2015

IF-MANET: Interoperable Framework for Mobile Ad Hoc Networks

Hamid Hassan[1]([⊠]), Philip Trwoga[1], and Izzet Kale[2]

[1] Mobile and Wireless Computing Group, University of Westminster,
115 New Cavendish Street, London, UK
Hamid.Hassan@my.westminster.ac.uk, Philip.Trwoga@westminster.ac.uk
[2] Applied DSP and VLSI Research Group, University of Westminster,
115 New Cavendish Street, London, UK
Izzet.Kale@westminster.ac.uk

Abstract. The rapid improvement in low power micro-processors, wireless networks and embedded systems has boosted the desire to utilize the very significant resources of mobile devices. The mobile ad hoc network (MANET), an infrastructure-less wireless network, is an emerging technology and is best suited to provide communication between wireless mobile devices. Due to the nature of MANET, the network topology changes frequently, unpredictably and has created the new challenges as traditional routing protocols are not suitable for multi-hop communication in mobile ad hoc environments. There are number of routing protocols proposed for MANET, and development is active in this area, however, there is no single routing protocol which is best suited to address all the basic issues of heterogeneous MANETs. This diverse range of routing protocols have created a new challenge as in general the heterogeneous mobile devices cannot communicate with each other and thus are unable to facilitate the full exploitation of mobile resources. To overcome the above mentioned issues, this paper has proposed an Interoperable Framework for MANET, called an IF-MANET, which hides the complexities of heterogeneous routing protocols and provides a homogeneous layer for seamless communication between these routing protocols. The IF-MANET resides in user space and it will be implemented at application layer to provide runtime interoperability and platform independence.

Keywords: Mobile application · Routing protocols · Ad Hoc networks · MANET · Mobile and wireless communication

1 Introduction

Mobile devices such as Smart Phones, PDA's, Laptops, and Sensors are gaining enormous processing power, storage capacity and wireless bandwidth. The advancement in wireless mobile technology has created a new communication paradigm via which an ad hoc wireless mobile network can be created without any priori infrastructure called mobile ad hoc network (MANET) [1]. While

© Springer International Publishing Switzerland 2015
P. Gaj et al. (Eds.): CN 2015, CCIS 522, pp. 54–68, 2015.
DOI: 10.1007/978-3-319-19419-6_6

progress is being made towards improving the efficiencies of mobile devices and reliability of wireless networks, the mobile technology is continuously facing the challenges of low power, limited resources, device heterogeneity, service discovery and routing protocol heterogeneity. In addition, the routing topology in a MANET changes frequently due to the movement of mobile nodes therefore; there are new challenges for routing protocols in MANETs. The traditional wired and wireless routing protocols are not suitable for MANET due to the unpredictable topology changes and heterogeneous environment.

The research community has developed and is busy developing different MANET routing protocols as surveyed in [2] e.g. DSDV, AODV, MAODV, DSR, OLSR, ZRP, TORA. But there is no single ad hoc routing protocol available which is best suited for MANETs to address all of their basic challenges like mobility, heterogeneity, dynamic resource discovery and QoS. This diverse range of ever growing protocols and heterogeneity of topologies have created barriers for nodes of different ad hoc networks to intercommunicate and hence wasting a significant amount of mobile resources. In order to utilize the mobile resources effectively and provide interoperability between heterogeneous routing topologies, this paper has proposed an interoperable framework for MANET called IF-MANET.

Unlike other proposed frameworks, discussed in [3], are implemented in the operating system kernel space, IF-MANET belongs to the user-space and will be implemented at application layer to provide platform and implementation independence. The framework will address the Interoperability of heterogeneous routing protocols and will provide a component based plug-in style adapters to communicate across different routing protocols. IF-MANET's abstraction layer will hide the complexities of heterogeneity and will provide a generic layer to access the mobile resources seamlessly and homogeneously. IF-MANET is a reusable artifact of software architecture, design and implementation for the development of middleware applications for MANETs. The key objectives of IF-MANET are:

1. Provide an interoperability between heterogeneous routing protocols but without modifying them.
2. Provide an interoperable route discovery protocol to discover MANET services in heterogeneous environments.
3. Achieve connectivity while providing comparable performance to their monolithic counterpart protocols.

The rest of the paper is structured as follows: Sect. 2 analyses the related work. Section 3 presents the IF-MANET design. Section 4 presents simulation and performance evaluation. Section 5 presents the conclusion and future work.

2   Related Work

The ad hoc routing protocols, based on their characteristics, are divided into three main categories [2] i.e. Proactive Routing (Table Driven), Reactive Routing (On Demand) and Hybrid Routing. There are enormous number of routing protocols already proposed and new ones are continuously arriving, under the above mentioned categories, as well as their variations which are best suited for different environments. However, there is no single routing protocol that can fulfil all the basic requirements of MANETs under heterogeneous environments. To address these issues of heterogeneity, there are different proposed research works, the main one are discussed in the following sub-sections.

2.1 Inter-domain Cluster Based Routing Approach

This approach addresses the issues of communication between heterogeneous routing protocols to provide one view of heterogeneous MANETs. It divides the nodes into different clusters called domains whereas each domain contains the same type of routing protocol. For routing within a domain it uses a proactive approach whereas a reactive approach is used for routing across different domains. It elects domain (cluster) head nodes from each cluster, which in turn is responsible of providing interoperable communication across clusters of different routing protocols. Following are several proposed protocols under this approach:

Classical Gateway routing protocol i.e. Border Gateway Protocol (BGP) [4] provides interoperability across heterogeneous nodes on internet where nodes are static but is not suitable for MANET environments where network topology changes randomly and have no physical boundaries due to node mobility.

Plutarch [5], TurfNet [6], CIDR [7], IDRM [8], ATR [9] and InterMR [10] have proposed inter-domain routing protocols specific to mobile ad hoc networks. These protocols provides the high level architectures and mainly deal with the intra (within Cluster) and inter (outside the Cluster) domain communication via cluster heads but they have not addressed the issue of heterogeneity i.e. how routing protocols of different taxonomies will communicate with each other.

Tarnoi et al. [11], PNCRM [12], Peppino Fazio et al. [13], SIR-AODV [14] have proposed protocols to reduce the transmission delays and increase the throughput in wireless networks. However they have not addressed the issues of dynamic mobility and protocols heterogeneity which are the main challenges in mobile ad hoc networks.

2.2 Middleware Framework Approach

Middleware is a software layer that lies between the operating system and the application layer. The purpose of this approach is to hide the complexities of heterogeneous resources, routing protocols and mobility from users and provide them a homogeneous environment. Traditional middleware frameworks like CORBA, Microsoft COM, Java/RMI, IBM Queries and Web Services requires powerful computing resources, fixed distributed systems, wired network topology and allow systems to be designed in advance in order to interoperate with each other [3]. However, where the topologies are heterogeneous and resources are limited, these traditional frameworks are not effective. To address the challenge of mobility, heterogeneity, scarce resources and interoperability across networks of different topologies, MANET research groups have proposed several frameworks and the key one are discussed below.

Mobile Gaia [15] addresses the pervasive and ad hoc computing environments and adopts a component-based approach. It decomposes application services into smaller components that can run on a cluster of different heterogonous devices. This yields to considerable memory and power savings, since the middleware allows only the required component to be loaded and unloaded to a device depending on its role. For communication between nodes, Mobile Gaia adopts the traditional event based model, namely, publish-subscribe. The framework is based on the "What You Need Is What You Get" model and focuses mainly on power, memory saving in heterogeneous mobile ad hoc networks.

ReMMoC [16] is an adaptive middleware framework, which is independent from particular discovery and interaction protocol. It allows client applications to be developed independently of service discovery and provides plug-in software to load the appropriate protocol at runtime to communicate with the encountered protocol. While suitable for systems that will interoperate with heterogeneous protocols, this approach cannot solve the problem of legacy platforms that are required to interoperate with one another.

PICA [17] provides multi-platform functionality for threading, packet queue management, socket-event notifications to waiting threads, and network device listing, as well as minimizing platform-related differences in socket APIs, and the kernel. It provides MANET specific APIs that can be used to developed components in user-space e.g. routing protocols. However, these systems are restricted to programming abstractions for operating system-level services only and ignore the generic routing protocol commonalities that could be reused across different implementations.

MESHMdl [18] proposed a middleware that uses mobile agents for logical mobility and tuple spaces to decouple applications components in order to address the dynamic mobility and frequent disconnections of mobile ad hoc networks. Lightweight mobile agents reduce the network overheads and tuple space decouples nodes to communicate across heterogeneous MANETs.

SELMA [19] presents a component based middleware platform for distributed applications in mobile multi-hop ad hoc networks. The middleware uses mobile agents to communicate through a "marketplace" pattern where mobile applications forward data or agents to specific geographical locations called marketplaces.

WARF [20] has proposed a component based middleware framework for IPV6 based Wireless Mesh Networks. It enables component based implementation of different routing protocols and provides cross-layer operations, multiple radio interfaces, real-time resource monitoring, dynamic resource allocation and multipath adaptive forwarding.

MANETKit [21] provides configurable component based framework that facilitates the support of multiple MANET protocols and accommodates pluggable protocol functionality. It allows protocols to be composed, decomposed and hybridized support for dynamic reconfiguration which can safely be executed at run-time. The main features of framework are: reduce MANET implementation effort, enhance the portability of protocol implementations, facilitate the exploration of protocol hybridization efforts, and seamlessly integrate MANET routing in a wider middleware framework. The main focus of this research is to address the mobile ad hoc networks in general and does not provide solution for communication between heterogeneous routing protocols.

Bluetooth JXME [22] implements a framework named JXBT (JXME over Bluetooth), which allows the JXME infrastructure to use Bluetooth as the communication medium. JXBT uses the basic features of JXME i.e. interoperability of binding peer-to-peer system to a single infrastructure, platform and programming language independence, and ubiquity. Its main focus is on Bluetooth based peer-to-peer ad hoc communication and does not address the heterogeneity of service discovery and routing protocols.

2.3 Summary

Table 1 compares the above mentioned related proposed solutions against the key MANET challenges i.e. limited resources, dynamic mobility, heterogeneity etc. From Table 1, it is evident that the proposed works have considered various issues regarding interoperation of multiple networks but none of them, except IF-MANET, have proposed a complete solution for interoperability of routing protocols in heterogeneous MANETs. The solutions presented in [9,10,20] and [21] are similar to our concept of interoperable MANET. However these approaches are specific to routing in hybrid networks of similar protocols or creating a new routing protocol using their framework. Whereas, the IF-MANET uses cluster based inter-MANET approach to support the routing in heterogeneous MANET and component based middleware approach to provide a seamless homogeneous platform to application developers. The detailed design of IF-MANET is presented in Sect. 3.

3   IF-MANET Novel Design

IF-MANET provides a novel component-based framework for MANET as shown in Fig. 1. The framework will overcome the heterogeneity of routing protocols which are the main barriers in achieving communication between heterogeneous resources. To provide interoperable communication between heterogeneous routing protocols it uses inter-domain cluster based approach to group nodes, elect cluster heads and provide communication between them. IF-MANET will provide an extensive API that can be used to develop different middleware applications without perturbing the complexities of heterogeneous resources. The key features of IF-MANET are:

- Provide communication between heterogeneous routing protocols.
- Provide route discovery across heterogeneous routing protocols.
- Special routing table to maintain the details of reachable gateways to pre-empt their routing path and protocol type.
- Inter-MANET cluster based routing for internal and external domains.
- Data transformation from source to destination protocols and vice versa.

Table 1. Comparison between MANET Frameworks

Research Works	Key Features	Mob	LR	RP	SB	SD	DH	PH	GS
ATR	Inter Domain Routing Inter/Intra Domain Communication	F	F	P	P	N	P	N	Y
InterMR	Inter Domain Routing, Attribute based addressing, Inter and Intra Gateway Election	F	F	P	P	N	P	N	Y
MESHMdl	Tuple Spaces, Mobile Agents	F	P	N	P	N	P	P	N
JXME	Peer2Peer, JXBT, Bluetooth	P	P	N	P	N	N	P	N
PICA	Component based, Service discovery	P	P	N	P	N	N	N	N
SELMA	Marketplace pattern, Component Based Mobile Agents, Neighbour discovery	F	P	P	N	P	N	N	P
Mobile Gaia	Component based, Clusters, Pub/Sub	F	F	P	N	N	N	P	P
ReMMoC	Multiple SDPs, Plugin Style protocols	F	F	P	P	N	N	N	N
MANETKit	Component based Framework, Runtime Configuration, Pub-Sub Event based	F	F	P	N	P	P	F	P
WARF	Component based, IPv6, Multi interfaces	F	F	P	N	P	P	P	P
IF-MANET	Component Based, Inter-Domain Routing Runtime Configuration, Event Driven Semantic Data Transformation	F	F	F	F	F	F	F	F

Keys: GS: Group Support, LR: Limited Resources, Mob: Mobility, SB: Scalability SD: Service Discovery, DH: Data Heterogeneity, PH: Platform Heterogeneity RP: Routing Protocol Interoperability, P: Partial, F: Full, N: Not Supported

3.1   Interoperable Routing Protocol Design

The IF-MANET provides a novel routing framework, as presented in Fig. 2, to overcome the heterogeneity of routing protocols and allow seamless communication between them. The key features of this framework are given below:

- Route Discovery in heterogeneous ad hoc networks.
- Control and Data transformation in a format compatible to source to destination routing protocols.
- Unique design such that gateway node can be used as a normal node within cluster to utilize its resources.
- A special routing table to maintain the type along with other details of all reachable nodes. It will allow packet transformation in-advance without broadcasting at different channels and hence reduce network overhead and response time. The fields of this routing table are: node Id, node type (Gateway or normal node), routing protocol type, MANET Network Type (e.g. Reactive, Proactive), lifetime (route expiry time).
- No change in existing routing protocols.

To support the above mentioned features and provide a communication between heterogeneous MANETs, IF-MANET provides a novel interoperable routing framework as shown in Fig. 2 below. The framework components are explained in the following sub-sections.

Gateway Engine. This component provides interoperability between heterogeneous routing protocols running in a diverse ad hoc networks (e.g. Pro/Reactive). It will maintain the state of the routing protocols and communicate with different components to fulfil the operation required. For Example, if a node

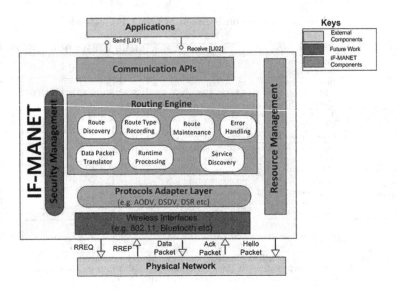

Fig. 1. IF-MANET logical architecture

sends a RREQ and the receiving node is running different routing protocols then the receiving node discards all the RREQ messages and cannot communicate with each other. To overcome this problem i.e. provide interoperability without changing the existing routing protocols, the Gateway Engine will maintain the RREQ counter. When the counter reaches RREQ Threshold (i.e. network wide search), it will assume that no node has replied because the packet is for different type of routing protocol. The gateway node will first search its routing table for the type of destination or next hop node. If found it will then load the relevant routing protocol adapter and transform the packet similar to the destination routing type, otherwise it will broadcast the RREQ message at different channels. On receiving RREP, it will update its routing table, convert the packet and RREP back to source node. For data transfer, the Routing Engine keeps track of source and destination's protocol type in an extended routing table to pre-empt the transformation the data from source to destination and vice versa.

Route Discovery Service (RDS). This component is responsible of discovering a routing path from source to destination node. If a node needs to send a packet and if it is a Reactive MANET, then the RDS will first discover the route by broadcasting a RREQ message. For Proactive MANET, the routes are maintained by period RREQ messages, and RREQ is broadcasted only when route to destination is stale in routing table.

Route Reply Service. This component implements the logic of Route Reply. The node will reply back with RREP message if it is a destination node or an intermediate node with fresh route to the destination.

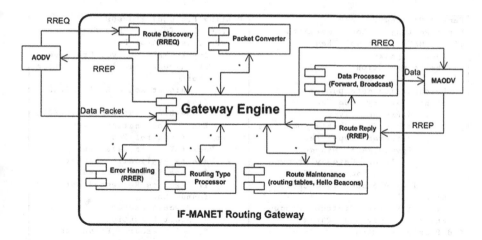

Fig. 2. IF-MANET routing gateway architecture

Message Converter. It will de-capsulate the incoming message and evaluate the type of interface whether its normal node or Gateway node. It will also check whether the source and destination nodes are same or not in order to avoid circular traffic.

Dispatcher (Data Processor). Evaluate the Packet type whether it is RREQ or Data packet. Process the Request and RREP or Broadcast RREQ. It will also semantically process the requested resource on current node.

Route Maintenance. Due to dynamic nature of MANET the network topology changes continuously and hence the nodes join and leave different MANET domains. To maintain the cluster heads routing information, IF-MANET provides a light weight sub-protocol to maintain the Gateway Nodes information via periodic beacons. Failure to receive a beacon indicates that a Gateway is lost or out of range and hence new Gateway Head will be selected.

Algorithm for IF-MANET Interoperable Routing Protocol. The pseudocode, presented in Fig. 3, is the core logic of IF-MANET routing framework. It shows the algorithms used for different components and services presented in Fig. 3.

Illustration of IF-MANET Routing Engine. Figure 4 illustrate a routing operation of IF-MANET routing engine between heterogeneous MANETs. The Cluster-A consists of nodes running Reactive Routing Protocol e.g. AODV whereas Cluster-B is using Pro-active routing protocol e.g. DSR. Due to difference in routing protocols, nodes in Cluster-A cannot communicate with nodes in Cluster-B without IF-MANET.

```
Packet Parsing Process:                     Routing Protocol Process:
Parse Routing Protocol Metadata               If Packet Type is RREQ Then
If Not MANET Protocol Then                    //RREQ is Control Packet
   Destroy packet                                Invoke RREQ Arrival Processor
   Aborts execution                              If Current Node is Destination
Else                                          Then
   De-capsulate data packet                         Unicast Route Reply (RREP)
   If not Gateway node Then                    Else
      Find Protocol Type                            Re-Broadcast packet (RREQ)
// Compare source and this node type          End // Destination check
   If Different Types Then                  End // If Control Packet
      Invoke MANET Manager                  If Packet Type is Data Then
         Invoke Gateway Process             // Data is Application Packet
         Send packet to Gateway                Process Packet Arrived
      End // Protocol Type Check               If Resource Requested Not
   Else                                     found Then
      Invoke Routing Protocol Process            Discard Packet
   End // Gateway Node Check                     Abort Execution
End Parsing Process                            Else If Resource Found Then
                                                 Process the Request
Gateway Process:                                 Create RREP
   Find Destination Protocol Type              If Destination Route Found
   Create packet of destination Type        In Routing Table Then
   Load Protocol of Destination Type                Unicast RREP
   Handover packet to loaded protocol        Else
   Invoke Routing Protocol Process                  Broadcast new RREQ
   Convert Packet to Originator Type         End //If Route to Destination
   Unicast or Broadcast RREP               End If Application Packet
End Gateway Process                         End Routing Protocol Process
```

Fig. 3. Algorithm for interoperable routing protocol

It is assumed here that the nodes in same cluster are running the same routing protocol as well as nodes of different clusters that cannot communicate directly. If a source node "Node-A1" wants to send message to destination node "Node-B3", then due to reactive (on-demand) nature, it will first initiates a route discovery by broadcasting RREQ messages. It will periodically send RREQ messages until it receives RREP or reaches a RREQ Threshold (also called network wide search). Source Node-A1 will not receive any RREP because the destination node is of a different routing type, and hence Node-A1 route discovery reaches RREQ Threshold. At this point, the IF-MANET Gateway node, which keeps count of RREQs, will assume that the destination node is running a different routing protocol. IF-MANET Gateway then stores the details of sender (type, routing path etc.), converts the protocol packet and re-broadcast RREQ.

The node (Node-B1, Node-B2), receive the RREQ from Gateway, processes the RREQ message and reply back with RREP. The Gateway node (Node-A3), on receipt of RREP will seek its routing table for source type, converts the packet compatible to source node type and send route reply (RREP) to node A-1. The source node then forward the data packet using routing path received. The Gateway node on receipt of data packet will find the type of next hop or destination node, transform the data packet and forward to next node along the path. Hence, interoperability between heterogeneous routing protocols is achieved.

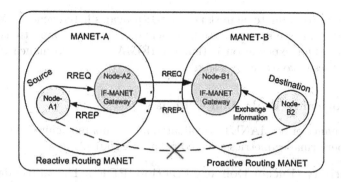

Fig. 4. Communication in heterogeneous MANETs

4   Simulations and Performance Evaluation

The IF-MANET has been implemented in NS-2.35 [23] using C++ and TCL (Tool Command Language) to conduct simulations and evaluate the performance of IF-MANET in heterogeneous MANET environment. It uses IEEE 802.11b PHY/MAC protocol which has CSMA/CA, 11 Mbps channel capacity and transmission range up to 250 m. For performance evaluation it uses two different routing protocols i.e. AODV and its variant MAODV along with IF-MANET. Figure 5 shows the deployment scheme of routing protocols under different clusters. AODV nodes are grouped in Cluster-A and MAODV in Cluster-B whereas IF-MANET nodes are distributed dynamically such that they can communicate with both type of routing protocols. The transmission range of AODV and MAODV nodes is configured such that they cannot communicate across the clusters except via IF-MANET nodes.

4.1   Evaluation Expectations

The main goal of this research is to achieve connectivity across heterogeneous routing protocols at the cost of low performance overheads. There are two

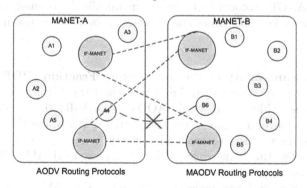

Fig. 5. IF-MANET simulation field map

extreme limits to evaluate against i.e. case-1: provides heterogeneous MANETs but without connectivity and case-2: provides maximum connectivity but no heterogeneity. Our expectation is that the IF-MANET performance will stand in-between these two extreme cases.

4.2 Performance Evaluation Criteria's

The performance of IF-MANET is evaluated and compared against the following three key performance metrics selected from RFC2501 [24]:

- **Connectivity (Packet Delivery Fraction [PDF]).** The ratio of data packets received by destinations to those generated by sources.
  PDF = (Total Packets Received/Total Packets Sent) × 100
  (Where Total = Sum of all packets).
- **Normalized Routing Load (NRL).** Total number of routing packets transmitted per data packet delivered at the destination. Each hop-wise transmission of a routing is counted as one transmission.
  NRL = Routing Packets Transmitted/Received Packets.
- **Average Packets End-to-End Delay.** Average time data packets takes to be transmitted across a MANET from source to destination. It includes all possible delays caused by buffering during the route discovery latency, queuing at the interface queue, retransmission delay at the MAC, propagation and transfer time.
  Delay = Total Receive Time – Total Sent Time
  (Where Total = Sum of all average times).

4.3 Simulation Results

The three performance criteria's, mentioned in Sect. 4.2, were evaluated against the combination of different parameters. The relationship of scenarios (legends) against set of parameters used is presented in Table 2. Each simulation scenario contains two cases: (1) a MANET with N nodes each running only AODV protocol, (2) two MANETs with N nodes each running AODV and MAODV protocols whereas IF-MANET protocol nodes were dynamically distributed. Each simulation was executed for three times and their mean values were taken for optimal results.

Scenario 1: Connectivity or Packet Delivery Fraction (PDF %). Simulations in Fig. 6 were executed at 5 m/s whereas 10 m/s in Fig. 7. From these simulations it is evident that single AODV (AODV-R100, AODV-R250), as expected, has achieved the maximum connectivity. The connectivity of heterogeneous MANETs was less at low node density irrespective of the speed and IF-MANET nodes. But it increases with the increase in IF-MANET Gateways and transmission range.

It implies that, at fewer nodes e.g. 25 nodes, the MANET has to create and maintain long multi-hop routing paths to reach the destination nodes via

Table 2. Simulation scenarios and parameters

Legend	Protocols	Mobile Nodes	IF-MANET Nodes	Speeds [m/sec]	Area [m x m]	Range [m]
AODV-R100	AODV	25, 50, 100	0	5, 10	500 x 500	100
AODV-R200	AODV	25, 50, 100	0	5, 10	500 x 500	250
GW10%-R100	AODV, MAODV	25, 50, 100	3, 5, 10	5, 10	500 x 500	100
GW20%-R100	AODV, MAODV	25, 50, 100	5, 10, 20	5, 10	500 x 500	100
GW10%-R250	AODV, MAODV	25, 50, 100	3, 5, 10	5, 10	500 x 500	250
GW20%-R250	AODV, MAODV	25, 50, 100	5, 10, 20	5, 10	500 x 500	250

IF-MANET Gateways and hence connectivity decreases. Similarly, there was a slight dip in connectivity when transmission range was 250 m, node density was 100 and IF-MANET Gateways were 10, 20. It is because the high density and transmission range has produced too much noise and hence overloaded the network bandwidth. It is evident that the IF-MANET has achieved a connectivity higher than 80 percent at a node density of 80 nodes which is far better than our expectations in Sect. 4.1. Hence, it proves that under optimal number of gateways, node density and transmission range, the IF-MANET will outperform the connectivity in heterogeneous MANETs.

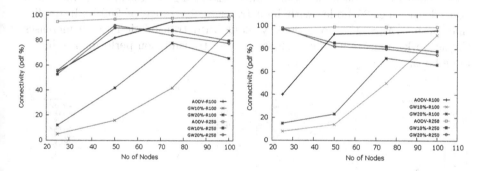

Fig. 6. Connectivity at 5 m/s   **Fig. 7.** Connectivity at 10 m/s

Scenario 2: Normalized Routing Load (NRL). Figures 8, 9 illustrates that a MANET with single AODV protocol is generating minimum routing packets for the transmission of data packets irrespective of speed and node density. However, IF-MANET was generating higher NRL at low node density i.e. when connectivity (pdf %) was less whereas it was decreased significantly with increase in node density and number of IF-MANET gateways. In contrast to our expectation of 50 % NRL, IF-MANET has outperformed and achieved up to 80 % reduction in NRL between node densities 40 to 80 irrespective of node movement speed. It implies that the IF-MANET has achieved high connectivity across heterogeneous protocols at less NRL.

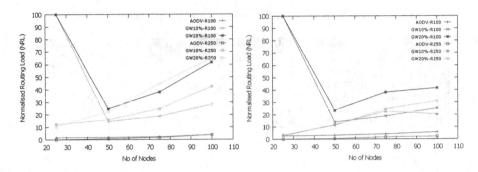

Fig. 8. NRL at 5 m/s   **Fig. 9.** NRL at 10 m/s

Scenario 3: Average Packets End-to-End Delay. Figure 10 (speed 5 m/s),
Fig. 11 (10 m/s) shows that packet transmission delay in single AODV scenario
decreased from 65 % to 15 % when node density increased to 60 nodes irrespective
of the speed. It means that the AODV initially takes more time to create a connected graph and once established it uses shortest path to deliver the packets.
Whereas, MANETs with IF-MANET gateways showed optimal performance, by
taking minimum time to transmit the data packets to the destinations at medium
node density i.e. at 50 nodes with transmission range of 100 m or at minimum
node density i.e. 25 nodes with high transmission range of 250 m. It implies
that by increasing the node density along with transmission range increases
the network connectivity but decreases packet transmission performance due to
more hops.

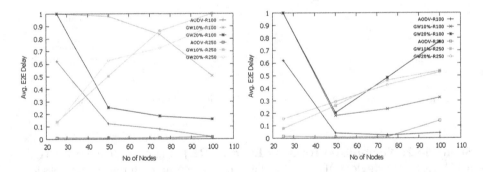

Fig. 10. Avg. E2E delay at 5 m/s   **Fig. 11.** Avg. E2E delay at 10 m/s

5   Conclusion and Future Work

This paper has investigated the challenges of communication in heterogeneous
mobile ad hoc networks and has proposed the IF-MANET for heterogeneous
MANETs to provide the interoperability between them. IF-MANET has addressed
the three key areas of heterogeneity i.e. service discovery, routing protocol and

data heterogeneity which are the main barrier in achieving the interoperability. It has then presented the novel algorithms to prove how service discovery and routing protocols will provide communication among heterogeneous resources.

To evaluate the performance of IF-MANET, this paper has performed the simulations using heterogeneous routing protocols i.e. AODV, MAODV and IF-MANET. The results were analysed against the three key performance metrics i.e. (1) Connectivity, (2) Normalized Routing Load (NRL) and (3) End to End (E2E) Delay using the two baseline use cases described in Sect. 4.2. The simulation results have confirmed that the heterogeneous routing protocols can communicate with each other through IF-MANET nodes and has outperformed the expectations by providing more than 50 % connectivity by keeping low NRL and E2E Delay. In future work, the IF-MANET will provide interoperability across different wireless technologies i.e. IEEE 802.11b, Zigbee and Bluetooth. To achieve this the IF-MANET will be extended to communicate over these different radio technologies.

References

1. Hong, X., Xu, K., Gerla, M.: Scalable routing protocols for mobile ad hoc networks. Network **16**(4), 11–21 (2002). IEEE
2. Abolhasan, M., Wysocki, T., Dutkiewicz, E.: A review of routing protocols for mobile ad hoc networks. Ad Hoc Netw. **2**(1), 1–22 (2004)
3. Hadim, S., Al-Jaroodi, J., Mohamed, N.: Trends in middleware for mobile ad hoc networks. J. Commun. **1**(4), 11–21 (2006)
4. Rekhter, Y., Li, T.: A border gateway protocol 4 (BGP-4) (1995). http://tools.ietf.org/html/rfc1771
5. Crowcroft, J., Hand, S., Mortier, R., Roscoe, T., Warfield, A.: Plutarch: an argument for network pluralism. ACM SIGCOMM Comput. Commun. Rev. **33**(4), 258–266 (2003). ACM
6. Schmid, S., Eggert, L., Brunner, M., Quittek, J.: *TurfNet*: an architecture for dynamically composable networks. In: Smirnov, M. (ed.) WAC 2004. LNCS, vol. 3457, pp. 94–114. Springer, Heidelberg (2005)
7. Zhou, B., Cao, Z., Gerla, M.: Cluster-based inter-domain routing (CIDR) protocol for MANETs. In: Sixth International Conference on Wireless On-Demand Network Systems and Services, WONS 2009, pp. 19–26. IEEE (2009)
8. Chau, C.K., Crowcroft, J., Lee, K.W., Wong, S.H.: IDRM: Inter-Domain Routing Protocol for Mobile Ad Hoc Networks. University of Cambridge Technical report UCAM-CL-TR-708 (2008)
9. Fujiwara, S., Ohta, T., Kakuda, Y.: An inter-domain routing for heterogeneous mobile ad hoc networks using packet conversion and address sharing. In: 32nd International Conference on Distributed Computing Systems Workshops (ICDCSW), pp. 349–355. IEEE (2012)
10. Lee, S.H., Wong, S.H., Chau, C.K., Lee, K.W., Crowcroft, J., Gerla, M.: InterMR: Inter-MANET routing in heterogeneous MANETs. In: MASS, pp. 372–381 (2010)
11. Tarnoi, S., Kumwilaisak, W., Saengudomlert, P., Ji, Y., Kuo, C.C.J.: QoS-aware routing for heterogeneous layered unicast transmissions in wireless mesh networks with cooperative network coding. EURASIP J. Wirel. Commun. Networking, 1–18 (2014)

12. Tan, G., Peng, X., Ni, X., Lin, B.A.F., Liu, X.: PNCRM: a novel real-time multicast scheme in MANETs based on partial network coding. J. Netw. **8**(10), 2414–2421 (2013)
13. Fazio, P., De Rango, F., Sottile, C., Santamaria, A.F.: Routing optimization in vehicular networks: A new approach based on multiobjective metrics and minimum spanning tree. Int. J. Distrib. Sens. Netw. (2013)
14. Fazio, P., De Rango, F., Sottile, C.: A new interference aware on demand routing protocol for vehicular networks. In: International Symposium on Performance Evaluation of Computer Telecommunication Systems (SPECTS), pp. 98–103. IEEE (2011)
15. Chetan, S., Al-Muhtadi, J., Campbell, R. Mickunas, M.D.: Mobile Gaia: a middleware for ad-hoc pervasive computing. In: 2005 Second IEEE Consumer Communications and Networking Conference, CCNC (2005)
16. Grace, P., Blair, G.S., Samuel, S.C.: ReMMoC: a reflective middleware to support mobile client interoperability. In: Meersman, R., Schmidt, D.C. (eds.) CoopIS/DOA/ODBASE 2003. LNCS, vol. 2888, pp. 1170–1187. Springer, Heidelberg (2003)
17. Calafate, C.M.T., Manzoni, P.: PICA: A multi-platform programming interface for protocol development. In: Parallel, Distributed and Network-Based Processing, pp. 243–249 (2003)
18. Herrmann, K., Mühl, G., Jaeger, M.A.: MESHMdl event spaces - a coordination middleware for self-organizing applications in ad hoc networks. Pervasive Mobile Comput. **3**(4), 467–487 (2007)
19. Görgen, D., Frey, H., Lehnert, J.K., Sturm, P.: SELMA: A middleware platform for self-organizing distributed applications in mobile multihop ad-hoc networks. In: Western Simulation MultiConference WMC, vol. 4 (2004)
20. Kukliński, S., Radziszewski, P., Wytrębowicz, J.: WARF: component based platform for wireless mesh networks. SmartCR **1**, 125–138 (2011)
21. Ramdhany, R., Grace, P., Coulson, G., Hutchison, D.: MANETKit: supporting the dynamic deployment and reconfiguration of ad-hoc routing protocols. In: Bacon, J.M., Cooper, B.F. (eds.) Middleware 2009. LNCS, vol. 5896, pp. 1–20. Springer, Heidelberg (2009)
22. Blundo, C., De Cristofaro, E.: A bluetooth-based JXME infrastructure. In: Meersman, R., Tari, Z. (eds.) OTM 2007, Part I. LNCS, vol. 4803, pp. 667–682. Springer, Heidelberg (2007)
23. The Network Simulator NS-2. http://www.isi.edu/nsnam/ns/
24. Corson, S., Macker, J.: Mobile Ad hoc Networking (MANET): Routing Protocol Performance Issues and Evaluation Considerations. RFC 2501 (1999). http://tools.ietf.org/html/rfc2501

Attractiveness Study of Honeypots and Honeynets in Internet Threat Detection

Tomas Sochor[✉] and Matej Zuzcak

University of Ostrava, Ostrava, Czech Republic
tomas.sochor@osu.cz
http://www1.osu.cz/home/sochor/en/

Abstract. New threats from the Internet emerging every day need to be analyzed in order to prepare ways of protection against them. Various honeypots combined into honeynets are the most efficient tool how to lure, detect and analyze threats from the Internet. The paper presents recent results in honeynet made of Dionaea (emulating Windows services), Kippo (emulating Linux services) and Glastopf (emulating website services) honeypots. The most important result consists in the fact that the differentiation among honeypots according to their IP address is relatively rough (usually two categories, i.e. academic and commercial networks, are usually distinguished, but the type of services in commercial sites is taken into account, too). Comparisons of results to other similar honeynets confirms the validity of the paper main conclusions.

Keywords: Computer attack · Dionaea · Glastopf · Honeynet · Honeypot low-interaction · Internet threat · Kippo · SSH server emulation · Windows services emulation · Website emulation

1 Introduction

Recent research in honeypots [1] showed that honeypots, especially when combined into honeynets, could be extremely beneficial for Internet threat assessment. Recent results [2] showed that a honeynet composed of low-interaction honeypots (emulating Windows and Linux services) captures primarily automated threats. Among them older malware (e.g. Conficker) prevails. The honeynet used for obtaining the previously published results consisted of three Dionaea[1] honeypots (sensors) emulating Windows services and one sensors emulating Linux[2]. The cited paper also showed that more sensors connected into various networks should be used to confirm the results and extend them into more general scope. This is important for better understanding of attackers' practices. The primary aim is to find how attractive for attackers are sensors according to their connection to various types of networks – commercial, academic etc. Also the influence of the domain associated to a specific network and port (well-known vs. registered [3]), where the specific service is available, was analyzed.

---

[1] http://dionaea.carnivore.it.
[2] http://code.google.com/p/kippo.

© Springer International Publishing Switzerland 2015
P. Gaj et al. (Eds.): CN 2015, CCIS 522, pp. 69–81, 2015.
DOI: 10.1007/978-3-319-19419-6_7

2   Honeypot Classification

Honeypot is a system serving as a lure for potential attackers. Its aim is to attract an attack and record it subsequently [4]. Honeypots are classified according to the level of interaction with an attacker into two categories [5–7]. Low-interaction honeypots provide services for an attacker as emulated in an isolated environment (without other additional services available) while high-interaction honeypots offering the whole operating system to an attacker. The operating system used in high-interaction honeypots is usually virtualized and easily restorable. It must be well secured against its abuse (e.g. for participating in DDoS). This paper analyses the results obtained from the honeynet made of low-interaction honeypots Dionaea, Kippo, and Glastopf. The results were compared to the data obtained courtesy to the Institute of Informatics of the Silesian University of Technology.

Dionaea emulates specific services of Windows operating system (namely SMB, MySQL etc.). Kippo honeypot provides an environment for SSH shell emulation that is widely used for remote administration of Linux systems and therefore it is often attacked. Glastopf honeypot[3] represents a system emulating a dynamic website. Adding web-emulating honeypots is due to the increasing importance of web services not only in providing information for public use but also in implementation of enterprise systems where various new methods of remote invocation are used [8]. A honeynet is usually understood as a combination of more honeypots into a single logical entity [9]. Honeynet allows obtaining more data and preparing more complex view on captured attacks.

3   Topology of the Research Honeynet

The honeynet is composed of 4 servers (Fig. 1) serving as low-interaction honeypots with all sensors described above (Dionaea, Kippo and Glastopf) implemented. Kippo honeypot emulates SSH service at well-know port 22 (Kippo default), and in one case at registered port 2222. The latter port 2222 was used instead of the well-known port in order to find the decrease in incoming connections comparing to the well-known port. Dynamic ports were used in other studies, too [1]. Other non-standard situations could be studied using registered or dynamic ports, too. Dionaea honeypot emulates SMB, MSSQL, MySQL, FTP services at well-known ports. Glastopf honeypot uses ports 80 and 8080.

The first server was connected into the Czech academic network CESNET and is located in Ostrava. The IP address does not have any Internet domain assigned. The second server was connected into the Slovak academic network SANET. The IP address does not have any Internet domain assigned, neither. The third server uses a commercial private hosting in the Czech Republic. The used IP address belongs to the scope that is used by so-called "greyzone" of the Internet (e.g. adult contents). The IP address has an .eu domain assigned. The well-known port 22 was used in all three above servers. The fourth server

---

[3] http://glastopf.org/.

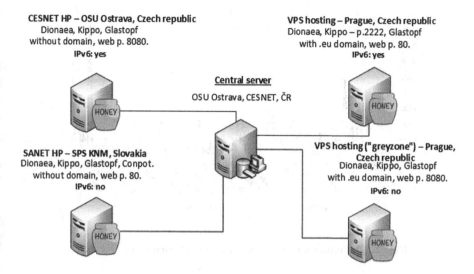

Fig. 1. Schematic layout of research honeynet

is also connected to a commercial VPS hosting in the Czech Republic but it is not available for adult contents unlike the third server. The registered port 2222 was used here. The IP address an .eu domain assigned, too. Both servers in commercial hostings are located in Prague, Czech Republic. All IP addresses belong to the scope assigned to the Czech Republic by RIPE. Only IP version 4 addresses are assigned to the honeypot servers.

4   Analysis of Attacks Against SSH Emulating Sensors

Detailed local statistics about individual Kippo sensors are presented here in order to allow the comparison of their attractiveness and evaluate attacks against individual sensors. Three Kippo sensors were the subject of the comparison, namely honeypots in academic networks CESNET (OSU) and SANET (SPS) and VPS hosting belonging to Internet greyzone (VPSgrey). The data were collected from July till November 2014. The data from another Kippo sensor were analyzed separately because the registered port 2222 was used there instead of well-known one, and its operation started later due to technical issues. Figure 2 shows the number of incoming connections to the sensor. All those connections could not be considered as attacks, however, because logging into the system did not happen in all cases [10].

Figure 2 clearly demonstrates that attackers interest is drawn the most to honeypots with IP addresses from commercial scopes. This observation could also be affected by the technique used for network traffic filtration that is under responsibility of individual network administrators and their agreed transit partners. An interesting observation consists in the fact that sudden ascents and

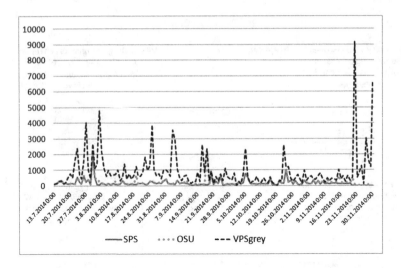

Fig. 2. The number of attacks tried against each of the servers

decays of numbers of connections usually reflect with the same trend on all sensors, however, with various scale. The recorded data led the authors to the hypothesis that just the filtering methods applied to specific networks and their transit partners had the most significant influence on attackers' attraction. This conclusion is supported by the data obtained from separate evaluation of IP addresses attacking against individual sensors. Another indicator for attackers is likely the IP address pertinence, i.e. its enlistment into certain autonomous system, scope of services provided, and its promotion. The difference in attackers' interest related to end stations is significant only in the case of the comparison between IP addresses belonging to academic networks (where the attraction for attackers is lower in the order of magnitude) and on the other hand the IP addresses commonly used by commercial subjects (where the attackers' interest is significantly higher).

The success ratio of connections (Fig. 3) clearly shows that significant part of attacks is fully automated because only around 300 out of thousands connection result in successful logging in, and this is the most favorable case for attackers, usually the success ratio is lower.

On the other hand, as Fig. 4 shows, once a successful logging in happens, certain activity identifiable as "human" usually occurs. Such activity represents an interaction of an attacker with the system (command or script execution, etc.). Therefore, an assumption could be made that robots that penetrate the system try to make use of the newly possessed system properly. Their activity consists the most frequently in downloading various, usually very simple, scripts. Those scripts either will allow an easier access to the system later, or try to execute similarly simple single-purpose malware. Sometimes also script-kiddies activities could be detected here.

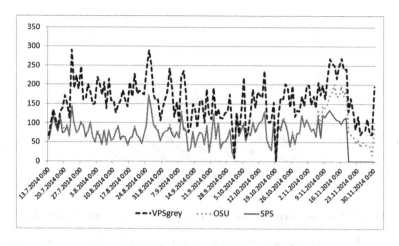

Fig. 3. Daily numbers of successful logins

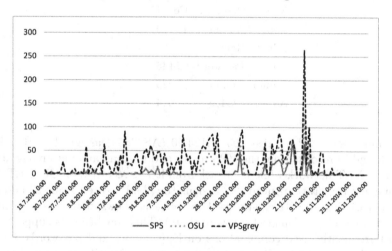

Fig. 4. Daily numbers of "human" activities in attacks per day

Obtained results showed that changing the well-known port 22 into another (registered) port represents very good way of protection. Despite the fact that the registered port 2222 used here is used for this purpose frequently, it is obvious that the daily numbers of incoming connections was very low. It did not exceed 500 connections per day (average was 12 connections) – Fig. 5. It should be noted that the IP address used for the sensor had a web domain associated in order to be more attractive. During the whole period of operation no one accepted connection, i.e. "human activity" was recorded. The most active IP address was 85.206.53.226 (AS8764) from Lithuania. This address occurs, however, only in connections to the VPS hosting. Therefore, it is highly likely that academic networks or their transit partners block this IP address. Similarly the

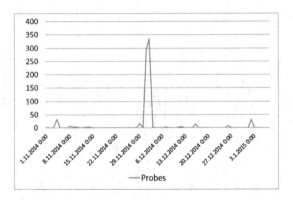

Fig. 5. Number of connection to Kippo sensor at registered port 2222

Table 1. TOP5 combinations of login names and passwords tried at Kippo honeypots

Name	Password	Count
Root	Admin	15 110
Admin	Root	3 506
Root	Password	2 145
Admin	Admin	1 925
Admin	123456	1 614

second most active IP address 103.41.124.50 (AS63854) assigned to Hongkong can be found in records from the VPS server and in small amount among attack against SANET network while no attack from it was recorded on CESNET. The WatchGuard ReputationAuthority[4] assigns the "bad" score to this IP address.

The statistics of SSH clients used for attacks shows that the majority of attacks is made by automated software (bots). This conclusion is supported by 10 most frequently used combinations on login names and passwords. On the other hand, also popular Putty SSH client used primarily by administrators (and human attackers, too) was recorded in relatively high number of attacks. This could indicate especially the activity of beginner attackers (so-called script-kiddies) that are unable to recognize the high level of emulation used by the sensor in short time after the connection to the system. The primary motivation of their activity is usually just verification of their capabilities. Most of such attacks are originated in Asia, especially China. Regarding the most frequently attacked combinations of login name and password in SSH shells (see Table 1) it is apparent that traditional root/admin combination prevails above all others. Comparing the results from smaller sample published last year [2] minor differences are observable. The overall results show that it is necessary to choose sufficiently strong password and avoid keeping default login setting on any machine that is connected to the Internet anyway.

---

[4] https://www.watchguard.com/products/reputation-authority.asp.

5 Analysis of Attacks Against Emulated Windows Services

Unlike previous studies, the primary focus was given here to honeypot attractiveness. We tried to find how attractive are Windows services emulating honeypots connected to various autonomous systems. Previous measurements confirmed that attackers are rarely attracted by honeypots at IP addresses from academic scopes. Therefore, a decision was made to add more sensors connected to commercial networks. In the past the single sensor with IP address from the scope assigned to a commercial VPS hosting offering among others also adult hosting ("greyzone") was used. Recently, another sensor connected to a commercial network (running on another VPS hosting not offering "adult hosting") was added. Figures 6 through 8 provide a comparison of results from all four sensors for the period from Oct till the end of 2014.

Figure 6 shows the number of connections to individual honeypot sensors. The significantly highest number of connections was recorded on the VPSgrey sensor connected to commercial network with greyzone services (incl. adult content hosting). It is apparent that in comparison with another commercial sensor (without greyzone services) the numbers of connections here were higher by the order of magnitude. While the average daily number of connection on VPSgrey was 11 497, the number on the other VPS hosting was only 347. However, significant fluctuations in number of connections are reflected in very high standard deviation values that are 14 091 and 1 132, respectively.

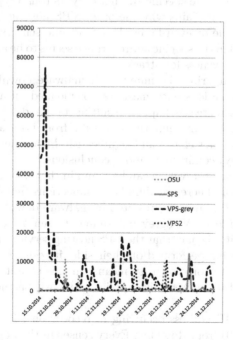

Fig. 6. Number of attacks where a malware was downloaded

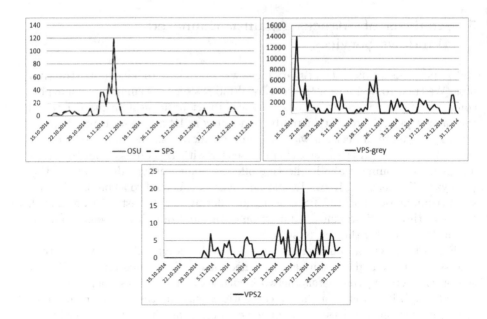

Fig. 7. Number of connections to all Dionaea sensors

This observation implies an assumption that attackers or botnets analyze the IP address pertinence to the specific AS, not only its rough type (e.g. academic/commercial) [11]. This explains why the other VPS hosting has the number of connections comparable or even lower than academic sensors. The information about services offered in the specific address range seems to be an important factor for the address attractiveness for attackers.

Figure 7 allow comparing the numbers of malware download. The average daily number of connections with malware downloaded on academic networks was 6.25 but sudden increases happen resulting in standard deviation as high as 16.4. Again, the apparent similarity of results from both academic locations (OSU and SPS in the first chart) and significantly higher numbers in the chart in the right (VPSgrey) confirm the above conclusions.

Another interesting observation can be found in the malware download activity at the VPS2 sensor. There are abundant connections (hundreds per day) but only a few of them (2 in average) resulted in malware download. This behavior made VPS2 different from VPSgrey but similar to sensors connected to academic networks. Despite the fact that the VPS hosting service provider guarantees that there is no filtering performed on their side, it seems to be apparent that this difference is due to some security means. In this case it is likely that such security means could be indirect, like applied in transit partner networks of the provider.

The most active IP address occurring in records on Dionaea honeynet was 78.97.199.63 (AS6830) from Romania. Every sensor in the honeynet has its own

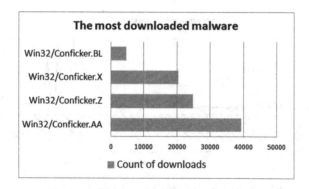

Fig. 8. The most frequently downloaded malware

Table 2. Top 10 most attacked ports

Port number	445	80	1 433	21	3 306	135	5 060	42
Number of connections	355 877	4 386	2 603	1 119	1 008	98	27	5

list of TOP10 attackers significantly differing from others', however. The attackers do not belong to the same AS and usually not even into the same country. For instance, the highest number of attack attempts against Dionaea honeypot in SANET academic network were recorded from the IP address 5.79.80.74 (AS16265) belonging to the Netherlands, that is quite paradoxical. It is rather an exception in statistics because usually the most active attackers' addresses belong to China, Romania, Bulgaria, Taiwan, Russia or Singapore.

The significant difference in numbers is observed both for number of connections and for number of malware downloads, too. Regarding the frequency of malware occurrence, the situation did not change significantly comparing results published in 2014. Win32/Conficker (Kiddo) worm in various mutations continue to dominate completely.

In addition to Conficker worm also other malware (e.g. Win32/Pepex.F and various variants of Win32/AutoRun.IRCBot, Win32/Agent.UOT, Win32/De-borm.NAB, Win32/TrojanDownloader.Tiny.NFP, Win32/Virut.NBP were recorded). The same as above (almost no change since last year) holds for the most frequently attacked ports. Here the port 445 utilized by SMB protocol dominates. Just SMB protocol is mostly used for Conficker worm dissemination. The detailed figures are shown in Table 2.

6 Attacks Against Emulated Web Services

Glastopf honeynet has been in operation since recently, namely from July 2014. The data presented here cover the period till November 2014. The trend of recorded attacks against emulated web applications is shown in the diagram in Fig. 9. About 100 interactions occurred every day with the exception of the

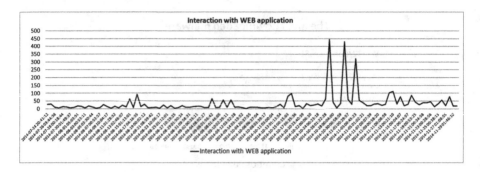

Fig. 9. Number of attacks on Glastopf sensors

period between Oct 26, and Nov 5. 2014 when the activity of attackers increased significantly. It is probably due to the increased interest of attackers to abuse the Shellshock vulnerability [12].

Most frequently attacked file types were SQL, LOG, and ASP but there are minor differences with other file types like REG, INI, DAT, CFG etc. probably due to relatively low figures (maximum 11 files of LOG type). Low figures are due to the fact that not each interaction results in uploading a file, such situations were relatively rare.

Several types of attacks were performed against web applications emulated by Glastopf honeypot. Most of these attacks had to be analyzed manually because the attack is detectable just based on the contents of HTTP GET and POST requests that can differ. This is the reason why Glastopf stores the requests in raw form into its database. The only automated part of the request analysis has been URL de-encoding (for URL encoding principle see RFC3986) yet.

The following types of attacks were detected so far:

- Attack against Cisco Linksys routers. The file named tmUnblock.cgi is vulnerable in certain cases facing Remote Command Execution (RCE); in this specific case The Moon attack[5] was recorded.
- Attack against Apache Struts 2 framework for Java applications.
- Shellshock [12] type attack.
- In general many attempts to execute a server shell and subsequent uploading of other malicious scripts or backdoors.

An example of an attack recorded on Glastopf honeypot containing RCE exploit – The Moon is shown below. The captured request tries to execute a script for downloading a file from attacker's server:

```
POST /tmUnblock.cgi HTTP/1.1

Content-Length: 1033
submit_button=&change_action=&action=&commit=&ttcp_num=2&
```

---

[5] http://packetstormsecurity.com/files/125252.

```
ttcp_size=2 &ttcp_ip=-h 'cd /tmp;echo"#!/bin/sh" > .893659fa.sh;
echo "wget -O .893659fa http://219.77.96.244:3200" >> .893659fa.sh;
echo "chmod +x .893659fa" >> .893659fa.sh;
echo "./.893659fa" >> .893659fa.sh;echo "rm .893659fa" >> .893659fa.sh;
chmod +x .893659fa.sh;./.893659fa.sh'&StartEPI=1
```

7   Comparison of Data from Different Honeynets

The authors' observations were compared to the results from cooperating institutions from Poland. First comparison was made with results from Silesian University of Technology (SUT) with data collected from Dec 2012 till Nov 2013. Only a single Dionaea sensor emulating Windows services was used, connected to the academic network SILWEB-AS-EDU. Table 3 shows that the numbers of attacks are similar to the authors' ones in academic networks (lines 1, 2 and 5 in the table). The most frequently attacked port in SUT was port 1433 (third in authors' honeynet) followed by port 445, the most frequently attacked in authors' honeynet. The most frequently downloaded malware was Conficker worm that conforms to authors' measurements.

Table 3. Attacks and malware download numbers compared to other honeypots

Dionaea honeypot	Connections per day avg./std. dev	Downloads per day avg./ std. dev.
CESNET honeypot (OSU)	840/1214	6/16
SANET honeypot (SPS)	774/1271	0.1/0.3
VPS-grey	11497/14091	2827/3333
VPS2	347/1132	1/2.4
SUT honeypot (SILWEB-AS-EDU)*	316/550	2/2

*different timeframe of taking measurements

Further comparison has been done with measurements from Polish Chapter of The Honeynet Project (with significant role of CERT-PL) for the period identical to author' measurements. Summarized comparison of Dionaea sensors is demonstrated in Table 4. It should be emphasized that sensors belonging to this honeynet are connected to various types of networks (academic, commercial etc.). Therefore, only 3 general indicators were compared, namely the frequently most attacked ports, the most frequently downloaded malware, and the most active (attacking) ASes.

Table 4 shows that the majority of attacks is attempted against port 445 on all Windows systems. An exception on a single academic honeypot on SUT does not matter because even here 445 was among the most frequent ports. The dominance of Conficker is apparent, too. On the other hand the AS with the most active attackers vary. This is likely due to differences in honeynets promotion

Table 4. Attacked ports and malware comparison across Dionaea honeynets

Honeypot	Top attacked ports	Top malware	Top attacker' AS and country
Authors' honeypots	445	Variants of	AS6830 – Romania
	80	Win32/Conficker	AS48161 – Romania
	1443		AS48475 – Russia
Silesian University of Technology	1433	Win32/Conficker	AS12741 – Poland
	445	Win32/Rbot	AS8508 – Poland
	80	Win32/AutoRun.IRCBot	
Polish Chapter Honeynet Project	445	Variants of	AS22822 – USA
	139	Win32/Conficker	AS8551 – Israel
	5060		AS24651 – Latvia

among potential attackers. As mentioned in previous sections, gaining attraction from attackers is one of crucial factors form honeypot success. Because of the fact that there is no united approach how to do so, ever honeynet administrator makes their measures. Therefore the differences in attacking IP addresses and ASes are ascribed to the differences in the specific honeypot promotion. What is a bit surprising, this is the occurrence of developed EU countries among the attacking AS.

On the other hand, the influence of previous use of honeypot IP addresses was not observed. Moreover, most of IP addresses used for the research honeypot were not used before at all.

Kippo results from the Polish Chapter of The Honeynet Project confirm our results, too. For the same period as in Sect. 4, the most active attacking AS was AS8708 – Romania. The most frequent passwords were only slightly different comparing to Table 1 (namely password, 1234 and toor). Regarding the ssh clients used for attacks the same library SSH-2.0-libssh prevailed.

8   Conclusions and Future Research

The results showed above may be concluded by the fact that attackers distinguish between sensors according to their IP addresses, namely whether the specific IP address belongs to a range interesting for them or not. Moreover it seems that attacker distinguish also services that are associated to specific address range, not just their pertinence to AS. Moreover it is highly likely that security measures used in the network play an important role in the attractiveness for attackers. For Glastopf the activity of attackers seems to respond to currently spreading threats. For Dionaea sensors (Windows services), the Conficker worm spreading continues to prevail. This indicates neglecting of updates by administrators not only in Asia but also in Europe where such behavior should have been rare.

Sensor attractiveness is an important topic for the future and it should be elaborated in details in the future. In addition to that, the future research will be focused on SCADA honeypots emulating industrial computers as well as to the implementation of highly sophisticated high-interaction honeypot with the aim of catching more elaborate (especially human-made) attacks could be captured.

Acknowledgment. The publication was supported by *Fuzzy modeling tools for adaptive search burdened with indeterminacy and system behavior prediction* project of the Student Grant Competition of the University of Ostrava. Thanks belong to the Center of Information Technologies of the University of Ostrava, and to Spojena skola in Kysucke Nove Mesto for providing the connection for research honeypots. Thanks are expressed to the Institute of Informatics of the Silesian University of Technology and Polish Chapter of The Honeynet Project and CERT-PL for providing the data, and The Honeynet Project, Czech Chapter, for consulting.

References

1. Kheirkhah, E., et al.: An experimental study of SSH attacks by using honeypot decoys. Indian J. Sci. Tech. **6**(12), 5567–5578 (2013)
2. Sochor, T., Zuzcak, M.: Study of internet threats and attack methods using honeypots and honeynets. In: Kwiecień, A., Gaj, P., Stera, P. (eds.) CN 2014. CCIS, vol. 431, pp. 118–127. Springer, Heidelberg (2014)
3. Reynolds, J., Postel, J.: Assigned numbers. IETF. RFC 1340 (1992). http://www.rfc-editor.org/rfc/rfc1340.txt
4. Spitzner, L.: Honeypots: Tracking Hackers. Addison-Wesley, Boston (2002)
5. Joshi, R.C., Sardana, A.: Honeypots: A New Paradigm to Information Security. Science Publishers (2011)
6. Grudziecki, T., et al.: Proactive detection of security incidents honeypots. In: ENISA (2012). https://www.enisa.europa.eu/activities/cert/support/proactive-detection/proactive-detection-of-security-incidents-II-honeypots/at_download/fullReport
7. Pisarcik, P., Sokol, P.: Framework for distributed virtual honeynets. In: Proceedings of the 7th International Conference on Security of Information and Networks, p. 324. ACM (2014)
8. Zacek, J., Hunka, F.: CEM: class executing modeling. Procedia Comput. Sci. **2011**, 1597–1601 (2011)
9. Sokol, P.: Legal issues of honeynet's generations. In: IWSSS 2014. Bucharest (2014)
10. Sokol, P., Zuzcak, M., Sochor, T.: Definition of attack in the context of low-level interaction server honeypots. In: Park, J.J.J.H., Stojmenovic, I., Jeong, H.Y., Yi, G. (eds.) Computer Science and Its Applications. LNEE, vol. 330, pp. 499–504. Springer, Heidelberg (2015)
11. Pomorova, O., Savenko, O., Lysenko, S., Kryshchuk, A., Nicheporuk, A.: A technique for detection of bots which are using polymorphic code. In: Kwiecień, A., Gaj, P., Stera, P. (eds.) CN 2014. CCIS, vol. 431, pp. 265–276. Springer, Heidelberg (2014)
12. Wheeler, D.A.: Shellshock (2015). http://www.dwheeler.com/essays/shellshock.html

User Trust Levels and Their Impact on System Security and Usability

Henryk Krawczyk[1] and Paweł Lubomski[2]([⊠])

[1] Faculty of Electronics, Telecommunications and Informatics,
Gdańsk University of Technology, Gdańsk, Poland
hkrawk@eti.pg.gda.pl
[2] IT Services Centre, Gdańsk University of Technology, Gdańsk, Poland
lubomski@pg.gda.pl

Abstract. A multilateral trust between a user and a system is considered. First of all we concentrate on user trust levels associated with the context-oriented CoRBAC model. Consequently, there were computed user profiles on the basis of its implementation in the information processing system "My GUT". Furthermore, analysis of these profiles and the impact of user trust levels on system security and usability have been discussed.

Keywords: Context · System security · CoRBAC model · Trust levels · User profile · System usability

1 Introduction

Let us consider two situations with respect to multilateral trust, i.e. from the system to the user and vice versa (see Fig. 1). From the system side the trust is estimated on the basis of the observed user behavior. Generally typical user behavior satisfies some rules, which mostly do not change over time. If the system detects a change, it is likely that somebody pretends to be a valid user. In such a situation the system trust to the user is reduced. In this case some extra actions must be taken by the system to verify the user. For example, the system may ask the user for some more detailed information or fire some stronger security mechanisms. This increases the chance that the system becomes more secure. However, frequent user verification takes more time and therefore the system usability and user satisfaction decreases. It is important to find some balance between the increase of system security, and system usability.

Similarly, let us consider the parallel situation when the user has some doubts concerning trust in the system's security. Then the user needs some tests to verify the system. These tests should be periodic and should be executed by an external auditor. In addition, the system must be reliable and produce the correct results of computations. High usability of the system also increases the user trust in the system.

In the paper we consider a system with three options of system security:

© Springer International Publishing Switzerland 2015
P. Gaj et al. (Eds.): CN 2015, CCIS 522, pp. 82–91, 2015.
DOI: 10.1007/978-3-319-19419-6_8

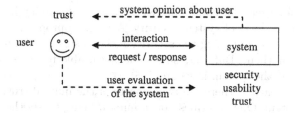

Fig. 1. The considered situation in the paper

1. a security scenario $secs_1$ with basic security mechanisms,
2. a security scenario $secs_2$ with strong security mechanisms,
3. a security scenario $secs_3$ with context-oriented security CoRBAC mechanisms [1].

In other words, we consider three variants of security scenario for the system, while functionality is the same. The first one (labeled $secs_1$) is a traditional security scenario with no context-oriented mechanisms. Let us assume there is only a basic security mechanism (e.g. password check on access). The second one (labeled $secs_2$) is the opposite case – it has strong security mechanisms, so that the security level of the system is much higher. As an example of these mechanisms, there could be a CAPTCHA mechanism [2], two-factor authentication or cryptographic signing, but the mechanisms are still not context-oriented. Let us assume that $secs_2$ has cryptographic signing, because it is the strongest security mechanism. The third one (labeled $secs_3$) is a system scenario with context-oriented security mechanisms implemented. These mechanisms could be similar to the second ones, but they are fired only in some situations, determined from the context.

We introduce the security levels of these variants of the system security – labeled respectively $secs_1$, $secs_2$, $secs_3$. According to [3] the following inequalities apply:

$$sec_1 < sec_3 < sec_2. \tag{1}$$

As an example we consider the IT system implemented in our university called "My GUT". Using the information about behaviour of over 46 thousand "My GUT" system users we introduced a four-level user trust model, and on the basis of that we tested both sides of the trust relationship. We assess that the increase of security in the third type of system does not decrease significantly the system usability. This property is the most important in the practical use of the system with CoRBAC mechanisms.

2   User and System Trust

It is a known fact that a human being is the weakest point in the security of a system [4,5]. On the other hand, the higher the level of user confidence in the system, the more secure the system is [6]. Pahnila, Siponen and Mahmood argue that sanctions for users not complying with the security policy have a insignificant impact on the level of system security. A more appropriate direction is to build user awareness, and using the security mechanisms which are possibly the least intrusive for them. But they should be noticeable to users so that users, could believe that the system is secure [7,8].

As mentioned earlier, there are two aspects of the mutual trust between the user and the system. One describes how a human being trusts the system as its user, the other one shows how the system trusts its user.

The first aspect of trust is how the system trusts the user. On the basis of the level of this trust, the system enforces weaker or stronger security mechanisms in CoRBAC security model [3]. The model assumes that the more trustworthy the user is the more she/he can do in the system. As a result, the level of system security is growing up, because only highly trusted users can interact with the system [9]. This fact also affects the user's opinion of the system security, and finally the user trust in the system.

The second aspect of trust is how the user trusts the system. A user opinion is based on the following aspects [10]:

- the functionality level of the system,
- the quality level of the system (meant as error prone level),
- the responsiveness of the system interface (especially during high load),
- the quality and reliability of the data being processed and stored in the system,
- human imagination about the system security [8].

There is one more parameter which we want to take into account for the considered system: usability (labeled, for introduced security scenarios, respectively $usab_1$, $usab_2$, $usab_3$). It can be defined as a human being's convenience in using a system. As is mentioned above, this parameter has also a big impact on the user trust in the system.

Implemented security mechanisms impact on usability in the following ways:

- the more, or the stronger security mechanisms involved, the lower the usability, as a user has to do some extra activities (e.g. typing a password multiple times),
- the more security mechanisms, the lower usability, because the responsiveness of the system decreases.

On the basis of these assumptions the following inequalities are true for the considered system:

$$usab_1 > usab_3 > usab_2. \tag{2}$$

Introducing the context-oriented CoRBAC model, and user trust levels, we claim that:

$$usab_1 \approx usab_3. \tag{3}$$

3   Context-Oriented Security Model

Users interact with the system through an interface located in an environment. This environment is described by many parameters which define the context of this activity [11–13]. The context may be considered as a certain period of time in which the user action is performed, a physical or logical localization of the user, a general state of the system, a relation between the user and the utilized data, a user interaction scenario, the kind of the device being used, etc. [14]. The ability to integrate contextual information makes the role-based security model flexible.

3.1   Context Definition

According to [1] context (C) can be defined as a set of different context parameters (CP):

$$c_i \in C = CP_1 \times CP_2 \times CP_3 \times \cdots \times CP_z. \tag{4}$$

Context parameter (CP_j) is a finite, discrete set of possible values of the parameter:

$$CP_j = \left\{ cp_{j1}, cp_{j2}, cp_{j3}, \ldots, cp_{jk_j} \right\}, \quad k_j = |CP_j|, \quad j = 1, 2, \ldots, z. \tag{5}$$

There are some assumptions:

1. The set of context parameter CP_j contains all possible values of this parameter.
2. Sets of context parameters CP_j can be of different size:

$$|CP_a| = k_a, \quad |CP_b| = k_b, \quad k_a = k_b \vee k_a \neq k_b. \tag{6}$$

3. The values of the context parameter (the elements of CP_j) are separable – in any given situation only one element of the set describes the context parameter:

$$\forall cp_{ja} \in CP_j, \quad \forall cp_{jb} \in CP_j, \quad cp_{ja} \neq cp_{jb}. \tag{7}$$

Some context parameters have a continuous character (e.g. time). During the context analysis they are clustered into certain groups, e.g. days of the week. The same situation applies to context parameters with discrete but numerous values, e.g. set of IP addresses. They are clustered into subsets of a specific netmask.

An example of the context parameters considered in this paper and their possible values is as follows:

- CP_1 – logical localization of user – a set of the following elements: $cp_{11} =$ internal network, $cp_{12} =$ campus network, $cp_{13} =$ external network (Internet),
- CP_2 – time of user's activity in the system – a set of elements: $cp_{21} =$ weekday, $cp_{22} =$ Saturday, $cp_{23} =$ Sunday.

Then $z = 2$, $|CP_1| = 3$, $|CP_2| = 3$, $|C| = 9$, e.g. $c_1 = (cp_{11}, cp_{21}) =$ (internal network, weekday).

3.2 Idea of CoRBAC Model

The CoRBAC model has been presented and discussed at an international congress of IT security experts and practitioners [1,3]. It is based on wide context acquisition and its analysis. To do that, we arbitrarily adopted 4 trust levels of the system towards the user: TL1, TL2, TL3, TL4, where the highest level is TL4 (user is the most trustworthy) and the lowest level is TL1 (user is the least trustworthy). The user trust level is determined on the basis of a user profile built on the history of the user interaction in the system, pointing to a current context value at each access request. This process is precisely explained in the next section. The initial trust level is TL4. If we want stronger security at the beginning of the user interaction in the system there should be assumed a lower initial trust level (e.g. TL1).

Figure 2 presents a trust level calculation mechanism of each user access request verification. Let us assume that for user u the current context accompanying the x-th access request is $c_x(u)$.

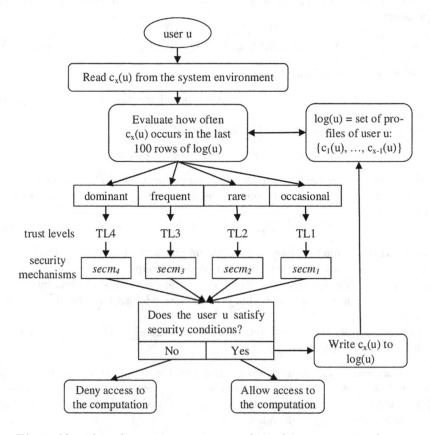

Fig. 2. Algorithm of security scenario $secs_3$ for x-th access request of user u

During the user interaction in the system, before each user access request in the system, the context is analyzed and the appropriate trust level is chosen. Next, the adequate security mechanism is fired.

The security mechanisms (*secm*) are some extra actions which the user is forced to do to prove that she/he is authorized. For TL4 it could be a password-based authentication. For TL3 it could be authentication based on both a password and CAPTCHA [2]. Two-factor authentication could be a security mechanism being forced on TL2. Computation on TL1 could be guarded by cryptographic signing of the user login request.

We can conclude that the security scenario $secs_1$ is similar to security scenario $secs_3$ working on trust level TL4. The second considered security scenario $secs_2$ with very strong mechanisms is similar to security scenario $secs_3$ working on trust level TL1. Security scenario $secs_3$ dynamically chooses the trust level and the corresponding security mechanism on the basis of a current context value $c_x(u)$ and user profile $\log(u)$. It is done for each user (u) and for each user access request separately.

4 User Profiles

We have analyzed the activity of over 46 000 users of "My GUT" system [15] within a nearly 4-year period. We took into account mainly security logs, whose first goal is to allow forensic analysis in case of any possible security incidents.

The analysis is based on the statistically unusual behavior of a user, taking into account context consisting of two context parameters. We have taken two assumptions – two limits of the frequency metric: 1 % and 5 % of the user activity. Below the first mentioned limit an activity is very suspicious, so the trust level is the lowest (TL1) and, according to the CoRBAC model, a user has to prove strongly that she/he is the right one. These mechanisms are the most inconvenient to the user. Between this and the next limit there is the second trust level (TL2). The second limit is assumed also arbitrarily, but we also consider one more limit – at the level of 10 %. We assume that activity performed more often than one out of ten activities is not suspicious enough. This way, we assume three limits of frequency metric (defined later), and on the basis of them, four trust levels (TL1 to TL4).

We have adopted one simplification: we have split user interaction into smaller scenarios corresponding to the user sessions from login to logout request. We have analyzed only login access requests.

The analysis consists of two aspects:

– how many users are under suspicion,
– how many times the trust level is reduced – how often the security mechanisms need to take action.

Each profile is computed in the following way:

1. For each user u, for each of their x-th access requests we assign a frequency metric describing the percentage of the previous access requests taking place

with given context value c_x (given type of localization and given type of day of week); we analyze only the 100 previous access requests of user u:

$$freq\left(access\_request_{u,x,c_x}\right) = \frac{\sum_{s=w}^{x-1} access\_request_{u,s,c_x}}{\sum_{s=w}^{x-1} access\_request_{u,s}} \cdot 100\,\%, \quad (8)$$

where:

$access\_request_{u,x,c_x}$ – x-th access request of user u, taking place with accompanying context value c_x,

$$w = \begin{cases} x - 100, & x > 100 \\ 1, & x \leq 100 \end{cases}.$$

2. If the frequency is lower than assumed limits, then the access request is marked as suspicious, and the trust level is reduced. In consequence this triggers stronger security mechanisms.

3. The first ten access requests of a user in the system do not change the trust level.

To clarify that, let us analyze the following example. Table 1 shows the log of access requests of one user pointing to their current context. There is assumed only one limit of mentioned-earlier frequency metric: "below $20\,\%$". This limit is assumed only for explanation of user profile building process. The profile is consequently updated after each positive access request.

Let us analyze a few rows of Table 1. In the first row there is the first access request in the system. There were no requests of the considered user earlier, so the request in context c_1 amounts to $0\,\%$ of previous requests. It exceeds the

Table 1. Example sequence of access requests for user u for x numbered from 1 to 13

x	c_x	Current frequency metric [%]	Below the 20 % limit
1	c_1	0	Yes (omitted)
2	c_1	100	No
3	c_1	100	No
4	c_2	0	Yes (omitted)
5	c_1	75	No
6	c_1	80	No
7	c_1	83.3	No
8	c_1	85.7	No
9	c_1	87.5	No
10	c_1	88.9	No
11	c_2	10	Yes
12	c_2	18.2	Yes
13	c_2	25	No

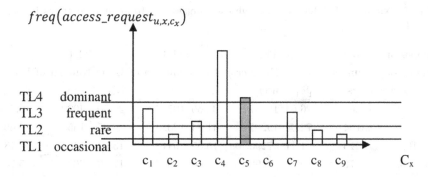

Fig. 3. Example user profile of x-th access request of user u, with accompanying current context $c_x = c_5$

"below 20 %" limit, but this value is omitted because of the rule that the first ten access requests do not change the trust level. Row 3 shows: access request in context c_1 in two access requests previous amounts to 100 % and this does not exceed the limit. In row 4 there is the first access request from the context c_2. The access request in context c_2 amounts to 0 % of three previous access requests. That exceeds the limit but it is also omitted because of the rule of the first ten. Row 11 shows the first interesting case. It is the second time that the context c_2 occured. Access request in this context value amounts to 10 % of ten previous access requests. That of course exceeds the limit and it is taken into account (triggers the stronger security mechanism). In row 13 the context c_2 occured for the fourth time. Access request in this context amounts to 25 % of twelve previous access requests. That does not exceed the limit.

Figure 3 presents a graphical representation of the profile of user u, on x-th access request, taken in context value c_5. This profile is computed later than the requests presented in Table 1.

We analyzed each user interaction log separately. For each user access request we built a user profile and assigned a corresponding trust level. Next we added the numbers of access requests taken in each trust level.

The result of this analysis is presented in Table 2. The numbers of requests belonging to intervals $(m, n]$ when users were classified from m to n times in trust level TL_i. For each trust level there is a number and percentage of users that were classified in this trust level.

To summarize this analysis, about 70 % of users were under suspicion. But it happened mostly from one to five times that the user was forced to take extra security actions, so these stronger security mechanisms could be slightly inconvenient to them. We have also analyzed the extreme cases (where required number of security actions amounted over 100). Those were the people who worked for a very long time in the administration section (which uses an internal network) and then changed their workplaces and positions, which affected the change of the network they used.

Table 2. Number of different users who were classified from m to n times in trust level TL_i

Number of requests of a single user	TL1 Number of users	Users [%]	TL2 Number of users	Users [%]	TL3 Number of users	Users [%]	TL4 Number of users	Users [%]
0	12265	26.64	15037	32.66	10727	23.30	4922	10.70
(0 – 5]	33164	72.04	19261	41.84	12085	26.25	1655	3.60
(5 – 10]	513	1.11	7296	15.85	6418	13.94	1463	3.18
(10 – 20]	72	0.16	3498	7.60	7824	17.00	2660	5.78
(20 – 30]	15	0.03	643	1.40	4111	8.93	2324	5.05
(30 – 40]	3	0.01	176	0.38	2207	4.79	2105	4.57
(40 – 50]	0	0	66	0.14	1155	2.51	1832	3.98
(50 – 100]	2	0.01	51	0.11	1376	2.99	7250	15.75
over 100	0	0	6	0.01	131	0.28	21823	47.41

5   Conclusions

The analysis shows that for most of the users (about 70 %) the forced stronger security mechanisms were almost invisible – max 10 times within a nearly 4-year period. So the usability can be considered as $usab_3 \approx usab_1$. However, it is true only when users do not change their behavior radically. The users who changed context very rarely were checked by stronger security mechanisms, which made them consider the system really secure (similar to [8]).

Context-oriented security scenario will not change the trust level when an attacker will make requests typical of user u (in the same values of context), which is (practically) unlikely.

We have also noticed that there was no significant difference in the analysis of 100 previous or all the previous to x-th access requests during the user profiles calculation.

In this paper we analyzed only two context parameters. There should be used a more complex context consisting of more context parameters. Then, user profiles will have a more uneven distribution of context values.

References

1. Krawczyk, H., Lubomski, P.: CoRBAC - kontekstowo zorientowany model bez-pieczeństwa. Stud. Informatica **34**(3), 185–194 (2013). (in polish)
2. Bursztein, E., Martin, M., Mitchell, J.: Text-based CAPTCHA strengths and weaknesses. In: Proceedings of the 18th ACM Conference on Computer and Communications Security - CCS 2011, vol. 2011, p. 125. ACM Press, New York (2011)
3. Lubomski, P.: Context in security of distributed e-service environments. In: Proceedings of the Chip to Cloud Security Forum 2014, Marseille, France (2014)

4. Stanton, J.M., Stam, K.R., Mastrangelo, P., Jolton, J.: Analysis of end user security behaviors. Comput. Secur. **24**(2), 124–133 (2005)
5. Balcerek, B., Frankowski, G., Kwiecień, A., Meyer, N., Nowak, M., Smutnicki, A.: Multilayered IT security requirements and measures for the complex protection of polish domain-specific grid infrastructure. In: Bubak, M., Kitowski, J., Wiatr, K. (eds.) PLGrid Plus. LNCS, vol. 8500, pp. 61–79. Springer, Heidelberg (2014)
6. Pahnila, S.P.S., Siponen, M.S.M., Mahmood, A.M.A.: Employees' behavior towards IS security policy compliance. In: 2007 40th Annual Hawaii International Conference on System Sciences (HICSS 2007) (2007)
7. Furnell, S.: Usability versus complexity – striking the balance in end-user security. Netw. Secur. **12**, 13–17 (2010)
8. Crawford, H., Renaud, K.: Understanding user perceptions of transparent authentication on a mobile device. J. Trust Manage. **1**, 1–7 (2014)
9. Dimmock, N., Belokosztolszki, A., Eyers, D., Bacon, J., Moody, K.: Using trust and risk in role-based access control policies. In: Proceedings of the Ninth ACM Symposium on Access Control Models and Technologies - SACMAT 2004, p. 156. ACM Press, New York (2004)
10. DeLone, W.H., McLean, E.R.: Information systems success: the quest for the dependent variable. Inf. Syst. Res. **3**(1), 60–95 (1992)
11. Cuppens, F., Cuppens-Boulahia, N.: Modeling contextual security policies. Int. J. Inf. Secur. **7**(4), 285–305 (2007)
12. Maamar, Z., Benslimane, D., Narendra, N.C.: What can context do for web services? Commun. ACM **49**(12), 98–103 (2006)
13. Mayrhofer, R., Schmidtke, H.R., Sigg, S.: Security and trust in context-aware applications. Pers. Ubiquit. Comput. **18**, 115–116 (2012)
14. Sliman, L., Biennier, F., Badr, Y.: A security policy framework for context-aware and user preferences in e-services. J. Syst. Architect. **55**, 275–288 (2009)
15. Gdańsk University of Technology: My GUT (2013). https://my.pg.gda.pl

Deploying Honeypots and Honeynets: Issues of Liability

Pavol Sokol[✉] and Maroš Andrejko

Faculty of Science, Institute of Computer Science,
Pavol Jozef Safarik University in Kosice, Jesenna 5, 040 01 Kosice, Slovakia
{pavol.sokol,maros.andrejko}@upjs.sk

Abstract. Honeypots and honeynets are common tools for network security and network forensics. The deployment and usage of these tools is affected by a number of the technical and legal issues. It is very important to consider both issues together. Therefore, paper outlines technical aspects of honeynet and discusses the liability of honeypot's and honeynet's administrator. Paper deals with civil and criminal liability. Also paper focuses on cybercrime and liability of attackers.

Keywords: Honeypot · Honeynet · Legal issue · EU law · Liability · Cybercrime

1 Introduction

Due to rapid growth of information and transfer messages, network security has become an increasingly important part of modern society. Traditional security tools, methods and techniques applied in protection are currently becoming increasingly ineffective. It is due to the fact that hackers' communities are several steps ahead of security mechanisms (firewalls, sandboxes etc.). Therefore it is necessary to collect and investigate as much information as possible about these communities.

From this point of view, honeypot seems to be very useful tool [1]. It can be defined as "a system that has been deployed on a network for the purpose of logging and studying attacks on itself" [2]. The most widespread classification is classification based on the level of interaction. There are low-level interaction and high-level interaction honeypots. On one hand, low-level interaction honeypots emulate the characteristics of network services or a particular operating system. On the other hand, a complete operating system with all services is used to get more accurate information about attacks and attackers. This type of honeypot is called high-level interaction honeypot.

Concept of honeypot is extended by a special kind of high-level interaction honeypot – honeynet. It is a highly controlled network of honeypots [3]. The honeynet can be also referred to as "a virtual environment, consisting of multiple honeypots, designed to deceive an intruder into thinking that he or she has located a network of computing devices of targeting value" [4]. The honeynet is composed of four core elements [2,4,5]:

© Springer International Publishing Switzerland 2015
P. Gaj et al. (Eds.): CN 2015, CCIS 522, pp. 92–101, 2015.
DOI: 10.1007/978-3-319-19419-6_9

- *Data control* – purpose of this element is to control the attackers' activities,
- *Data capture* – monitors and logs all of the attackers' activitiesthe,
- *Data collection* – purpose of this element is to capture and collect all information from multiple honeynets,
- *Data analysis* – purpose of this element is to analyse, understand and track the attacks and their activities.

1.1 Motivation

Deployment and usage of honeypots and honeynets (hereinafter honeypots) brings a lot of advantages [6], such as an easy deploying, low operating costs etc. On the other hand, there are a number of issues that need to be addressed in their deployment and usage. The most frequent problems are [6,7]:

- *limited view* – honeypots capture the only data collected from the interaction with the attackers,
- *inaccurate results* – in some cases, data obtained from the honeypots lead to poor result due to a limited amount of data and
- *detectability* – the attackers can detect the honeypots.

Abuse of honeypots belongs to the problems associated with their deployment and usage. They can be used for attack against servers or whole computer networks. The *risk of* honeypots' *abuse* is relatively high [8]. Therefore, the primary motivation for elaborating this paper is fact that administrator of honeypots or honeynets (hereinafter administrator) must *be aware of their liability*. *The liability* is referred as "a legal duty or obligation [9]. It represents one of the most significant words in the field of law and it can be defined as legal responsibility for one's acts or omissions". "Failure of a person to meet that responsibility leaves him open to a lawsuit for any resulting damages or a court order to perform" (as in a breach of contract or violation of statute) [10].

1.2 Presented Contributions and Organization of the Paper

There are several scientific papers dealing with the legal aspects of honeypots. Paper discusses these papers in more details in the following section. Unfortunately, these *papers deal* with liability in context of honeypots *only marginally*. Therefore, primary contribution of this paper is discussion of liability in context of honeypots. Paper also focuses on civil and criminal liability of administrator and attacker. Based on this analysis the paper outlines data control, which meets the legal requirements.

Paper discusses the European Union (EU) regulations, EU directives and international agreements. National legislation of member states of EU are based on these legal documents (EU directives, international agreements) or legal documents are an integral part of national legislation (EU regulations, international agreements). Therefore, some native legislation may be slightly different from the concept of EU law or international law.

This paper is organized as follows: In Sect. 2 paper outlines the review of literature in field of legal aspects of honeypots. Section 3 focuses on civil liability in field of honeypots including liability for omissions to protect others, liability for spurious content and liability in testbeds. In Sect. 4 paper provides an overview of criminal liability in field of honeypots including issue of elements of criminal offence and issue of offender of criminal offence. Section 5 outlines proposed data control taking into account the issue of liability. The last section contains conclusions and author's opinion on the future research.

2   Review of Literature

There are few papers that are dealing with legal aspects of honeypots. Most of these papers focus on *U.S law perspective*. These authors jointly agree that if honeypots are deployed in the United States, three legal issues are considered (entrapment, privacy, and liability).

Salgado [2] outlines the legal framework of honeypots' usage. If a honeypot is used, three legal issues are taken into consideration. Firstly, it is recommended to take into account the laws that restrict monitoring users' activities. Secondly, attackers may misuse the honeypots to harm others. Thirdly, if the honeypot is used for prosecuting and catching attackers, it is possible that attackers (defendants) might argue that the undercover computer entrapped them. Salgado extended his analysis in [3].

Scottberg [11] considers the honeypot an entrapment and outlines the privacy issues of the attackers' files. According to his opinion these files are not protected. He also deals with liability issues. According to him, honeypot might be used to attack to other systems and administrator may be liable for lacking in due diligence of corporate assets.

Mokube [12] focuses on the aspects of deployment and usage of honeypots in the United States. One of these aspects are legal issues. According to him, attackers may argue entrapment. In the privacy issue the laws might restrict right to monitor users on system. The attackers may misuse honeypot to harm others. Therefore, they may be liable.

Important analysis of the legal aspects of honeypots from the U.S. perspective is in [2]. *Spitzner* discusses the same legal issues as in previous papers. According to him, honeypots are not a form of entrapment. In issues of privacy he distinguishes two types of information that is being collected: transactional and content. He also thinks that administrator may be sued if the honeypot is used to harm others. *Rubin* [13] focuses on legal and ethical aspects researchers' and professionals' security methods in field of honeypots.

There are a few papers that would at least outline the legal issues of the honeypots from *the perspective of EU law*. Example of this legal analysis is in [14]. Dornseif et al. focus on vulnerability assessment, using honeypots. They outline criminal and civil liability concerning German laws, but they discuss these issues only marginally.

3 Civil Liability

Civil liability is based on the fact that a person to whom damage to another is legally attributed is liable to compensate that damage. Central term in civil liability is term *damage*. It can be defined as "a material or immaterial harm to a legally protected interest" [15]. In the case that the damage is not caused there is no civil liability. Interesting problem in this issue is the issue of damage in honeypot. Can the attacker directly cause damage by a honeypot? The primary role of the honeypots is to attract attackers and to be threatened by the attacks. It may occur damage in honeypot during an attack. According to author's opinion the administrator is aware of the risk of damage. He sets limits for attacker taking into account the corresponding risk of damage. In following text if paper discusses the damage it means damage of the no-honeypot's system. This paper focuses on the following aspects of civil liability:

- liability for omissions to protect others,
- liability for spurious content and
- liability in testbeds.

3.1 Liability for Omissions to Protect Others

In perspective of honeypots the *liability for omissions to protect others* is important issue. A duty to act positively to protect others from damage may exist [15] if:

- law so provides, or
- the actor creates or controls a dangerous situation, or
- there is a special relationship between parties or
- the seriousness of the harm on the one side and the ease of avoiding the damage on the other side point towards such a duty.

Since the first two cases are relevant for administrator, paper focuses on them in more details in the following text. There are a several examples of duty to act positively to protect others from damage may exist if *law so provides*. The first example is *Article 13a of EU Directive 2009/140/EC* that outlines the security and integrity of networks and services. According this article the administrators take appropriate technical and organisational measures to appropriately manage the risks posed to security of networks and services. Having regard to the state of the art, these measures shall ensure a level of security appropriate to the risk presented. In particular, measures shall be taken to prevent and minimise the impact of security incidents on users and interconnected networks.

In [16] authors outlines technical guidance on the security measures in Article 13a of EU Directive 2009/140/EC. According to this guide the minimum security measures have been derived from a set of international standards (e.g. ISO 2700x standards) and national standards. Administrator should establish and maintain an appropriate information security policy for his honeypot, especially for data control. Also administrator should ensure his sufficient security knowledge and regular security training in field of honeypots. According to author's opinion the critical security measure is incident detection and incident response capabilities. These capabilities should be fast as possible.

The second example is *Article 4(2)* of EU Directive 2002/58/EC on privacy and electronic communications (*e-privacy EU Directive*). In case of a particular risk of a breach of network security, the administrator must inform the production network's users concerning such a risk and, where the risk lies outside the scope of the measures to be taken by the administrator, of any possible remedies, including an indication of the likely costs involved.

The second case is duty to act positively to protect others from damage may exist if the *actor creates or controls a dangerous situation* [15]. It is most relevant case for administrator. The honeypot's deployment presents the dangerous situation for the production network and for the no-honeypot's systems and administrator controls this situation. Based on this the administrator has a duty to act positively to protect others from damage.

Liability for omissions to protect others is closely connected to administrator's care to *ensure the security of the honeypots*. This care consists of the reactive care and proactive care [17]. In *reactive care* the duty of administrator to respond to the statement is a consequence of the obligation to intervene in a particular case. *Particular case* is the situation, when attacker takes over the server and attacks to no-honeypot's systems. In this case administrator has duty to respond to this attack (e.g. attack redirecting, connections blocking).

On the other hand, *proactive care* is ensuring certain security of honeypot's usage for the future. It this case administrator has duty to create security measures to prevent successful attacks. Important question is what measures is required to create. In author's opinion, the administrator should minimum measures against already carried out attacks. The maximum limit of proactive care is found in Article 15 of EU Directive on electronic commerce. Although this provision is relating to information society services, it can be used to determine *maximum limit of proactive care* in EU law. Within the meaning of Article 15 of EU Directive on electronic commerce, no general obligation to monitor the information which is transmitted or stored, nor a general obligation actively to seek facts or circumstances indicating illegal activity.

This implies the boundary between what can be considered as possible in proactive care and what not. For example of the crossing the boundary it can be considered the zero-day vulnerabilities, which represent the attacks or threats is attacks that exploit a previously unknown vulnerability in a computer applications. Other example of the crossing the boundary includes specific vulnerabilities, which are not known honeynet's community (e.g. honeynet.org).

At the last the liability for omissions to protect others is closely connected to *culpable breach of administrator's duty*. In a number of national legal orders, fault is presumed. The possible forms of fault are direct intention, indirect intention, conscious negligence and unconscious negligence. In the case that administrator do not respond to the attack and his omission reinforce the damage, he is usually solidary liable for the entire implementation of harm.

3.2   Spurious Content

In certain situations, the civil liability issue is linked to liability for *spurious content*. There are situations where there is a general liability and obligation of the owner of such a member's account to keep its data under lock that no one of them acquired knowledge [18]. Example of this liability is deployment and usage of spam honeynet, which consists of several email servers. Administrator provides email accounts the attacker. The aim is to detect attacker's methods, attacks goals etc. If administrator poorly set up or use the data control (e.g. allow an attacker to send e-mail – spam), he is liable for spurious content.

3.3   Testbeds

Interesting issue is situation when honeypot can be used as a testbed (e.g. cuckoo sandbox [19]). *Testbed* can be defined as "controlled environment to analyse vulnerabilities in new applications, operating systems, or security mechanisms" [4]. In case that administrator uses honeypot as testbed he should be liable for infiltrated code based on civil law. One potential tort-based theory is negligence, which is the doctrine that courts apply to compensate injured parties after accidents. In contrast to negligence, which requires proof that a defendant failed to take precautions appropriate to prevent harm (discounted by the probability of harm), strict liability does not involve any notion of fault: if strict liability applies to an activity (a big if) and an accident occurs, the person conducting the activity is liable for harm to others [20].

4   Criminal Liability

Criminal liability is linked to the concept of cybercrime. The *cybercrime* can be defined as "a crime that utilizes a computer network during the committing of crimes such as online fraud, online money laundering, identity theft, and criminal uses of Internet communication" [21].

The Council of Europe's Cybercrime Treaty (Convention on Cybercrime) uses the term cybercrime to refer to offences ranging from criminal activity against data to content and copyright infringement [22]. The national cybercriminal law of a many countries is based on Convention on Cybercrime. It follows that attack from honeypot's to non-honeypot's system placed in country has acceded to the Convention on Cybercrime, will be considered an *criminal offense*. On the other hand, in countries that did not join this convention, these criminal offenses may have or may have not regulated in their legal orders. In this case, it is not possible to state that countries which did not join the Convention on Cybercrime do not have such an acts considered to be criminal offenses. The examples are Greece and Ireland, which haven't ratified the Convention on Cybercrime, but it considers these actions as offenses.

In analysis of criminal liability, it is necessary to distinguish between several kinds of criminal acts – attacks targeted at honeypot and attacks from honeypot.

Attacks targeted at honeypot – it means illegal activity, which is malicious and directed against honeypot. In this case the attackers attack against honeypot to order to:

- access to the whole or any part of honeypot without right (i.e. without authorization),
- damaging, deletion, deterioration, alteration or suppression of any data in honeypot without right,
- the serious hindering without right of the functioning of honeypot by inputting, transmitting, damaging, deleting, deteriorating, altering or suppressing any data in honeypot.

Attacks from honeypot – if honeypot is penetrated or invaded, it can serve for further criminal activity. It includes also establishing honeypot, which is created intentionally to pursue further criminal activity. The attackers attack from honeypots to no-honeypot's systems to harm them. Administrator is liable in case that he intentionally makes available of honeypot to attacker (misuse of devices). Examples of these attacker's activities are offenses, such as computer-related offences (e.g. computer-related forgery, computer-related fraud) and content-related offences (e.g. offences related to child pornography).

New EU Directive *2013/40/EU on attacks against information systems* (hereinafter EU Directive on attacks) will replace Council Framework Decision 2005/222/JHA. This Directive establishes minimum rules concerning the definition of criminal offences and sanctions in the area of attacks against information systems. From the perspective of criminal liability, it is necessary to adress issue of elements of criminal offence and issue of offender of criminal offence.

4.1   Elements of Criminal Offence

The first aspect is what action constitutes the above mentioned offences (*elements of offense*). Important issue is a question, whether the attack to honeypot can be considered as criminal offence. According to author, the attack can be considered as criminal offence within the meaning of EU Directive on attacks. This attack satisfies the elements of some criminal offense, especially illegal access to information systems, illegal system interference and illegal data interference. Attack to no-honeypot's system from honeypot can satisfy the elements of any criminal offense referred in EU directive against attacks. Equally important issue is whether there is the criminal offense if *the attack fails*. The answer to this issue is found in Article 8 of EU Directive on attacks, under which the attempt to commit illegal system interference and illegal data interference is punishable as a criminal offence.

4.2   Offender of Criminal Offence

The second aspect is who can be considered as offender of above mentioned offences (*offender of criminal offence*). From the perspective of honeypot the

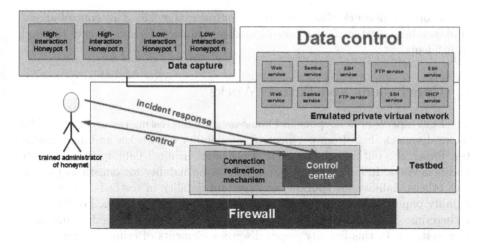

Fig. 1. Scheme of data control

attacker and the administrator can be considered as offender. Attacker can be statue of the offender in two cases, such as attack to honeypot itself and attack to no-honeypot's systems. Criminal liability of administrator is more complex. Administrator could be liable in some cases. The first case is that he uses his honeypot to attack no-honeypot's systems (like attacker). The second case is to meet expectations set out in Article 8 of the EU Directive on attacks. It would be particularly the case for aid an attacker by the administrator. The definition and substance of aid will be determined by the laws of the member states of EU.

5   Proposed Data Control Based on Issue of Liability

The core element of honeynet, which is closely linked to liability, is data control. For example, if attacker takes over the honeypot, he may attempt to launch web exploitation against a no-honeypot's system. In this case, the data control has limitations. After several successful attempts all further activity including any exploits would be blocked. In such case the attack is carried out. Therefore, concept of data control is an essential issue. As we shown in previous sections, adequate level of security is required by EU law.

In this section the data control meeting the requirements of EU law is proposed (Fig. 1). This honeypot consists of five parts, which enhance security in honeynet. Basic part is *firewall* with restrict white list (minimal opened network ports). It is standard part of data control in honeynet. Second part is a *dynamic connection redirection mechanism* [23]. This mechanism redirects connections from/to data capture to *testbed* (e.g. cuckoo sandbox [19]). If there is a trusted connection, it is allowed to outside of the honeynet. Otherwise it is redirected to testbed or to *emulated private virtual network* (e.g. honeyd [24]). The reason for using this private network is the fact that the system restricts attackers. Honeynet collects data mainly from the services (e.g. web service, ftp service) of this private network. The last part is *control center* [25]. This control allows trained administrator of honeynet to monitor connections and quickly respond to incidents.

6   Conclusions and Future Works

The legal aspects of honeypots and honeynets are interesting research topic. The paper discusses them from perspective the liability of attacker and administrator. Paper also outlines the concept of civil and criminal liability in honeypots and honeynets. In civil liability paper focuses on liability for omissions to protect others, liability for spurious content and liability in testbeds. In criminal liability paper focuses on criminal offences considering The Council of Europe's Cybercrime Treaty and EU Directive 2013/40/EU on attacks against information systems. In this liability paper discusses elements of criminal offence and offender of criminal offence. Paper also outlines proposed data control taking into account the issue of liability. In the future research, author will focus primarily on related problems, such as privacy, jurisdiction and applicable law and legal aspects of distributed honeynets.

Acknowledgment. Paper is the result of the Project implementation: University Science Park TECHNICOM for Innovation Applications Supported by Knowledge Technology, ITMS: 26220220182, supported by the Research & Development Operational Programme funded by the ERDF. This paper is partially funded by the VVGS grant under contract No. VVGS-PF-2015-472.

References

1. Sochor, T., Zuzcak, M.: Study of internet threats and attack methods using honeypots and honeynets. In: Kwiecień, A., Gaj, P., Stera, P. (eds.) CN 2014. CCIS, vol. 431, pp. 118–127. Springer, Heidelberg (2014)
2. Spitzner, L.: The honeynet project: trapping the hackers. In: IEEE Security and Privacy Conference, 2004, pp. 15–23. IEEE (2004)
3. The Honeynet project: Know Your Enemy: Learning about Security Threats, 2nd edn. Addison-Wesley, Boston (2004)
4. Abbasi, F., Harris, R.: Experiences with a generation III virtual honeynet. In: Telecommunication Networks and Applications Conference (ATNAC) 2009, pp. 1–6. IEEE, Australasian (2009)
5. Balas, E., Viecco, C.: Towards a third generation data capture architecture for honeynets. In: Proceedings from the Sixth Annual IEEE SMC Information Assurance Workshop, IAW 2005, pp. 21–28. IEEE (2005)
6. Mairh, A., Barik, D., Verma, K., Jena, D.: Honeypot in network security: a survey. In: Proceedings of the 2011 International Conference on Communication, Computing and Security (ICCCS 2011), pp. 600–605. ACM (2011)
7. Vrable, M., et al.: Scalability, fidelity, and containment in the potemkin virtual honeyfarm. ACM SIGOPS Operating Syst. Rev. **39**(5), 148–162 (2005)

8. Husak, M., Vizvary, M.: POSTER: reflected attacks abusing honeypots. In: Proceedings of the 2013 ACM SIGSAC Conference on Computer and Communications Security, pp. 1449–1452. ACM (2013)

9. Martin, E.A., Law, J. (eds.): A Dictionary of Law. Oxford University Press, Oxford (1997)

10. Black, H.C., Garner, B.A.: Black's Law Dictionary. West Publishing Company, USA (1999)

11. Scottberg, B., Yurcik, W., Doss, D.: Internet honeypots: protection or entrapment? In: International Symposium on Technology and Society, (ISTAS 2002), pp. 387–391. IEEE (2002)

12. Mokube, I.: Honeynets - concepts, approaches and challenges. In: Proceedings of the 45th Annual Southeast Regional Conference, pp. 321–326. ACM (2007)

13. Rubin, B., Cheung, D.: Computer security education and research: handle with care. In: Security and Privacy, pp. 56–59. IEEE (2006)

14. Dornseif, M., Gärtner, F.C., Holz, T.: Vulnerability Assessment using Honepots. K.G. Saur Verlag, München (2004)

15. Koch, B.A.: The principles of european tort law. ERA Forum 8, 107–124 (2007). Springer-Verlag

16. Enisa: Technical Guideline for Minimum Security Measures. Guidance on the security measures in Article 13a (2011)

17. Haapio, H.: Introduction to proactive law: a business lawyer's view. Stockholm Institute for Scandinavian Law 1999 (2010)

18. Decision BGH, Halzband, sp. zn. I ZR 114/06 (2006)

19. Cuckoo Sandbox project (2015). http://www.cuckoosandbox.org/. Accessed 28th January 2015

20. Burstein, A.J., Aaron, J.: Conducting cybersecurity research legally and ethically. In: LEET, vol. 8, pp. 1–8 (2008)

21. Kshetri, N.: The simple economics of cybercrimes. IEEE Secur. Priv. 4(1), 33–39 (2006)

22. Krone, T.: High Tech Crime Brief. Australian Institute of Criminology, Canberra (2005). ISSN 1832-3413

23. Alata, E., et al.: Internet attacks monitoring with dynamic connection redirection mechanisms. J. Comput. Virol. 4(2), 127–136 (2008)

24. Honeyd project (2015). http://www.honeyd.org/. Accessed 20th Febrary 2015

25. Pisarcik, P., Sokol, P.: Framework for distributed virtual honeynets. In: Proceedings of the 7th International Conference on Security of Information and Networks, p. 324. ACM (2014)

On the Balancing Security Against Performance in Database Systems

Damian Rusinek[1], Bogdan Ksiezopolski[1,2(✉)], and Adam Wierzbicki[2]

[1] Institute of Computer Science, Maria Curie-Sklodowska University, Lublin, Poland
[2] Polish-Japanese Academy of Information Technology, Warsaw, Poland
bogdan.ksiezopolski@acm.org

Abstract. Balancing security against performance for IT systems is one of the most important issues to be solved. The quality of protection of systems can be achieved on different levels. One can choose factors which have a different impact on the overall system security. Traditionally, security engineers configure IT systems with the strongest possible security mechanisms. Unfortunately, the strongest protection can lead to unreasoned increase of the system load and finally influence system availability. In such a situation the quality of protection models which scale the protection level depending on the specific requirements can be used. In the article, we present the approach which enables balancing security against performance for database systems. The analysis is performed by Automated Quality of Protection Analysis (AQoPA) tool which allow automatic evaluation of system models which are created in the Quality of Protection Modelling Language (QoP-ML).

Keywords: Quality of protection · Security economics · Modelling and protocol design · Cryptographic protocols · Security protocol analysis · Network security

1 Introduction

The simultaneous analysis of the security and its influence on the system performance in IT systems is one of the most important topics to be solved. On one hand, the traditional approach assumes that implementation of the strongest security mechanisms makes the system as secure as possible. Unfortunately, on the other hand, such reasoning can lead to the overestimation of security measures which causes an unreasonable increase in the system load [1–4]. The system performance is especially important in the systems with limited resources such as the wireless system networks or the mobile devices [5,6]. Another example where such analysis should be considered is cloud architecture. The latest research indicates three main barriers for using cloud computing which are security, performance and availability [7]. Unfortunately, when the strongest security mechanisms are used, then it will decrease system performance and further influence system availability. This tendency is particularly noticeable in complex

© Springer International Publishing Switzerland 2015
P. Gaj et al. (Eds.): CN 2015, CCIS 522, pp. 102–116, 2015.
DOI: 10.1007/978-3-319-19419-6_10

and distributed systems. Among other applications where balancing security against performance are most demanding are those which are using Peer-to-Peer (P2P) protocols, due to their decentralized and distributed nature (for a good overview, see [8]). Another example are application-layer multicast protocols that are a practical solution for streaming media delivery in the absence of network-layer support [9]. These practical applications are vulnerable to man-in-the-middle attacks, require specialized authentication and access control [10], and have strong requirements for fairness [11]. The latest results show [12,13] that in many cases the better way is to determine the required level of protection and adjust security measures to these security requirements [14]. Such approach is achieved by means of the Quality of Protection models where the security measures are evaluated according to their influence on the system security.

2 Related Work

In the literature the Quality of Protection (QoP) models were created for different purposes and have different features and limitations. Below the related research in this area is presented.

S. Lindskog and E. Jonsson attempted to extend the security layers in a few Quality of Service (QoS) architectures [15]. Unfortunately, the descriptions of the methods are limited to the confidentiality of the data and based on different configurations of the cryptographic modules. C.S. Ong et al. in [16] present the QoP mechanisms, which define security levels depending on security parameters. These parameters are as follows: key length, block length and contents of an encrypted block of data. The article [17] presents an adaptable protocol concentrating on the authentication. By means of this protocol, one can change the version of the authentication protocol which finally changes the parameters of the asymmetric and symmetric ciphers. The article [18] introduces mechanisms for adaptable security which can be used for all security services. In this model the quality of protection depends on the risk level of the analysed processes. A. Luo et al. present the quality of protection analysis for the IP multimedia systems (IMS) in [19]. This approach presents the IMS performance evaluation using Queuing Networks and Stochastic Petri Nets. E. LeMay et al. create in [20] the adversary-driven, state-based system security evaluation, the method which quantitatively evaluates the strength of system's security. In the article [21] authors present the performance analysis of security aspects in the UML models. This approach takes as an input a UML model of the system designed by the UMLsec extension [22] of the UML modelling language. This UML model is annotated with the standard UML Profile for schedulability, performance and time and then analysed for performance.

In the article [23] the Quality of Protection Modelling Language (QoP-ML) is introduced. It provides the modelling language for making abstraction of cryptographic protocols that put emphasis on the details concerning quality of protection. The intended use of QoP-ML is to represent the series of steps which are described as a cryptographic protocol. As mentioned above, during the QoP analysis one can not consider only primary cryptographic operations or basic communication steps. This analysis must be multilevel. The QoP-ML introduces the multilevel protocol analysis that extends the possibility of describing the state of the cryptographic protocol. Every single operation defined by the QoP-ML is described by the security metrics [24] which evaluate the impact of this operation on the overall system security [25].

The article presents an analysis of performance versus security in data storage system. First, the QoP model of system which realize the service at different quality of protection level was created. Then, performance analysis has been conducted for various scenarios and the results were verified experimentally. The model, which was created in QoP-ML, is the first model that presents the use of language QoP-ML and analysis by AQoPA (Automated Quality of Protection Analysis) tool of the IT system using database systems.

In presented case study performance refers to the time behaviour from performance efficiency category and to the availability from reliability category from ISO/IEC 25010 [26]. The metrics for performance are the response times and throughput rate which must meet the availability requirements (e.g. response time under 500 ms, ability to handle 100 sessions per hour).

The quality of protection (QoP) is understood as the set of security properties (security category in ISO/IEC 25010) ensured by the system. Its quality is measures as the complexity of performing a successful attack on the cryptographic mechanisms ensuring selected properties. The complexity can be measured as the number of brute force checks in order to find the key (assuming that cryptographic primitives are ideal) or as the complexity of the best known attack on particular primitive (more practical approach). The level of protection can be different for the same set of properties depending on the cryptography mechanisms used and its parameters (e.g. algorithm: AES, 3DES, key size: 128, 256, mode: ECB, CBC).

3   Modelling of Storage Center in QoP-ML

Detailed and precise examination of different parameters and comprehensive analysis of IT systems is a challenging task. To solve a problem, it should be formulated by means of standardized methodologies and all relevant information should be collected. As the decision making action is a reproducible process, a more comprehensive view should be taken into account. The decision cycle includes four phases: problem definition (a decision situation dealing with difficulty or with an opportunity), model construction (for description of the real-world problem using specialized tools), identification and evaluation (identification and evaluation of possible solutions) and recommendation and implementation (examination, comparison and choice of potential solutions).

Different solutions of a given problem can be considered and evaluated by means of QoP-ML approach and AQoPA evaluation. Such decision support system enables defining distinct scenarios and examination the obtained results. The scenarios also referred to as versions are possible sets of input parameters and characteristics of the modelled environment which allow analysis of consequences of decision alternatives with past experience in a context being described. The decision making cycle is useful in proper security management. Properly designed, specialized, dedicated support system allows defining and modelling various scenarios differing in utilized security mechanisms and assessing its performance.

The modelling of IT systems in the Quality of Protection Modelling Language consists of four successive stages presented on Fig. 1: stage 1 – model creation, stage 2 – security metrics definition, stage 3 – scenarios definition and stage 4 – simulation. These stages refer to the methodology of creating QoP-ML models defined in the article [23] where the details about syntax and semantics can be found.

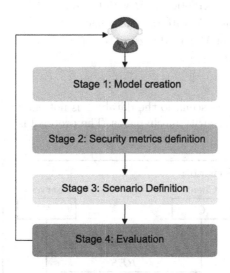

Fig. 1. The four stages of QoP-ML modelling

In this section we present the QoP modelling and AQoPA analysis for storage center as the example of IT system. We analyse different configurations and scenarios of protocol retrieving and searching information from the database. One can imagine that the database stores the details of company customers (corporate network) while employees use devices (public network) with thin clients to obtain information about particular customers. In the analysis process we prepare different scenarios for searching and retrieving data from the database. The differences in scenarios come from different configurations of security mechanisms and assuring different sets of security attributes. We also examine the server load, depending on the number of concurrent client connections. Each scenario is tested for 100 simultaneous clients.

3.1 Scenarios

In the description of scenarios (Figs. 2, 3, 4, 5) we use the following notation:

ID_A – identification of site A,

Q – search query,

RES – search results,

$REQ(K_Q)$ – request for the key of the part of database selected
           by main query Q,

K_Q – the key that particular part (selected by query) of
           database is encrypted with,

$K_{S,SKS}$ – symmetric key shared by the S and SKS,

PK_A – public key of A,

SK_A – secret key of A,

$\{M\}_K$ – symmetric encryption of M with the key K,

$\{M\}_{PK_A}$ – encryption of M with the public key of A,

$\{M\}_{SK_A}$ – signature of M with the secret key of A,

$H(M)$ – hash of M.

Scenario 1. In the first scenario the database is not encrypted and we analyse the protocol without security mechanisms. The protocol is shown in Fig. 2.

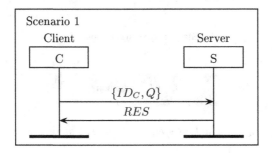

Fig. 2. Protocol in Scenario 1

In the first step the client sends his/her identifier and search query to the server. The server finds the part of database that query refers to and executes the SELECT query using the remaining part of the query. The selected rows are sent back to the client in the second step.

Scenario 2. In the second scenario we assume that the whole database is encrypted in order to protect it against data theft and leakage. However, the server must have access to the plaintext when executes search operations. We assume that the server can obtain the encryption key temporarily (for the time of one search operation) from Secure Keys Storage Server to decrypt the data and select a subset of records. The decryption of the whole database may take a long time therefore we assume that the database is indexed and divided into small parts of data (e.g. customers are divided based on the first letter of the last name). However, on the other hand, the database cannot be indexed in too much detail, because an attacker may distinguish records according to the indexing information (e.g. there may be only one customer on the particular street).

In this scenario the database was encrypted by means of the AES-128-CBC cipher. The protocol is presented in Fig. 3.

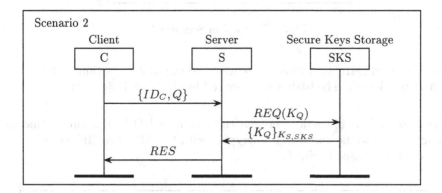

Fig. 3. Protocol in Scenario 2

In the first step the client sends his/her identifier and search query to the server. The server, in the second step, sends the request for the database key with the obtained query to the Secure Keys Storage Server. Secure Keys Storage Server selects the part of the database according to the subset of query containing the information about indexing (e.g. all customers with the last name staring with A). In the third step Server sends back the key, encrypted with the shared key $K_{S,SKS}$, required for decrypting the appropriate part of the database. Server decrypts the selected database part and executes the database query using the rest of the search query. In the fourth step, the selected rows are sent back to the client.

Scenario 3. The third scenario is the same as the second one, but different configuration is used. The database is encrypted with the AES-256-CBC cipher.

Scenario 4. In the fourth scenario the database is not encrypted, but the cryptographic modules which ensure integrity and authentication are implemented. The protocol is presented in Fig. 4.

In the first step the client sends his/her identifier and search query to the server. The server finds the part of database that search query refers to and executes the database query using the remaining part of the search query. Afterwards, the server calculates the hash of results and signs it with his/her private key. The selected rows and signed hash are sent back to the client in the fourth

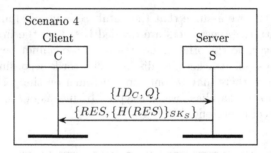

Fig. 4. Protocol in Scenario 4

step. For asymmetric operations there was used the RSA algorithm with the key of 2048 bits length. The hash was generated by the SHA-1 algorithm.

Scenario 5. The difference between the fourth and fifth scenarios is that in the fifth scenario the database is encrypted with the AES-256-CBC cipher. The protocol is presented in Fig. 5.

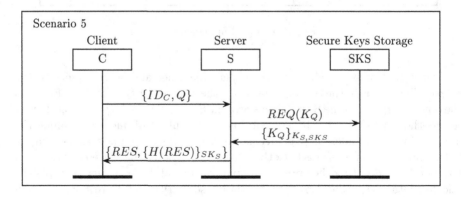

Fig. 5. Protocol in Scenario 5

In the first step the client sends his/her identifier and search query to the server. The server, in the second step, sends the request for the database key with obtained query to the Secure Keys Storage Server. Secure Keys Storage Server selects the part of the database according to the subset of query containing the information about indexing. In the third step Server sends back the key, encrypted with the shared key $K_{S,SKS}$, required for decrypting the appropriate part of the database. The server decrypts the selected database part and executes the database query using the remaining part of the search query. Afterwards, the server calculates the hash of results and signs it with his/her private key. The selected rows and signed hash are sent back to the client in the fourth step.

In Table 1 a summary of security properties ensured by scenarios is presented.

Table 1. Security properties ensured by analysed scenarios

Scenario	Database encryption	Authentication	Integrity
1	—	—	—
2	AES-128-CBC	—	—
3	AES-256-CBC	—	—
4	—	RSA-2048	SHA-1
5	AES-256-CBC	RSA-2048	SHA-1

3.2   Stage 1 – Model Creation Stage

This stage is divided into 4 phases: functions definition (Phase 1), equations definition (Phase 2), channels definition (Phase 3) and protocol flow definition (Phase 4). Below we are defining this structure for analysing the case study in this paper.

Phase 1 – Functions Definition. In order to model the presented scenarios we define the functions which refer to the cryptographic operations required in the protocol. These functions are presented below. In the round bracket the description of these functions is presented.

```
functions {
  fun nonce() (create nonce);

  fun s_enc(data, key) (symmetric encryption of data with key);
  fun s_dec(data, key) (symmetric decryption of data with key);
  fun sign(data, s_key) (sign data with secret key s_key);
  fun a_enc(data, p_key) (asymmetric encryption of data with public key p_key);
  fun a_dec(data, s_key) (asymmetric decryption of data with secret key s_key);
  fun pkey(skey) (get public key for secret key skey);
  fun skey() (generate secret key);
  fun hash(data) (calculate hash of data);
  fun id_c() (generate identification of Client);
  fun query() (generate search query);
  fun database() (create non-encrypted database);
  fun key_request() (create key request);
  fun get_db_key(key) (retrieve key for database);
  fun select_rows(database, query) (get rows from database selected by query);
}
```

Phase 2 – Equations Definition. After defining the functions one can describe the relations between them.

```
equations {
  eq s_dec(s_enc(data, key), key) = data;
  eq a_dec(a_enc(data, pkey(skey)), skey) = data;
  eq get_db_key(key) = key;
}
```

The first equation says that the symmetric decryption of symmetrically encrypted data with the same key returns the encrypted data. The second equation is similar to the first one, but it concerns the asymmetric encryption and decryption.

The difference is in the key. Data must be encrypted with the public key corresponding to its secret key. The third equation says that the function *get_db_key* returns its parameter. This function was used to show that there is a process of obtaining the key that may have influence on the system (i.e. different mechanisms of retrieving and storing keys may cause differences in performance or quality of protection).

Phase 3 – Channels Definition. In the presented example we define four channels with unlimited buffer on each channel.

```
channels {
  channel ch1,ch2,ch3,ch4 (*);
}
```

Phase 4 – Protocol Flow Definition. The last and the most important operation during the modelling process is abstracting the protocol flow. In the presented example we are analysing five scenarios of the above presented architecture. The protocol flow is defined by hosts. In our case study we have three hosts: Client, Server and Secure Keys Storage, all containing one process. Below we will present the QoP-ML model of the analysed case study.

```
hosts {
 host Client(rr)(ch1, ch2) {
   #ID_C = id_c();
   #QUERY = query();

   process Client1(ch1, ch2) {
     M1 = (ID_C, QUERY);
     out(ch1: M1);
     in(ch2: M2);
   }
 }

 host Server(rr)(ch1, ch2, ch3, ch4) {
   #DB_KEY = nonce();
   #DB = s_enc(database(), DB_KEY);
   #SK_S = skey();
   #K_S_SKS = nonce();

   process Server1(ch1, ch2, ch3, ch4) {
     while(true) {
       in(ch1: M1);
       QUERY = M1[1];

     subprocess get_db_key(ch3, ch4) {
       R = key_request();
       out (ch3: R);
       in (ch4: E_TMP_DB_KEY);
       TMP_DB_KEY = s_dec(E_TMP_DB_KEY,
           K_S_SKS)[AES,256,CBC,32B,1];
     }

     subprocess decrypt_aes_128_sim_100() {
       DB_PLAINTEXT = s_dec(DB, TMP_DB_KEY)
           [AES,128,CBC,300MB,100];
     }

     subprocess decrypt_aes_256_sim_100() {
       DB_PLAINTEXT = s_dec(DB, TMP_DB_KEY)
           [AES,256,CBC,300MB,100];
     }
```

```
     subprocess get_db() {
       DB_PLAINTEXT = s_dec(DB, DB_KEY);
     }

     subprocess select_rows_100() {
       ROWS = select_rows(DB_PLAINTEXT,
           QUERY)[100];
     }

     subprocess get_rows() {
       M2 = ROWS;
     }

     subprocess
       get_rows_with_hash_and_signature_100
       () {
       H = hash(ROWS)[SHA1,1MB,100];
       SGN = sign(H, SK_S)[20B,RSA
           ,2048,100];
       M2 = (ROWS,SGN);
     }

     out(ch2: M2);

   }
 }
 host SecureKeysStorage(rr)(ch3, ch4) {
   #DB_KEY = nonce();
   #K_S_SKS = nonce();

   process Store1(ch3, ch4) {
     while (true) {
       in (ch3: Request);
       KEY = s_enc(get_db_key(DB_KEY), K_S_SKS
           )[AES,256,CBC,32B,1];
       out(ch4: KEY);
     }
   }
 }
}
```

3.3   Stage 2 – Security Metrics Definition

In the analysed case study we need security metrics for the following crypto-graphic operations: symmetric decryption of database (AES-128-CBC, AES-256-CBC), symmetric encryption and decryption of symmetric key (AES-256-CBC), hash function (SHA-1) and signing (RSA-2048). We also need metrics for executing the SELECT query on the database. The metrics are obtained by running listed operations on the same device for which the analysis will be performed. All metrics (except symmetric encryption and decryption of key) are calculated for the database of the size approx. 300 MB and the results size approximately 1 MB. We calculated the metrics for 100 simultaneous number of clients. We conducted 10 tests and average is presented in Table 2. Each value represents the time (in ms) that operation takes to finish for one client.

Table 2. Times of operations used in protocol (per one client)

Security operation	Execution time [ms]
Decryption AES-128-CBC (300 MB)	12132.50
Decryption AES-256-CBC (300 MB)	12239.87
Encryption AES-256-CBC (32 B)	8.54
Decryption AES-256-CBC (32 B)	8.30
RSA2048 Signature	0.28
SHA-1 Hash	0.18
SQL Query	2235.59

The security metrics are defined for each device that is analysed. We have defined metrics for the device which we call *Server* and all hosts will use these metrics. They are divided into *conf* and *data* structures. The first one contains the list of device hardware and software that is important for QoP evaluation. We have used a computer with twelve Intel Core i7 processors, Debian 7 as operating system and openssl 1.0.1c library for cryptographic operations.

3.4   Stage 3 – Scenarios Definition

In the third stage we define scenarios which will be analysed. In our case study we have defined five scenarios which were described in the previous section. In this section the QoP-ML construction for this stage is presented. Due to page limit we present definition only of the Scenario 5, the definition of the rest scenarios can be found on [27].

```
versions {
  version scenario_5 {

    set host Server(Server);
    set host SecureKeysStorage(Server);
```

```
run host SecureKeysStorage(*) {
  run Store1(*)
}
run host Server(*) {
  run Server1(get_db_key,decrypt_aes_256_sim_100,select_rows_100,
      get_rows_with_hash_and_signature_100)
}
run host Client(*){100}[ch1, ch2] {
  run Client1(*)
}
}
}
```

3.5 Stage 4 – Evaluation

The last step in the QoP-ML modelling and AQoPA analysis process is evaluation. It depends on the analysis modules which were enabled during the Scenarios Definition stage. In the article which introduces QoP-ML [23] one module was described and is called Time Analysis Module. This module is implemented in the AQoPA and can be used for the analysis of availability security service. By means of the Time Analysis Module one can calculate the total execution time of the selected host (T_{Total}).

In the presented paper the time analysis was performed for five previously described scenarios. After the AQoPA estimation the runtime of an actual implementation was conducted. We have created scripts for clients and a server that execute the selected scenario. We have used a computer with twelve Intel Core i7 processors, Debian 7 as operating system and openssl 1.0.1c library for cryptographic operations (the same as for the security metrics generation). Afterwards, each scenario was conducted for 100 simultaneous clients. The results of analysis are presented in Table 3. In this table we present the estimated execution time of the scenarios which were modelled in the AQoPA and the runtime of an actual implementation (experiment results) of these scenarios. All the scenarios were executed 10 times and the average was calculated. Comparing the results of the protocol runtime estimated in the AQoPA to the runtime of an actual implementation one can conclude that the protocol runtime estimated in the AQoPA is in the range specified by the standard deviation. These results confirm the correctness of the modelling based on the AQoPA approach.

During the analysis, one can notice that the most time consuming scenario is the fifth one where the data is encrypted by AES-256-CBC and the hash is generated and digitally signed. One can notice that similar results give second and third scenarios. These scenarios differ from each other only by hash and its digital signature generation, therefore one can conclude that adding these security mechanisms is less computationally expensive.

The least computationally expensive is scenario number one where the database is not encrypted. The similar results are obtained from the fourth scenario which adds only hash and digital signature compared to the first scenario. One can conclude that comparing the first and fourth scenarios, the hash and its digital signature do not increase the system load significantly.

Table 3. Total times of executing of the scenarios

Scenario	Estimated time in AQoPA [ms]	Runtime of an actual implementation [ms]	Standard deviation [ms]
1	223559.03	230145.65	29390
2	1438493.29	1523261.45	104389
3	1449230.09	1527315.33	76762
4	223605.39	238870.36	27846
5	1449276.45	1534928.44	107408

The significant difference can be found comparing the first and the third scenarios. When the database is not encrypted, the server will generate the response for 100 simultaneous clients in 223559.03 ms. In the third scenario, where the database is encrypted with AES-256-CBC, the response for the clients will be generated in 1449230.09 ms. The difference between these scenarios is significant and the usage of the third scenario should be adapted to the specific security and functional requirements.

3.6 Server Load

We estimated the hourly server load of the storage center being proceeded by different scenarios. Utilizing results obtained by AQoPA along with those that have been estimated, we evaluated the number of clients (sessions) with different scenarios, the server is able to handle within an hour. Our assessment is quite straightforward: knowing that for the existing server, it takes 223.55 s to handle data in Scenario 1, and using the simplest possible formula, (1 h = 3600 s, so 3600 s/223.55 s ≈ 16) one can say that within an hour server is able to deal with approximately 16 sessions (Fig. 6).

Analysing obtained results one can easily notice, that the storage server is able to handle data in scenario 1 and 4 faster than the same number of clients permitted to perform actions in scenarios 2, 3 and 5. Gathered results clearly indicate the relationship between the quality of protection of scenarios and consumption of server resources: the longer time the action needs to accomplish, the more server resources are going to be used. A server, which works longer, utilizes more resources, thereby consuming a greater amount of energy.

3.7 Scalability

The analysis allows us to create scenarios to cope with a situation that will require greater efficiency. Such events may include a sudden and significant increase in requests to the system or type of denial of service attack. In this case, the system may switch to the operation mode where the efficiency is greater than the security.

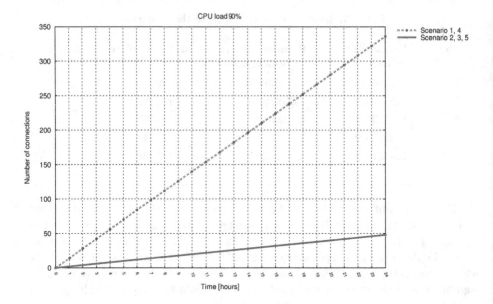

Fig. 6. Server's performance in our scenarios

The other solution to the need of high efficiency is scaling. However this can entail high costs while the idea of QoP is to extend the requirements list in SLA (Service-Level Agreement). To cost and quality of service (QoP) requirements one introduces the third which is the quality of protection. As the result, the trade-off between these three requirements is achieved by negotiations.

On the other hand some security requirements may refer not only to the database but also to integrated systems (e.g. authorization server). In such situation these systems can be the bottle-necks and the scaling of database would not solve the problem. Of course, the integrated systems could also be scaled (e.g. by mirroring) but it would increase management complexity and even more the costs.

4   Conclusions

In the article we proposed the approach for balancing security against performance for IT systems. The proposed approach can automatically answer the question what is the influence of the ensured security properties on the overall system performance. Through this analysis you can make a trade off between the means of information protection and the required performance.

Presenting high level of abstraction, having the possibility of maintaining processes and communication steps consistently, QoP-ML provides a flexible approach for modeling complex systems and performance of multilevel security analysis. The analysis from [28] showed the flexibility of the quality of protection modeling language and confirmed the fact that the multilevel analysis is crucial in modeling security systems.

In the article the case study of the QoP-ML modelling and AQoPA analysis was presented. The analysis was performed for different versions of the cryptographic protocol which are searching and retrieving data from the database. We focus on one level of analysis which concerns the cryptographic functions related with the time analysis module. The evaluation was performed and the total protocols runtime was calculated. The protocols runtime estimated in the AQoPA was validated by the actual implementation of the protocols. This confirms the correctness of the modelling with QoP-ML and evaluating with AQoPA tool.

References

1. Ksiezopolski, B., Kotulski, Z., Szalachowski, P.: Adaptive approach to network security. In: Kwiecień, A., Gaj, P., Stera, P. (eds.) CN 2009. CCIS, vol. 39, pp. 233–241. Springer, Heidelberg (2009)
2. Ksiezopolski, B., Kotulski, Z., Szalachowski, P.: On QoP method for ensuring availability of the goal of cryptographic protocols in the real-time systems. In: European Teletraffic Seminar, pp. 195–202 (2011)
3. Stubblefield, A., Rubin, A.D., Wallach, D.S.: Managing the performance impact of web security. Electron. Commer. Res. 5, 99–116 (2005)
4. Sklavos, N., Kitsos, P., Papadopoulos, K., Koufopavlou, O.: Design, architecture and performance evaluation of the wireless transport layer security. J. Supercomputing 36(1), 33–50 (2006)
5. Ksiezopolski, B., Kotulski, Z.: On scalable security model for sensor networks protocols. In: 22nd CIB-W78 Conference Information Technology in Construction, Dresden, pp. 463–469 (2005)
6. Szalachowski, P., Ksiezopolski, B., Kotulski, Z.: On authentication method impact upon data sampling delay in wireless sensor networks. In: Kwiecień, A., Gaj, P., Stera, P. (eds.) CN 2010. CCIS, vol. 79, pp. 280–289. Springer, Heidelberg (2010)
7. Jürjens, J.: Security and compliance in clouds. In: IT-Compliance 2011, Berlin, 4th Pan-European Conference (2011)
8. Khan, J.I., Wierzbicki, A.: Foundations of Peer-to-Peer Computing (2008)
9. Wierzbicki, A., Szczepaniak, R., Buszka, M.: Application layer multicast for efficient peer-to-peer applications. In: Proceedings of the Third IEEE Workshop on Internet Applications, WIAPP 2003. IEEE (2003)
10. Wierzbicki, A., Zwierko, A., Kotulski, A.: Authentication with controlled anonymity in P2P systems. In: Sixth International Conference on Parallel and Distributed Computing, Applications and Technologies, PDCAT 2005. IEEE (2005)
11. Wierzbicki, A.: The case for fairness of trust management. Electron. Notes Theoret. Comput. Sci. 197(2), 73–89 (2008)
12. Ksiezopolski, B., Rusinek, D., Wierzbicki, A.: On the modelling of kerberos protocol in the quality of protection modelling language (QoP-ML). Ann. UMCS Inf. AI XII 4, 69–81 (2012)
13. Ksiezopolski, B., Rusinek, D., Wierzbicki, A.: On the efficiency modelling of cryptographic protocols by means of the quality of protection modelling language (QoP-ML). In: Mustofa, K., Neuhold, E.J., Tjoa, A.M., Weippl, E., You, I. (eds.) ICT-EurAsia 2013. LNCS, vol. 7804, pp. 261–270. Springer, Heidelberg (2013)

14. Lambrinoudakis, C., Gritzalis, S., Dridi, F., Pernul, G.: Security requirements for e-government services: a methodological approach for developing a common PKI-based security policy. Comput. Secur. **26**, 1873–1883 (2003)
15. Lindskog, S.: Modeling and Tuning Security from a Quality of Service Perspective. Ph.D. dissertation, Department of Computer Science and Engineering, Chalmers University of Technology, Goteborg, Sweden (2005)
16. Ong, C.S., Nahrstedt, K., Yuan, W.: Quality of protection for mobile applications. In: IEEE International Conference on Multimedia & Expo 2003, pp. 137–140 (2003)
17. Schneck, P., Schwan, K.: Authenticast: An Adaptive Protocol for High-Performance, Secure Network Applications. Technical report GIT-CC-97-22 (1997)
18. Ksiezopolski, B., Kotulski, Z.: Adaptable security mechanism for the dynamic environments. Comput. Secur. **26**, 246–255 (2007)
19. Luo, A., Lin, C., Wang, K., Lei, L., Liu, C.: Quality of protection analysis and performance modelling in IP multimedia subsystem. Comput. Commun. **32**, 1336–1345 (2009)
20. LeMay, E., Unkenholz, W., Parks, D.: Adversary-driven state-based system security evaluation. In: Workshop on Security Metrics, MetriSec (2010)
21. Petriu, D.C., Woodside, C.M., Petriu, D.B., Xu, J., Israr, T., Georg, G., France, R., Bieman, J.M., Houmb, S.H., Jürjens, J.: Performance analysis of security aspects in UML models. In: Sixth International Workshop on Software and Performance, Buenos Aires, Argentina, ACM (2007)
22. Jürjens, J.: Secure System Development with UML. Springer, Heidelberg (2007)
23. Ksiezopolski, B.: QoP-ML: quality of protection modelling language for cryptographic protocols. Comput. Secur. **31**(4), 569–596 (2012)
24. Mazur, K., Ksiezopolski, B., Kotulski, Z.: The robust measurement method for security metrics generation. Comput. J. (2014) (in press)
25. Ksiezopolski, B., Zurek, T., Mokkas, M.: Quality of protection evaluation of security mechanisms. Sci. World J. 2014, Art. ID 725279 (2014)
26. ISO: ISO/IEC 25010: Systems and Software Engineering - Systems and Software Quality Requirements and Evaluation (SQuaRE) - System and Software Quality Models (2011)
27. The web page of the QoP-ML project: http://www.qopml.org
28. Mazur, K., Ksiezopolski, B.: Comparison and assessment of security modeling approaches in terms of the QoP-ML. In: Kotulski, Z., Księżopolski, B., Mazur, K. (eds.) CSS 2014. CCIS, vol. 448, pp. 178–192. Springer, Heidelberg (2014)

Analysis of Objective Trees in Security Management of Distributed Computer Networks of Enterprises and Organizations

Olga Dolinina, Vadim Kushnikov, and Ekaterina Kulakova[✉]

Department of Applied Information Technologies, Yuri Gagarin State Technical University of Saratov, Politechnicheskaya 77, 410054 Saratov, Russia
{odolinina09,kulakovakm}@gmail.com, kushnikoff@yandex.ru
http://www.sstu.ru/

Abstract. The paper contains the formulation of the problem of situational safety management of distributed computer networks, mathematical models and algorithms, which can help to identify quickly and eliminate the causes of security breaches of these networks during their operation. The developed software allows to create a methodological basis for the creation of new information security management systems with cognitive functions, the use of which will improve the reliability and efficiency of the enterprises and organizations.

Keywords: Security · Distributed computer networks · Algorithms · Mathematical models · The automatized system

1 Introduction

"The only truly secure system is one that is powered off, cast in a block of concrete and sealed in a lead-lined room with armed guards – and even then I have my doubts" [1]. This phrase of Professor Eugene Spafford, one of the leading experts in the field of safety and security of operating systems, director of the Purdue University's Center for Education and Research in Information, Assurance and Security, can be considered the best at showing complexity, multidimensional and complex nature of the problems of information security management of distributed computer networks.

Methods of information security management in structural subdivisions of enterprises and organizations are well known today and elaborated, their requirements are detailed, for example, in the international standards ISO/IEC 27002, British Standard BS 7799, GOST R (Russian National Standard) ISO/IEC 15408-2002, and in many other documents [2–11].

Successful implementation of key provisions of these documents in practice is impossible without the development of a detailed activity plan for ensuring information security of distributed computer networks of an enterprise or organization, which affects the operation of all its structural subdivisions. These plans

© Springer International Publishing Switzerland 2015
P. Gaj et al. (Eds.): CN 2015, CCIS 522, pp. 117–126, 2015.
DOI: 10.1007/978-3-319-19419-6_11

are generally formed as objective trees; they have a complex structure and can consist of hundreds of individual activities performed in different time intervals at different levels of the controlled object hierarchy. At the same time each activity is influenced by various conditions that can limit or push forward its implementation, which should also be kept in mind when developing a systematic plan for information security management of distributed computer networks of an enterprise or organization.

Currently, numerous automatic operation control systems have been created and tested, in which one can arrange and control the execution of objective trees of information security in distributed computer networks of enterprises and organizations [6,7]. In the works of foreign and Russian researchers, such as E. Feigenbaum, D. Waterman, S. Russell, P. Norvig, D.A. Pospelov, O.I. Larichev, A.F. Rezchikov and others one can find elaborately developed general principles of these systems [12,13].

However, in the specialized literature, there are practically no reports of the software, which allow to confirm quickly or deny the ability to perform major objectives that are necessary, for example, for information security management of distributed computer networks of large enterprises and organizations, as well as to define the reasons which prevent their implementation.

This fact complicates greatly the implementation of the well-known concept of the Objective Management, according to which the objective is to be [14]:

- Specific;
- Measurable;
- Achievable;
- Relevant;
- Time-bounded.

As a result, even the most experienced, highly qualified professionals can get considerable difficulties in assessing the feasibility of complex objective trees consisting of several hundreds of leaf nodes, which often lead to the breakdown of the planned security management strategy of distributed computer networks, misallocation and, as a result, to a significant economic damage.

The above mentioned facts determine the topicality, economic efficiency and practical significance of the problems connected with the development of mathematical models and algorithms, which allow to evaluate quickly the feasibility of complex objective trees, implemented in the information security management of distributed computer networks of enterprises or organizations, and establish reasons prevented their implementation.

2   Problem Definition, Constraints and Assumptions

If there is a difficult situation in an industrial enterprise or organization $w(\overrightarrow{x}, \overrightarrow{u}) \in \{W(\overrightarrow{X}, \overrightarrow{U})\}$, involving a violation of the information security system of distributed computer networks, as a result of which the controlled object was moved to the condition $s_0(\overrightarrow{x}, \overrightarrow{u}) \in \{S(\overrightarrow{X}, \overrightarrow{U})\}$ and suffered substantial damage ($\{S(\overrightarrow{X}, \overrightarrow{U})\}$ – set of various possible states of the controlled object, $\{W(\overrightarrow{X}, \overrightarrow{U})\}$ – set of manageable situations of an enterprise or organization, environmental parameters vectors $\overrightarrow{x} \in \{(\overrightarrow{X})\}$ and management actions $\overrightarrow{u} \in \{(\overrightarrow{U})\}$, respectively).

Let also assume that the transition of the controlled object to the condition $s_k(\overrightarrow{x}, \overrightarrow{u}) \in \{S(\overrightarrow{X}, \overrightarrow{U})\}$ will lead to the elimination of a difficult situation and will minimize the damage caused.

In order to solve the difficult situation $w(\overrightarrow{x}, \overrightarrow{u}) \in \{W(\overrightarrow{X}, \overrightarrow{U})\}$ the management staff developed a set of measures, presented in the form of an objective tree $d \in \{D\}$, on the step-wise transfer of controlled object from condition $s_0(\overrightarrow{x}, \overrightarrow{u}) \in \{S(\overrightarrow{X}, \overrightarrow{U})\}$ to condition $s_k(\overrightarrow{x}, \overrightarrow{u}) \in \{S(\overrightarrow{X}, \overrightarrow{U})\}$, which marks the end of a difficult situation ($\{D\}$ – set of feasible objective trees). Hereafter we assume that the objective tree $d \in \{D\}$ consists of a finite set of the objectives $\{z_1, z_2, \ldots, z_n\}$, the implementation of each of them includes a transition of the controlled object from the condition $s_i(\overrightarrow{x}, \overrightarrow{u}) \in \{S(\overrightarrow{X}, \overrightarrow{U})\}$ to the condition $s_{i+1}(\overrightarrow{x}, \overrightarrow{u}) \in \{S(\overrightarrow{X}, \overrightarrow{U})\}$, $i = \overline{0, k-1}$.

Let assume that the execution of each of the objectives $z_i \in \{z_1, z_2, \ldots, z_n\}$ is affected by the conditions $B_i(\overrightarrow{x}, \overrightarrow{u})$, $i = \overline{1, g}$ arising from the performance features of the object and management system, as well as dependent on the environment. These conditions, which influence the execution of the objective $z_i \in \{z_1, z_2, \ldots, z_n\}$, generally can be formalized by means of the rules which have the form of the following expressions:

IF $<B_1(\overrightarrow{x}, \overrightarrow{u})R_1B_2(\overrightarrow{x}, \overrightarrow{u})R_2 \ldots R_{k-1}B_k(\overrightarrow{x}, \overrightarrow{u})>$ IS PERFOMED,

THEN THE OBJECTIVE $<z_i \in \{z_1, z_2, \ldots, z_n\}>$ WILL BE PERFOMED

$R_i \in \{\text{AND}, \text{OR}, \text{NOT}, \text{AND} - \text{NOT}, \text{OR} - \text{NOT}\}$, $i = \overline{1, 5}$.   (1)

If the number of test conditions is two or one, then in (1) they are recorded in the form of expressions $B_1(\overrightarrow{x}, \overrightarrow{u})ANDB_2(\overrightarrow{x}, \overrightarrow{u})$ or $B_1(\overrightarrow{x}, \overrightarrow{u})$, respectively.

The objective $z_i \in \{z_1, z_2, \ldots, z_n\}$ is considered feasible if:

– all objectives of the tree, prior to this objective, were performed;
– all rules $B_i(\overrightarrow{x}, \overrightarrow{u})$, $i = \overline{1, g}$, set in the form of expressions (1) and affecting the objective $z_i \in \{z_1, z_2, \ldots, z_n\}$ were performed.

If you violate at least one of these conditions, the objective $z_i \in \{z_1, z_2, \ldots, z_n\}$ cannot be performed.

Objective tree d will be feasible if all its objectives are achievable, and unfeasible if it contains at least one objective which is impossible to achieve under current conditions. In view of the above given definitions and assumptions the formalized statement of the problem has the following definition.

For automatized information security management systems of distributed computer networks of enterprises and organizations, it is necessary to develop mathematical models and algorithms allowing by formal methods in real-time to confirm or deny the feasibility of objectives, presented in the form of an objective tree d, to establish obstacles to achieving it, and to recommend ways to eliminate them.

3   General Characteristics of the Method of Solution

The proposed method of solving the problem is based on the known heuristic approach based on the objective tree in the form of a scheme of a digital discrete device DU, built on the basis of conjunctors, disjunctors and inverters. This approach to the analysis of the feasibility of objective trees and activity plans was first introduced and justified for dialogue systems of operational management of production processes in the works of A.F. Rezchikov and representatives of his scientific school [15,16].

For each activity of the plan, a system of conditions is formed, the execution of which, in the opinion of management personnel, directly affects the execution or non-execution of the corresponding activity. Conditions must be formed in a product; i.e. have the form of expression (1).

Using this system allows to formalize the knowledge of decision-makers about the cause-and-effect relationships, without which the implementation of an activity of the plan for the elimination of difficult production situation $w(\overrightarrow{x}, \overrightarrow{u}) \in \{W(\overrightarrow{X}, \overrightarrow{U})\}$ can lead to undesirable results.

Each input of a digital device DU in one-one mapping is put under objective $z_i \in \{z_1, z_2, \ldots, z_n\}$ or condition $B_i(\overrightarrow{x}, \overrightarrow{u})$. During the implementation of these measures and the conditions, a unit impulse is given to the corresponding input of a digital device DU, and if they are not implemented, zero signal is given.

Management personnel of the enterprise, forming binary signals on the input of digital device DU, which correspond to the fulfillment or non-fulfillment of conditions $B_i(\overrightarrow{x}, \overrightarrow{u})$, $i = \overline{1, g}$ and activities $z_i \in \{z_1, z_2, \ldots, z_n\}$ can quickly confirm, based on the output value, the feasibility of large objectives, characterized by the objective tree d, or determine the reasons for its failure.

4   Mathematical Models and Algorithms

Depending on whether during the formal description of the objective tree conditions are used or not (1), the analysis of the feasibility of objectives can include different formal models and algorithms, differing by complexity of their construction.

Model A. Let represent the objective tree d in the form of a direct graph $G(U, E)$ which node set U represents activities of the developed plan and the set E represents arcs connecting these nodes $u_i \in U$. In this case, the nodes $u_i, u_j \in U$ of the graph $G(U, E)$ are connected by the arc $e_{ij} \in E$ if and only if the two activities of the plan m_i, m_j $in M$, corresponding to these nodes, imply the execution of dependence R_1 – "execution intransitively depends on".

The process of forming a Model A with a graph $G(U, E)$ is described by the rule-based system formed by the following algorithm (Algorithm 1):

1. Begin.
2. Determine the node u^* with a zero in-degree $d^-(u^*) = 0$ on the graph $G(U, E)$. On the activity plan scheme M this node corresponds to the node M_1 – "Activity Plan is completed".
3. Determine all the nodes $u_{m0}, u_{k0}, u_{h0}, \ldots, u_{l0} \in U$ of graph $G(U, E)$ connected by arcs with a node u^*.
4. Write down the following condition into the formed rule-based model:
   PLAN M WILL BE EXECUTED IF THE FOLLOWING ACTIVITIES, CORRESPONDING TO THE GRAPH NODES u_{m0} AND u_{k0} AND u_{h0} AND ... AND u_{l0}, WILL BE EXECUTED.
5. For the top u_{m0}, determine all nodes $u_{m1}, u_{k1}, u_{h1}, \ldots, u_{l1}$ connected by arcs with a node u_{m0}.
6. Write down the following condition into the formed rule-based model:
   ACTIVITY u_{m0} WILL BE EXECUTED IF THE FOLLOWING ACTIVITIES, CORRESPONDING TO THE GRAPH NODES u_{m1} AND u_{k1} AND u_{h1} AND ... AND u_{l1}, WILL BE EXECUTED.
7. Continue the formation of the rule-based model until the end nodes of the graph $G(U, E)$ will be achieved, i.e. the nodes with a zero out-degree $d^+(u_k) = 0$.
8. End.

Then, according to certain rules used in the design of digital computers [17, 18], the rule-based system is assigned a logic function $f(u_{1k}, u_{2k}, \ldots, u_{vk})$ where $u_{1k}, u_{2k}, \ldots, u_{vk}$ – are the end nodes of the graph $G(U, E)$, i.e. the nodes with a zero out-degree. In addition, each argument u_{ij} of the function $f(u_{1k}, u_{2k}, \ldots, u_{vk})$ should be set to 1 or 0, which would mean the execution or non-execution of activity m_{ij} $in M$ corresponding to the node u_{ij} of the graph $G(U, E)$. Logic function generated at the output DU, takes the following values:

$$f(u_{1k}, u_{2k}, \ldots, u_{vk}) = \begin{cases} 1 & \text{if plan is completed} \\ 0 & \text{if plan is not completed} \end{cases}.$$

By assigning different combinations of arguments $u_{1k}^0, u_{2k}^0, \ldots, u_{vk}^0$ and by determining the appropriate values $f(u_{1k}^0, u_{2k}^0, \ldots, u_{vk}^0)$ of these arguments, it is possible to analyze the degree of completeness of the activity plan; to determine the activities which were not executed and which constrain the successful implementation of the developed plan, and gain new knowledge about the conditions of its implementation.

Model B. This model is based on the Model A and differs from it in that Model B takes into account the impact of conditions on the execution of individual activities of the plan $m_i \in M$ (1) defined by management personnel on the basis of personal experience, intuition, good knowledge of the operation of the facility and management system and so on. These conditions can be of necessary or sufficient character and relate to the implementation of various measures of the plan M. In this case, these conditions must be formed in advance at the stage of initial development of an activity plan, or added to the plan during its implementation. Conditions should be presented in the form of rules and combined by logic operations AND, OR, NOT. In particular, they have the following form:

$$\text{THE ACTIVITY } m_i \text{ WILL BE EXECUTED,} \tag{2}$$

$$\text{IF (H1} \leq \text{N1) OR ((H2} \leq \text{N2 AND H3} \geq \text{N3) AND H4} \leq \text{N4),} \tag{3}$$

where:

H1 :: = <number of dead computers>;
H2 :: = <duration of virus attack, h>;
H3 :: = <the intensity of the impact on the information system,%>;
H4 :: = <expected recovery period time, h>

(I – the serial number of the activity in the plan M; H1, H2, H3, H4 – logical-linguistic variables whose values are defined by the above relations; N1, N2, N3, N4 – known constants).

Using the conditions (1) enables to improve the quality of planning of activities for difficult situations $w(\overrightarrow{x}, \overrightarrow{u}) \in \{W(\overrightarrow{X}, \overrightarrow{U})\}$, as well as to pre-test the feasibility of the developed plan in evolving conditions and, if necessary, to carry out its correction. In addition, these rules can be entered into the information system database together with the documents governing the behavior of the operating and dispatching personnel of the enterprise or organization and can be used both in the process of preparation and decision making in the eliminating of situations $w(\overrightarrow{x}, \overrightarrow{u}) \in \{W(\overrightarrow{X}, \overrightarrow{U})\}$, and at the stage of training the personnel on effective actions for information security violations of distributed computer networks.

In order to construct the model B, as in the case with the model A, a graph $G_1(U, E)$ is formed under the block diagram of the activity plan M in accordance with the Algorithm 1. The difference is the addition of rules (1) into paragraphs 4, 6 and 7 of Algorithm 1, which link the realization of one or another activities of the plan $m_i \in M$ with the implementation of pre-defined conditions. Then according to the graph $G_1(U, E)$ a logical function $f(u_{1k}, u_{2k}, \ldots, u_{vk})$ is formed (where $u_{1k}, u_{2k}, \ldots, u_{vk}$ are end nodes of the graph $G_1(U, E)$) and digital devices are synthesized based on conjunctors, disjunctors and inverters having the same state table as the logic function $u_{1k}, u_{2k}, \ldots, u_{vk}$. The logical functions formation and synthesis of the digital device DU is carried out in the same manner as in model A.

With the help of the digital device DU, the implementation of which can be carried both with the development or installation of the software and hardware, the possibility of the plan realization under various conditions is estimated.

Rapid Assessment of the Feasibility of an Activity Plan. At the current time the execution of all activities $m_i \in M$ is analyzed, as well as the conditions of those activities that are the part of rules (1). The input to the digital device

DU receives individual signals corresponding to the already executed activities and conditions (1), and zero signals, corresponding to those, which were not executed.

The output signal of the DU is determined. If it equals 1, then at the current time the objective tree $d \in \{D\}$ on elimination of the difficult situation $w(\overrightarrow{x}, \overrightarrow{u}) \in \{W(\overrightarrow{X}, \overrightarrow{U})\}$ that has arisen as a result of violations of information security of distributed computer networks can be realized. Otherwise, the correction of conditions (1) and verifiable activity plan is carried out.

5   Application of the Proposed Algorithms

The top level of the objective tree that must be reached to achieve the required level of security of distributed computer networks of two schools of Yuri Gagarin State Technical University of Saratov is shown in the left part of the Fig. 1, where some of the quantities are defined:

$M1$ – assuring the required level of computer network security of the University;
$U1, U2$ – network security of International School of Applied Information Technology and School of Electrical and Mechanical Engineering, respectively;
$U3, U4$ – installation of Kaspersky or Dr.Web anti-virus software, respectively;
$U5$ – physical protection of the network;
$U6$ – network security at the software level;
$U7$ – safe performance mode of operating systems;
$U8$ – development of the concept of granting data ownership;
$U9$ – organization of efficient access to data for different groups of network users;
■ – the symbol of the conjunction;
■ – a symbol of disjunction.

The right side of the Fig. 1 shows a diagram of a discrete device used to test the feasibility of this objective tree fragment according to the algorithms described above. Conditions (1) are formulated in the following expressions.

Condition f_1: IF passwords on the computers of School of Electrical and Mechanical Engineering are not changed at least 1 time per year, THEN the execution of goal $U4$ is impossible.
Condition f_2: IF the warranty period of Dr.Web anti-virus software system is less than 1 year, THEN the execution of objectives $U1$ and $U6$ is not possible.
Condition f_3: IF the discount on the installation of Dr.Web anti-virus system on all computers in the network is less than 15 %, THEN the execution of objectives $U1$ and $U6$ is not possible.
Condition f_6/f_7: IF system administrators of International School of Applied Information Technology/School of Electrical and Mechanical Engineering were not retrained over the past 2 years, the execution of the target $M1$ is not possible.

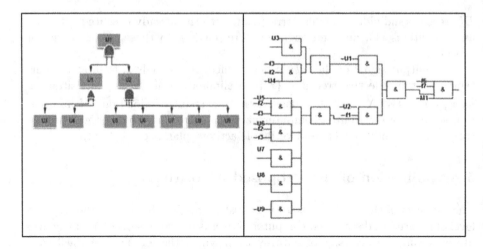

Fig. 1. A fragment of the upper level of the objective tree for ensuring computer network security of Yuri Gagarin State Technical University of Saratov and the scheme of the discrete device used at its feasibility check

6   Advantages of the Proposed Way of Ensuring Distributed Computer Networks Safety

1. Availability check of the required level of security of distributed computer networks at Yuri Gagarin State Technical University of Saratov is reduced to a feasibility check of the respective objectives tree.
2. Feasibility of the objectives tree consisting of 678 items is checked in a fraction of a second with the help of the program that emulates the work of the discrete device, the upper level of which is shown in the right part of the Fig. 1.
3. The manager can independently verify the security level, without involvement of the specialists responsible for the implementation of specific activities to ensure security.
4. An algorithm for the security check of computer networks can be quickly and simply rebuilt when changing objectives tree and conditions (1).
5. When achievability check of the required level of security through the use of conditions (1) is used, it is quite easy to register the impact of a large number of different factors that affect the security of a computer network, without the development of complex mathematical models and costly experimentation.

7   Conclusion

The approach to the analysis of the feasibility of the objective trees, developed in this paper, allows to create a methodological basis for the formation of new information security of management systems with cognitive functions that enhance the reliability and efficiency of the distributed computer networks of enterprises and organizations. The proposed approach is based on the formal verification of the set of objectives feasibility which ensures the security of distributed computer networks of enterprises and organizations. The advantage of this approach, in our opinion, is connect with the possibility of formal conditional test of safe operation of computer networks which do not require construction of complex mathematical models and costly experiments.

Practical implementation of the developed software which allows to conduct the conditional test of safe operation of computer networks, was carried out under the modernization of the safety management system of distributed computer networks of the International School of Applied Information Technology of Yuri Gagarin State Technical University of Saratov. Practical use of the computer networks information security system of the university showed that the average number of emergencies and pre-emergency situations decreased by 19.5 %, and the degree of readiness of the computer network to work under extra strain significantly increased, for example, during preparation for the exams, tests, etc.

References

1. Spafford, E.H.: Cyber security: assessing our vulnerabilities and developing an effective defense. In: Gal, C.S., Kantor, P.B., Lesk, M.E. (eds.) ISIPS 2008. LNCS, vol. 5661, pp. 20–33. Springer, Heidelberg (2009)
2. BS 7799–1: 2005. Information security management. Code of practice for information security management
3. BS 7799–2: 2005. Information security management. Specification for information security management systems
4. BS 7799–3: 2006. Information security management systems. Guidelines for information security risk management
5. BS 7799–3: 2006. Information security management systems - Part 3: Guidelines for information security risk management
6. Information Security Management Handbook. 5th edn. CRC Press (2004)
7. ISO/IEC TR 13335–3: 1998. Information technology - Guidelines for the management of IT Security - Part 3: Techniques for the management of IT security
8. Risk Management Guide for Information Technology Systems. Recommendations of the National Institute of Standards and Technology. Special Publication 800–30 (2002)
9. GOST R (Russian National Standard) ISO/IEC TR 18044. Information technology. Security techniques. Information security management
10. GOST R (Russian National Standard) ISO/IEC 18045. Information technology. Security techniques. Methodology for IT security evaluation
11. GOST R (Russian National Standard) 50922–2006. Protection of information. Basic terms and definitions
12. Vasil'ev, S.N.: From the classic problems of regulation to intelligent control I: bulletin of the Russian academy of sciences. Theor. Manage. Syst. **1**, 5–22 (2001). Science Publishers, Moscow
13. Rezchikov, A.F., Kushnikov, V.A., Shlychkov, E.I., Boikova, O.M.: Models and algorithms of problem definition of ACS development by industrial facilities: devices and systems. Manage. Control Diagn. **9**, 64–68 (2006). NauchTekhLitIzdat, Moscow

14. Shlychkov, E.I., Pohaznikov, M.Y., Kushnikov, V.A., Kalashnikov, O.M.: Feasibility analysis of activity plans under the operational management in a machine-building enterprise. Bull. Yuri Gagarin State Tech. Univ Saratov **1**(1), 88–95 (2007). SSTU, Saratov
15. Avetisjan, J.A., Kushnikov, V.A., Rezchikov, A.F., Rodichev, V.A.: Mathematical models and algorithms for operative administration of processes dealing with elimination of emergencies. Mechatron. Autom. Control **11**, 43–47 (2009). New Technologies, Moscow
16. Kushnikov, V.A., Rezchikov, A.F., Tsvirkun, A.D.: Control in man-machine systems with automated correction of objectives. Meitan Kexun Jishu **26**, 168–175 (1998). Coal Science and Technology, Peking
17. Drucker, P.F.: Management: Tasks Responsibilities Practices. HaperCollins Publishers, New York (1993)
18. Anderson, J.A.: Discrete Mathematics with Combinatorics. Prentice Hall, New Jersey (2003)

A Technique for the Botnet Detection Based on DNS-Traffic Analysis

Oksana Pomorova, Oleg Savenko, Sergii Lysenko,
Andrii Kryshchuk[✉], and Kira Bobrovnikova

Department of System Programming, Khmelnitsky National University,
Instytutska, 11, Khmelnitsky, Ukraine
o.pomorova@gmail.com, savenko_oleg_st@ukr.net, sirogyk@ukr.net,
rtandrey@rambler.ru, kirabobrovnikova@mail.ru
http://spr.khnu.km.ua

Abstract. A technique for botnet detection based on a DNS-traffic is developed. Botnets detection based on the property of bots group activity in the DNS-traffic, which appears in a small period of time in the group DNS-queries of hosts during trying to access the C&C-servers, migrations, running commands or downloading the updates of the malware. The method takes into account abnormal behaviors of the hosts' group, which are similar to botnets: hosts' group does not honor DNS TTL, carry out the DNS-queries to non-local DNS-servers. Method monitors large number of empty DNS-responses with NXDOMAIN error code. Proposed technique is able to detect botnet with high efficiency.

Keywords: Botnet · Bot · DNS-traffic · DNS-query · Group activity

1 Introduction

During the last years, the most dangerous malware class are botnets. They perform dangerous acts such as DDoS attacks, malware distribution, theft of confidential data, organization of anonymous proxy servers etc. Botnets can be the mean of the corporate espionage and hiding phishing, spreading of the spam, provide remote machines services etc. [1,2].

Attackers can circumvent the traditional methods of botnets detection such as packet filtering, traffic analysis and signatures based method by dynamically changing the malicious code, system controls and ports, or by using standard HTTP/S ports. However, the complete analysis of the packets content demands high computational resources [3].

2 Related Works

In order to manage and control the infected hosts the vast majority of botnets use DNS [4–7]. A characteristic feature of the behavior of such type of botnets is a

© Springer International Publishing Switzerland 2015
P. Gaj et al. (Eds.): CN 2015, CCIS 522, pp. 127–138, 2015.
DOI: 10.1007/978-3-319-19419-6_12

coordination in the DNS-traffic. Known detection techniques, described in [8,9,11] are based on this feature and have following disadvantages: they rely on group queries only for the same domain names, have low accuracy of the detection the migrations of the command-and-control servers (C&C-servers) and bots' group activity. Mentioned methods also require large amounts of computing resources and considerable processing time for applying to large networks; short duration of monitoring; arbitrary division of the monitoring period into intervals within which the search for infected groups is being performed.

3   A Technique for the Botnet Detection Based on DNS-Traffic Analysis

The proposed method for the botnet detection in the corporate networks is based on the property of group bots activity in the DNS-traffic. Such traffic appears in group DNS-queries of hosts in a small period during trying to access the C&C-servers, during migrations, running commands or downloading the updates of the malware [3]. The method takes into account abnormal behaviors of the hosts' group, which are similar to botnets' behavior: hosts' group does not honor DNS TTL, that is they flush local DNS-cache and carry out repeated queries for domain names before TTL expiration [3,4], implement the DNS-queries to non-local DNS-servers [4]. Also proposed method monitors large number of empty DNS-responses with error code RCODE=3 (NXDOMAIN, domain name does not exist).

In [8–10] in order to compare the groups of hosts the symmetric similarity measures are used. They can be used only for equal groups. That is why in the article the asymmetric similarity measures, which are suitable for evaluation of the different size groups, are used.

The proposed method also takes into account the C&C-servers migration and other DNS-queries related to a botnet functioning. Method takes into account the possible change the size of the hosts' group, and does not require large amounts of computing resources and considerable processing time.

In contrast to [11] the developed method involves the gathering of DNS-traffic. In the developed method the division of the monitoring period into intervals for the searching of the infected groups is based on the TTL values, which are contained in DNS-messages.

A scheme of the botnet detection is shown in Fig. 1.

4   The Similarity Evaluation of Hosts' Groups

Gathering the Incoming Traffic. Incoming DNS-traffic is gathered by the set of network sensors $E = \{e_i\}_{i=1}^{N_E}$, where N_E – number of sensors, connected to the switch with port mirroring.

Comparison "White" and "Black" Lists of the Domain Names. In order to reject the legitimate queries and to reveal known malicious domain names, the comparison of the collected data from the "white" and "black" lists of the domain names is carried out.

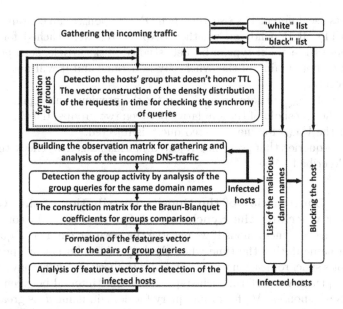

Fig. 1. Functioning principle of the DNS-method for botnet detection

Detection the Hosts' Group that Does not Honor TTL. A group flushing of the local hosts DNS-caches means that hosts' group does not honor TTL [10]. In order to detect that fact we build the observation matrix V_{MAC}, in which each row contains the hosts' MAC-addresses H_j that requested the specific domain name during the TTL.

Therefore, the rows of the matrix V_{MAC} contain the MAC-addresses of hosts that possibly perform a group activity.

If host's MAC-address is represented in the group, then the cell of the observation matrix V_{MAC} is marked as "1", otherwise – "0". If the host sent a repeated request for domain name d_i, then MAC-address of the host is marked as "1" in a row of the matrix V_{MAC} (a new row is created for repeated query).

Example of observation matrix V_{MAC} is presented in Fig. 2.

H_1	H_2	H_3	H_4	...	H_j
1	1	1	1	...	1
1	0	1	1	...	0

Fig. 2. Observation matrix V_{MAC}

The groups formation is performing till the expiration of the highest value of the TTL, received in the DNS-response for the repeated query.

Note. If hosts use different DNS-servers then DNS-responses can contain different values of TTLs. It depends on whether DNS-records were cached for domain name on these servers. For type of botnets studied in the article, such possibility is excluded as bots maintain the synchrony in actions and it will not affect the detection accuracy.

Let N_G and $N_{G_{rep}}$ denote groups' sizes for the previous and repeating group queries, δ – the threshold of the similarity between two groups. If $\delta \cdot N_G > N_{G_{rep}}$, then row of the V_{MAC} for the repeated query is rejected.

For group queries that were not rejected at this stage, we check their synchrony as described below.

The Vector Construction of the Density Distribution of the Requests in Time for Checking the Synchrony of Queries. We will consider the group of queries as synchronous if we observe the geratest number of queries for the domain name during the time when the bots of the botnet are performing queries – bot's synchronization time t_s. In order to check the synchrony of queries of the DNS-queries we perform such steps. If the time interval between the first and last DNS-responses Δt_q for group query for domain name d_i is greater than the duration of the time window t_s, then time interval Δt_q is divided into z intervals:

$$z = (t_{last} - t_{first}) / ((1/3) \cdot t_s), \tag{1}$$

where t_{last} and t_{first} – time of the last and first DNS-responses for domain name d_i within the TTL, during which the group activity is searching or the group flushing of local DNS-caches is fixed.

Such division makes it possible to minimize the number of queries that do not fall in the interval ts (Fig. 3a).

For group query we build the vector of density distribution of z-elements for queries in time $\overline{W_{d_i}} = (\Omega_j)_{j=1}^z$, where Ω_j – number of queries within the z-th interval. For the element of vector $\overline{W_{d_i}}$ with a maximum value Ω_{max} within $j = max \pm 2$, we find two adjacent elements with the largest values so that all three elements could describe the query distribution of continuous interval, and then we calculate their sum (Sum_s). If $(1 - \delta) \cdot Sum_s > Sum_r$, then the sets of MAC-addresses in the matrix V_{MAC} hosts groups are combined and the group of query is the subject, and is subject to further analysis, otherwise such group is discarded, where Sum_r – the sum of other vector elements $\overline{W_{d_i}}$ (Fig. 3b).

Building the Observation Matrix for Analysis the Incoming DNS-Traffic. In order to gather and analyze the incoming DNS-traffic during specified monitoring time t_k we build the observation matrix M_k, where k – number of observation iterations. It contains domain names d_i, requested by hosts' groups; MAC-addresses of hosts' groups H_j received from the matrix V_{MAC}; the sign of query to local/non-local DNS-servers S; the sign of repeated request within TTL-period, F; the sign of presence of the NXDOMAIN error code in DNS-responses, R; the sign of the hosts' group "infected" or "suspicious", obtained

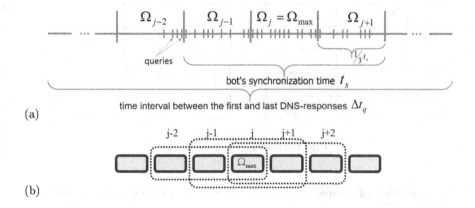

(a)

(b)

Fig. 3. Checking the synchrony of queries: (a) division the interval delta t into z intervals; (b) searching the greatest number of the DNS-queries during time t_s

at intermediate stages of analysis, M; number of observation iteration when the sign "suspicious" was fixed, N; number of hosts in the group, N_G.

If the requests' synchrony was observed, then the set of MAC-addresses H_j of the hosts' groups are transferred from the matrix V_{MAC} to the observation matrix M_k.

If the group flushing of the local hosts DNS-caches was observed, then the cell of the matrix observation $M_k(d_i, F)$ is marked as "1", otherwise – "0".

If the hosts' group have been requesting the domain name d_i to a local and other DNS-servers, then the cell of observation matrix $M_k(d_i, S)$ is marked as "0", if only to the local DNS-server – "0.5", if only to a non-local DNS-servers – "1". If the DNS-responses for this group contain NXDOMAIN error code, then the cell of observation matrix $M_k(d_i, R)$ is marked as "1", otherwise – "0". The cells of observation matrix $M_k(d_i, M)$ and $M_k(d_i, N)$ are filled with zeros. The number of MAC-addresses represented in the group is filled into the cell of observation matrix $M_k(d_i, N_G)$. The cell $M_k(d_i, F)$ will be filled at a next stage. Example of the observation matrix M_k is presented in Fig. 4.

5   The Similarity Evaluation of Hosts' Groups

In order to compare *two* groups of hosts G_1 and G_2, that sent the DNS-queries for two domain names d_1 and d_2 at time intervals Δt_1 and Δt_2 respectively, we used the Braun-Blanquet coefficient [5], which is asymmetric similarity measure and enables to define group similarity with high accuracy:

$$K_B(G_1, G_2) = N_o/(\max[N_{G_1}, N_{G_2}]),\qquad(2)$$

where N_o – the number of common elements in groups G_1 and G_2; N_{G_1} and N_{G_2} – the number of hosts in groups G_1 and G_2, respectively; $K_B(G_1, G_2) \in [0, 1]$.

	H₁	H₂	...	Hj	N_G	S	F	R	M	N
d₁	1	1	...	1	5	1	1	0	0	0
d₁	1	0	...	1	4	1	0	0	0	0
d₂	1	0	...	1	6	0.5	0	0	0	0
...	...	...	...	...	...	...	...	...	...	...

Fig. 4. Observation matrix M_k

If the number of compared groups is *more than two* we used the Koch index of dispersity [6]:

$$K_K(G_1, \ldots, G_q) = \frac{C - A}{(q - 1) \cdot A}, \tag{3}$$

where $G_1, \ldots, G_q$ – comparable groups of hosts; q – the number of comparable groups; $C = \sum_{i=1}^{q} N_{G_i}$ – total number of MAC-address in all groups; A – number of different MAC-addresses presented in groups; $K_k(G_1, \ldots, G_q) \in [0, 1]$.

We consider the group of hosts as infected if the similarity coefficient for group exceed the threshold $K_B \geq \delta$ or $K_K \geq \delta$, where δ – the similarity threshold. Also we apply an additional threshold similarity δ', which indicates a suspiciousness of the hosts' group, $\delta' \leq K_B < \delta$ or $\delta' \leq K_K < \delta$. We use MAC-address as the ID of host in the network.

Detection the Group Activity by Analysis of the Group Queries for the Same Domain Names. At this stage, we carry out an analysis of the observations matrix M_k in order to detect the groups' queries for the same domain name. For this purpose, we compare the groups by MAC-adresses. Depending on the number of group queries for a specific domain name d_i we choose the Braun-Blanquet coefficient to compare *two* groups or dispersion index Koch for *3 or more* groups (Fig. 5a). If the result of comparison exceeds the threshold $K_B \geq \delta$ or $K_K \geq \delta$, the hosts' group are considered as infected.

If the result of the comparison is $\delta' \leq K_B < \delta$ or $\delta' \leq K_K < \delta$, then we carry out an additional analysis of the observation matrix M_k whether group queries do not honor TTL $M_k(d_i, F) = 1$; and non-local whether group use non-local DNS-servers $M_k(d_i, S) = 1$. If *any* of group queries does not honor TTL, or *all* queries to non-local DNS-servers were observed, then the hosts' group is considered as infected.

If groups that queried the same domain name are defined as infected or suspicious, we combine the set of MAC-addresses into one row for the domain name d in the matrix M_k (Fig. 5b). It will be used in further search for related to the group queries. If the hosts' group is identified as infected the cell of the observation matrix $M_k(d, M)$ is marked as "1", and as "0.5" – if it was defined as suspicious.

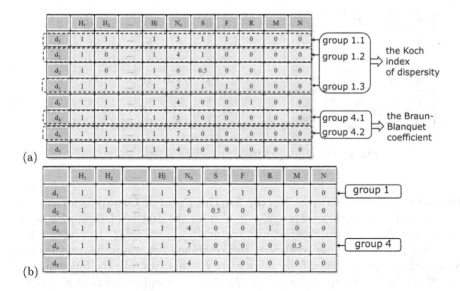

Fig. 5. Observation matrix M_k: (a) comparison the groups' queries for the same domain name d_i; (b) observation matrix M_k after the set of MAC-addresses is combine into one row for the domain name d_i

In these cases, the cells of the observation matrix $M_k(d, F)$ and $M_k(d, S)$ are marked as "1", domain name is entered into the list of malicious domain names. Otherwise, the group is considered as suspicious and the cell of observation matrix $M_k(d, F)$ is marked as "0". If the groups' queries were carried out to local and other DNS-servers the cell of observation matrix $M_k(d_i, S)$ is marked as "0"; if the groups' queries were carried out only to local DNS-servers the cell of observation matrix $M_k(d_i, S)$ is marked as "0.5". If the DNS-responses for one of the groups contained NXDOMAIN error code, then the cell of observations matrix $M_k(d, R)$ is marked as "1", otherwise – "0".

If none of these conditions is not satisfied, group queries for such the domain name are removed from the observation matrix M_k.

The Construction of the Lower Triangular Matrix for the Braun-Blanquet Coefficients. On the basis of the observation matrix M_k we built the lower triangular matrix for the Braun-Blanquet coefficients B_k. All features N_G, S, F, R, M, N from the matrix M_k are transferred to the matrix B_k. The rows of the matrix B_k are formed by increasing number of MAC-addresses in groups N_G columnwise. Also the Braun-Blanquet coefficients are filled in the matrix, which were calculated for pairs of hosts' groups (Fig. 6). Calculation of the values for the column cells is terminated if $N_{G_i}/N_{G_{i+1}} < \delta'$. Empty cells of the matrix B_k shown in Fig. 6 demonstrate, that the calculation is terminated. It will reduce time and computational resources required for analysis.

	d_3	d_5	d_1	d_2	d_4	...	N_Q	S	F	R	M	N
d_3	1					...	4	0	0	1	0	0
d_5	1	1				...	4	0	0	0	0	0
d_1	0.8	0.8	1			...	5	1	1	0	1	0
d_2			0.5	1		...	6	0.5	0	0	0	0
d_4			0.71	0.43	1	...	7	0	0	0	0.5	0

Fig. 6. The matrix of Braun-Blanquet coefficients B_k

Formation of the Features Vector for the Pairs of Group Queries. For each pair of group queries when $K_B \geq \delta'$ from the matrix of Braun-Blanquet coefficients B_k we form the features vector $\overline{W_{G_1,G_2}}$ (Fig. 7). It contains five elements: Braun-Blanquet coefficient and combined behavioral features for two compared groups, obtained from the matrix B_k that can take the following values: "Unusual" (not typical for bots), "Neutral" (typical for users as well as for bots), "Suspicious", "Dangerous" (typical for bots). The features vector $\overline{W_{G_1,G_2}}$ can be defined:

$$\overline{W_{G_1,G_2}} = (K_B(G_1, G_2), S_{G_1,G_2}, F_{G_1,G_2}, R_{G_1,G_2}, M_{G_1,G_2}), \qquad (4)$$

where $S_{G_1,G_2}, F_{G_1,G_2}, R_{G_1,G_2}, M_{G_1,G_2}$ – combined behavioral features for two compared groups.

Combined behavioral features S_{G_1,G_2} and M_{G_1,G_2} can be defined as follows:

$$S_{G_1,G_2} = \begin{cases} \text{Unusual,} & \text{if } B_k(d_1, S) = B_k(d_2, S) = 0, \\ \text{Neutral,} & \text{if } B_k(d_1, S) = B_k(d_2, S) = 0.5, \\ \text{Dangerous, if } B_k(d_1, S) = B_k(d_2, S) = 1, \\ \text{Suspicious} & \text{otherwise.} \end{cases} \qquad (5)$$

	d_3	d_5	d_1	d_2	d_4	...	N_Q	S	F	R	M	N
d_3	1					...	4	0	0	1	0	0
d_5	1	1				...	4	0	0	0	0	0
d_1	0.8	0.8	1			...	5	1	1	0	1	0
d_2			0.5	1		...	6	0.5	0	0	0	0
d_4			0.71	0.43	1	...	7	0	0	0	0.5	0

$K_b(G_1, G_4)$	S_{G_1,G_4}	F_{G_1,G_4}	R_{G_1,G_4}	M_{G_1,G_4}
0.71	Suspicious	Suspicious	Neutral	Dangerous

Fig. 7. Formation of the features vector $\overline{W_{G_1,G_2}}$ for the pairs of group queries from the matrix of Braun-Blanquet coefficients B_k

$$M_{G_1,G_2} = \begin{cases} \text{Neutral,} & \text{if } B_k(d_1, M) = B_k(d_2, M) = 0, \\ \text{Suspicious, if } ((B_k(d_1, M) = 0.5 \vee B_k(d_2, M) = 0.5) \wedge \\ \quad \wedge B_k(d_1, M) \neq 1 \wedge B_k(d_2, M) \neq 1) \wedge \\ \quad \wedge B_k(d_1, M) \neq B_k(d_2, M)), \\ \text{Dangerous, if } B_k(d_1, M) = 1 \vee B_k(d_2, M) = 1 \vee \\ \quad \vee (B_k(d_1, M) = B_k(d_2, M) = 0.5 \wedge B_k(d_1, N) \neq \\ \quad \neq B_k(d_2, N) \vee B_k(d_1, N) = B_k(d_2, N) = 0), \end{cases} \tag{6}$$

where $B_k(d_1, N)$ and $B_k(d_2, N)$ – the number of iterations k, which is filled only for groups identified as suspicious at the stage of the features vectors analysis during each iteration.

Combined features F_{G_1,G_2} and R_{G_1,G_2} are defined similarly, for example:

$$F_{G_1,G_2} = \begin{cases} \text{Neutral,} & \text{if } B_k(d_1, F) = B_k(d_2, F) = 0, \\ \text{Suspicious,} & \text{if } B_k(d_1, F) \neq B_k(d_2, F), \\ \text{Dangerous,} & \text{if } B_k(d_1, F) = B_k(d_2, F) = 1. \end{cases} \tag{7}$$

Analysis of Features Vectors for Detection of the Infected Hosts. Analysis of the features vectors $\overline{W_{G_1,G_2}}$ is performed by the following rules, where the output function f can take four values "Not_Infected", "Not_Suspicious", "Suspicious", "Infected":

$$f(\overline{W_{G_1,G_2}}) = \begin{cases} \text{Not\_Infected,} & \text{if } K_B(G_1, G_2) < \delta \wedge S_{G_1,G_2} = \text{Unusual} \wedge \\ & \quad \wedge \forall \overline{W_{G_1,G_2}}(j) \neq \text{Suspicious} \wedge \\ & \quad \wedge \forall \overline{W_{G_1,G_2}}(j) \neq \text{Dangerous,} \\ \text{Not\_Suspicious, if } K_B(G_1, G_2) < \delta \wedge S_{G_1,G_2} \neq \text{Unusual} \wedge \\ & \quad \wedge \forall \overline{W_{G_1,G_2}}(j) \neq \text{Suspicious} \wedge \\ & \quad \wedge \forall \overline{W_{G_1,G_2}}(j) \neq \text{Dangerous,} \\ \text{Infected,} & \text{if } \exists \overline{W_{G_1,G_2}}(j) = \text{Dangerous} \vee K_B(G_1, G_2) \geq \delta, \\ \text{Suspicious} & \text{otherwise.} \end{cases} \tag{8}$$

where $j = \overline{2,5}$ – number of the element in the features vector.

The same group within the iteration can get several different values. In this case, a highest value of danger is chosen. Hosts' groups that were identified as not infected are discarded. Concerning to hosts' groups which are identified as infected, the measures for infection elimination are applied (blocking, eliminating of the system vulnerabilities, installation (upgrade) antivirus software, etc.).

Hosts' groups from the observation matrix M_k, which were not added to the matrix of Braun-Blanquet coefficients B_k, and groups for which the condition $K_B \geq \delta'$ is not satisfied, and groups, that weren't identified as non-suspicious or suspicious will be analyzed with the data that will be received on the next iteration of observation (observation matrix M_{k+1}) in order to detect possible repeated group requests.

At the same time, if the group that queried the domain name d was defined as suspicious, for this group in the cell of observation matrix $M_{k+1}(d, M)$ is marked as "0.5", and the iteration number k is filled in the cell of observation matrix $M_{k+1}(d, N)$.

However, if the group that queried about domain name d, was defined as suspicious, for this group we put "0.5", and the iteration number k is filled in the cell of observation matrix $M_{k+1}(d, N)$.

6 Experiments

In order to determine the efficiency of the proposed method several experiments were held. For this purpose the bots of the SDBot family with the centralized architecture were generated. For experiments the university network with 100 hosts was used. Hosts were infected with bots.

Each experiment lasted 8 h. DNS-traffic of the local network was captured by means of tcpdump utility (Fig. 8).

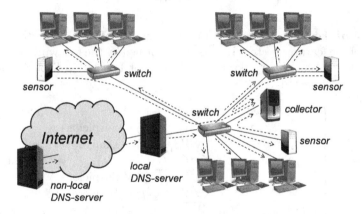

Fig. 8. Scheme of the network used for botnet detection

In order to determine the optimal parameters of the proposed method the experiments concerning to the synchronization time of the requests t_s and similarity threshold δ were held.

In order to identify the optimal value of the synchronization time of requests t_s (in seconds) the experiments in the range $t_s = \overline{0, 30}$ (in increments of 2 s) were conducted.

In order to identify the optimal value of similarity threshold δ, two groups of experiments for values of $\delta = 0.7$ and $\delta = 0.8$ were conducted.

Also, we investigated the efficiency of the method in situation when different number of hosts were infected (during 1st experiment 90 hosts were infected by bots, and during the 2nd experiment only 10 hosts were infected).

The results of experimental studies are presented in Fig. 9.

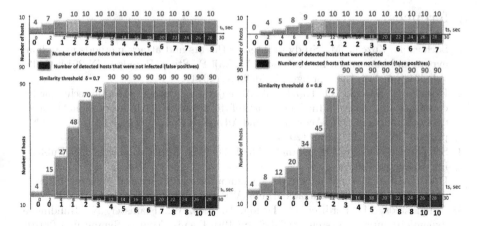

Fig. 9. The results of experimental studies: (a) similarity threshold $\delta = 0.7$; (b) similarity threshold $\delta = 0.8$

It was found that the decrease of value t_s leads to reduction of the botnet detection rate, while the increasing of value t_s leads to increasing of the detection rate, but also leads to increasing the false positives rate. Decrease of the similarity threshold δ also demonstrates the increasing of the detection rate, but it increases the false positives rate. Thus, the experimentally was found that the maximum detection rate with minimal false positives is achieved with values of synchronization time t_s from 12 to 16 sec and value of the similarity threshold $\delta = 0.8$.

It worth mentioning that a limitation of the method is the inability to detect the hosts' infections in situation when group contains less than 4 hosts.

7   Conclusions

A technique for botnet detection based on DNS-traffic is developed. Botnets detection is based on the property of bots group activity in the DNS-traffic, which appears in a small period of time in the group DNS-queries of hosts during trying to access the C&C-servers, migrations, running commands or downloading the updates of the malware. The method takes into account abnormal behaviors of the hosts' group, which are similar to botnets: hosts' group do not honor DNS TTL, carry out the DNS-queries to non-local DNS-servers. In addition, method monitors large number of empty DNS-responses with NXDOMAIN error code.

The method is able to detect unknown bots on initial stage of hosts infection in the corporate network. Method is applicable both to a small and to large networks and does not need high computational resources for data processing.

References

1. Sochor, T., Zuzcak, M.: Study of internet threats and attack methods using honeypots and honeynets. In: Kwiecień, A., Gaj, P., Stera, P. (eds.) CN 2014. CCIS, vol. 431, pp. 118–127. Springer, Heidelberg (2014)
2. Lysenko, S., Savenko, O., Kryshchuk, A., Kljots, Y.: Botnet detection technique for corporate area network. In: Proceedings of the 2013 IEEE 7th International Conference on Intelligent Data Acquisition and Advanced Computing Systems (IDAACS), pp. 363–368. IEEE, Berlin (2013)
3. Schiller, C., Binkley, J.R.: Botnets: The Killer Web Application. Syngress Publishing, Rockland (2007)
4. DAMBALLA. Botnet Detection for Communications Service Providers. https://www.damballa.com/downloads/r_pubs/WP_Botnet_Detection_for_CSPs.pdf
5. Antonakakis, M., Perdisci, R., Dagon, D., Lee, W., Feamster, N.: Building a dynamic reputation system for DNS. In: 19th Usenix Security Symposium (2010)
6. Bilge, L., Kirda, E., Kruegel, C., Balduzzi, M.: EXPOSURE: finding malicious domains using passive DNS analysis. In: NDSS (2011)
7. Villamarín-Salomón, R., Brustoloni, J.C.: Identifying botnets using anomaly detection techniques appliedto DNS traffic. In: Consumer Communications and Networking Conference (2008)
8. Choi, H., Lee, H., Lee, H., Kim, H.: Botnet detection by monitoring group activities in DNS traffic. In: Seventh IEEE International Conference on Computer and Information Technology, pp. 715–720 (2007)
9. Manasrah, A.M., Hasan, A., Abouabdalla, O.A., Ramadass, S.: Detecting botnet activities based on abnormal DNS traffic. Int. J. Comput. Sci. Inf. Secur. (IJCSIS) **6**(1), 97–104 (2009)
10. Choi, H., Lee, H.: Identifying botnets by capturing group activities in DNS traffic. Comput. Netw. **56**, 20–33 (2012)
11. Roshna, R.S., Vinodh, E.: Botnet detection using adaptive neuro Fuzzy inference system. Int. J. Eng. Res. Appl. (IJERA) **3**(2), 1440–1445 (2013)

SysML-Based Modeling of Token Passing Paradigm in Distributed Control Systems

Marcin Jamro and Dariusz Rzonca[⊠]

Department of Computer and Control Engineering, Rzeszow University
of Technology, al. Powstancow Warszawy 12, 35-959 Rzeszow, Poland
{mjamro,drzonca}@kia.prz.edu.pl
http://kia.prz.edu.pl

Abstract. Distributed control systems are often used in many branches
of industry and frequently replace standalone controllers. However, their
operation is more complex and include aspects of communication between
various devices. To operate correctly, it is crucial to ensure that timeliness
of communication is satisfied. In this paper, the approach to modeling of
the token passing paradigm, as well as the multi-master communication
with token exchange has been presented. The proposed models are based
on a few kinds of SysML diagrams, namely Block Definition, Internal
Block, State Machine, and Sequence Diagrams. The paper presents a set
of dedicating modeling rules together with their detailed explanation.

Keywords: Control systems · Communication · Modeling · Token
passing

1 Introduction

Distributed control systems (DCSs) are frequently used in industry. They perform various operations, such as controlling the process and devices. According to the distributed nature of DCSs, it is crucial to provide efficient and reliable communication between various elements inside the system. This assumption is important, because improper communication could lead to serious problems, including performing operations using incorrect or outdated values of variables, what in turn could have a negative impact on the whole automation system.

Various paradigms of access to the communication link may be used in DCSs, depending on a bus type and system architecture [1]. Typically, they are based on the Time Division Multiplexing (TDM) scheme, however, the Frequency Division Multiplexing method has been also discussed [2]. The commonly used TDM models are master-slave, token passing and producer-distributor-consumer [3]. Each of them assigns appropriate time slots to every device connected to the common bus, thus defines a scenario of data exchanges in a particular case. Such parameters are usually fixed, however, an interesting concept of changing them on-line to improve flexibility and reliability has been considered in [4]. Industrial applications typically involve a fieldbus [5] and one of the field protocol defined in the IEC 61158 standard [6]. Nonstandard solutions, such as using Hypertext Transfer Protocol (HTTP) as a field protocol, have been also described [7,8].

© Springer International Publishing Switzerland 2015
P. Gaj et al. (Eds.): CN 2015, CCIS 522, pp. 139–149, 2015.
DOI: 10.1007/978-3-319-19419-6_13

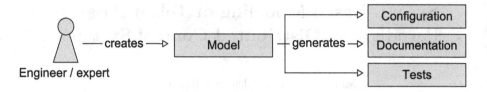

Fig. 1. Model as a source of data for configuration, documentation, and tests

In this paper, the authors present an approach to modeling of communication between devices in DCSs that uses the token passing. The model is created in the Systems Modeling Language (SysML) [9], which is a profile of the Unified Modeling Language (UML) [10]. For the purpose of this article, four kinds of SysML diagrams are used, namely Block Definition Diagram (BDD), Internal Block Diagram (IBD), State Machine Diagram (STM), and Sequence Diagram (SD). What is more, four dedicated modeling rules have been proposed.

Using this research, the model of communication could be included in the overall system model and parts of communication configuration could be generated automatically, together with documentation (Fig. 1). What is more, the model may be used as a source for tests verifying requirements of communication performance between particular devices in a DCS. Such tests could behave similarly to their versions for the master-slave paradigm, as described in [11,12].

The current work extends the overall methodology of modeling a control system, whose other parts are explained in [13–16]. Their application in the CPDev engineering environment for development of software for various programming controllers and DCSs is mentioned by the authors in [17].

The paper is organized as follows. In the next section, short information about related work is presented. Then, the modeling approach regarding the token passing communication is introduced. Such a topic is particularized in the two following sections. Modeling of the system structure is explained in Sect. 4, while modeling of communication in Sect. 5.

2   Related Work

The topic of modeling of control software is frequently analyzed by researchers. It is related to various aspects of systems, including their structure, behavior, requirements, as well as tests and configuration. The modeling is beneficial for the overall control software quality for many reasons [18], such as increasing the abstraction level, automatic generation of implementation or configuration, simplification of modification and maintenance, as well as presentation of the system in a more readable form, even for people who do not have specific expert knowledge in the software engineering domain.

Introduction of modeling into the software development process could be performed variously. This research assumes that engineers comply with the Model-Driven Development (MDD) paradigm [18], which uses the model as a primary artifact in the software development process. Such an artifact is later used to generate a part or whole implementation in an automatic way. The model contains data on various abstraction levels, both high-level with only the overall view of the system and low-level that presents various aspects with details. It is worth mentioning that MDD is not the only paradigm related to modeling. Among others, Model Driven Architecture (MDA) [19, 20], Model Driven Engineering (MDE) [21], or Model Based Testing (MBT) [22] exist. The model could be created using various languages and methods, among which formal (such as various classes of Petri Nets [23, 24]) and semiformal (UML and SysML) exist.

There are several papers regarding modeling of automation systems using semiformal languages. Some of them are also related to modeling of communication. For example, Wenger et al. [25] create the model to configure communication in networked industrial automation systems. They use two modeling languages, namely Architecture Analysis and Design Language (AADL) for a software part and Field Device Configuration Markup Language (FDCML) for hardware. The model transformation is used to integrate such models. It is worth mentioning that similar topic is also researched by Schimmel et al. [26].

Among other papers, the MDD processes for control and automation software are proposed and described by Hastbacka et al. [18], Thramboulidis et al. [27], as well as Zaeh and Poernbacher [28]. The overall MDD approach consistent with the solution from the current paper is explained in [13, 14].

3   Token Passing-Based Modeling Approach

The token passing is one of the commonly used paradigms defining a deterministic access to the communication link. The main idea is that the devices connected to the common bus share so-called "token", which can be understood as the right to control the bus and send messages. Only one station at the time, the one which currently holds the token, might transmit data over the network, while other devices are listening. After defined amount of time, the station with token passes it to another device. The token itself is a special message, also transmitted by the bus. Such an exchange of token is repeated periodically, thus every station has a deterministic communication time slot.

While such strictly token passing networks are indeed used in industry [29], sometimes this paradigm is combined with other ones, typically master-slave. In the master-slave network, one of the devices (master) controls data flow over the common bus. Other devices (slaves) transmit data only in response to messages sent by the master, but they cannot initiate data exchange themselves. The token-passing paradigm may be easily enhanced by adapting the master-slave mechanism, thus forms the multi-master communication network with token exchange between master stations. Such a solution is used e.g. in the Profibus protocol [30]. A similar generalized approach is considered in this paper. An exemplary network is shown in Fig. 2.

It is assumed that numerous devices are connected to the common data bus. Some of them are masters exchanging token according to token-passing rules.

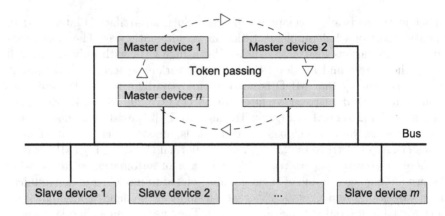

Fig. 2. Token-passing paradigm of medium access

During time slots defined by the token flow, the master currently having the token may communicate with other devices (slaves). Such repeated periodically data exchange between devices forms a communication task. Each master may serve numerous communication tasks during its time slot, if time permits.

The proposed approach to modeling of token passing-based communication in DCSs requires to prepare four diagrams in SysML. They present:

- a structure of the whole system (Sect. 4),
- physical connections between devices involved in communication (Sect. 4),
- a sequence of passing the token between devices (Sect. 4),
- specification of communication tasks (Sect. 5),
- assignment of communication tasks to connections between devices (Sect. 5),
- values exchanged during communication tasks (Sect. 5).

4   Modeling of System Structure

The first step of the proposed methodology is modeling of the system structure. It includes specification of bus and available devices, together with marking a role of each of them as a master or slave. The overall model is presented in Fig. 3. It is created according to the following modeling rule:

- The model is placed on the BDD diagram named *Overall*, located in the *Structure* package, as specified in the diagram header.
- The diagram contains the «*dcs*» block, named *SYSTEM*, representing a DCS.
- The diagram contains «*node*» blocks indicating devices involved in the token passing communication. They are connected with the «*dcs*» block via the composition relationship.
- Each «*node*» element contains the *address* parameter. It should be unique among all devices, either master or slave.
- Master devices are marked with additional stereotype «*master*» (together with «*node*»), while slave devices – with «*slave*».

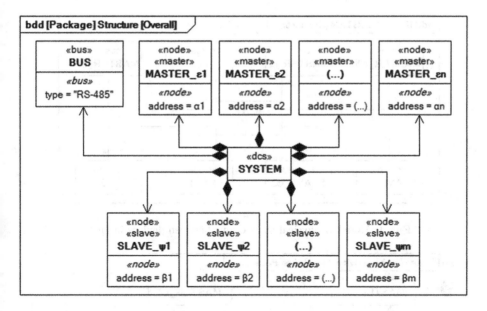

Fig. 3. Model of the DCS structure

- The bus is represented as a «*bus*» element. It contains the *type* parameter indicating a type of the communication link, such as RS-485.

In the next step, the prepared diagram is particularized by specification of physical connections between devices in a DCS. Such a diagram is created according to the overall model from Fig. 4 and the following modeling rule:

- The model is placed on the BDD diagram, named *Connections*, specifying the *SYSTEM* block from the previous modeling rule (related to Fig. 3).
- The diagram contains the «*bus*» element (representing the bus), as well as «*master*» and «*slave*» that indicate devices involved in the communication.
- Paths on the diagram specify physical connections in the DCS.

After specification of the DCS structure and physical connections between elements, it is necessary to provide information about a sequence of token passing between master devices. This task is performed using another kind of diagram, as presented in Fig. 5 and explained in the following modeling rule:

- The model is placed on the STM diagram, named *Token Passing*, located in the *Structure* package.
- The diagram contains states indicating all master devices that share the token. Each state has a name equal to the device name.
- States are connected using time-dependent transitions indicating expected value of how long the token should stay in a particular device.
- The diagram contains the initial state that is connected with a master device that starts the token passing procedure.

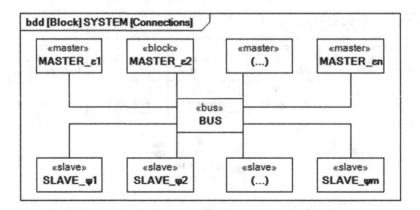

Fig. 4. Model of physical connections between devices in DCS

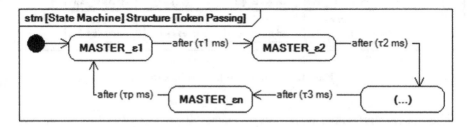

Fig. 5. Model of the token passing sequence

5   Modeling of Communication Tasks

Modeling of the system structure is not enough to create the overall model of communication using the token passing paradigm. Another part of the model involves specification of communication tasks and their assignment to particular connections between devices in a DCS. It is assumed that a *communication task* is a cyclical execution of a communication transaction between two devices – master and slave. The communication task is executed with a given cycle time.

At the beginning, communication tasks for each master device are specified. The overall diagram is presented in Fig. 6. It is created according to the following modeling rule:

- The model is placed on the BDD diagram located in the *Communication Tasks* package. The diagram name is equal to the master device name.
- The diagram contains elements with the *«comTask»* stereotype. Each represents a single communication task and contains a unique name.

The modeled communication tasks are used on the next diagrams, such as for assignment to a particular connection. It is accomplished in a way shown in the overall model from Fig. 7 and explained in the following modeling rule:

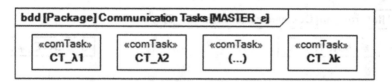

Fig. 6. Model of communication tasks assigned to a particular master device

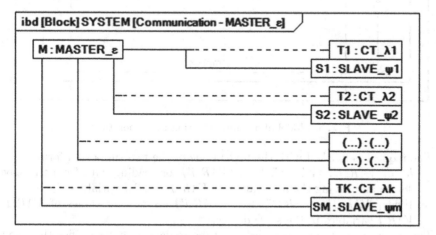

Fig. 7. Model of assignment of communication tasks to connections between devices

- The model is placed on the IBD diagram specifying the *SYSTEM* block. The diagram name is equal to *Communication - [master device name]*.
- The diagram contains an instance of the «*master*» block from Fig. 3.
- The diagram contains instances of «*slave*» blocks from Fig. 3 that represent devices communicating with the given master device.
- Instance of the master device is connected with instances of proper slave devices to indicate exchange of data between them.
- The communication task instance is assigned to each connection using the association block.

The last group of necessary diagrams specifies details of communication tasks in the system. For each of them, an engineer should define an operation (read or write), as well as indicate variables where communication results are stored or from where data are taken before communication. The overall diagram is shown in Fig. 8. It is created according to the following modeling rule:

- The model is placed on the SD diagram, located in the *Communication Tasks* package. The diagram name is equal to the communication task name.
- The diagram contains two lifelines. The first represents a master device, while the other – a slave device involved in the communication task.
- The diagram contains the *loop* interaction operator that indicate that the synchronous request and asynchronous response are performed in a loop.

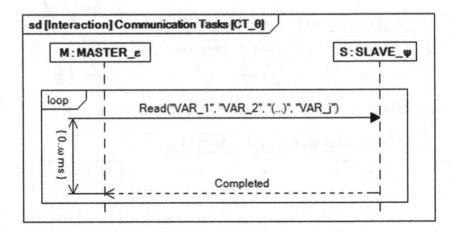

Fig. 8. Model of a communication task operation

- The request is marked with the text in one of the two supported forms:
  - *Read("VAR_1", "VAR_2", "...", "VAR_j")* for reading data from the slave device and saving them as values of *VAR_1–VAR_j* variables.
  - *Write("VAR_1", "VAR_2", "...", "VAR_l")* for writing values of *VAR_1– VAR_l* variables to the slave device.
- The diagram contains the additional time indicator that specifies the maximum waiting time between sending a request and receiving a response.

6   Applications and Limitations

The presented methodology has been designed especially for the CPDev engineering environment. However, the proposed approach is generalized to use similar concepts in other applications. One of the crucial goals to achieve is related to choosing a suitable semi-formal tool to support engineers in creating efficient communication subsystems for mini-sized DCSs. The SysML diagrams prepared during the modeling stage, according to the described set of rules, behave as a part of the system documentation, as well as are used as a source for generating the system configuration. By such a solution, a consistence between documentation and implementation is ensured.

The proposed approach allows to model numerous communication parameters directly on graphical SysML diagrams, including the communication tasks and the token passing sequence. Such parameters could be also used for further transformations and calculations. Finally, the generated communication settings are transferred to controllers as configuration values. Unfortunately, details of such implementation related to hardware possibilities of particular devices does not fit in the generalized approach and are out of the scope of this paper. Thus, the necessity of creating individual implementations for different hardware platforms is a significant limitation. However, the practical benefits from creating parts of documentation and configuration in a single common step are important, as well. Another possible limitation of the proposed methodology is related to their genesis. As stated before, the presented SysML-based approach has been designed especially for the CPDev engineering environment for DCSs based on CPDev and then generalized. Thus, it is possible that different engineering tools or DCSs may need slight changes in the way of modeling.

It is worth mentioning that SysML models can also be used for generation of test cases to check performance of communication between devices in a DCS. The parameters specified in the model are taken into account to check correctness and performance of a communication subsystem. With such a mechanism that is possible to verify whether the assumed parameters are obtainable, such as weather the times in token passing sequence are feasible in case of given communication tasks. What is more, tests could verify performance of the final prototype, similarly as described in [11,12]. Such a topic is currently under research and the next results will be described in the future papers.

7   Conclusion

Nowadays, more and more automation systems are developed as distributed control systems. Such solutions involve advanced communication mechanisms whose improper operation could have a negative impact on correctness and stability of the whole control system. For this reason, it is important to propose various approaches that simplify configuration of communication.

One of possible solutions involves modeling using a graphical modeling language, such as SysML, and is presented in this paper. It enhances the previous research, related to the master-slave protocols, by supporting the token passing paradigm, as well as multi-master communication with token exchange between masters. Such a solution complies with the Model-Driven Development approach, therefore simplifies creation of an efficient communication subsystem.

The planned future work includes a semi-automatic generation of test cases to verify communication performance in DCSs based on the token passing paradigm, similarly as it has been already done for the master-slave models.

References

1. Silva, M., Pereira, F., Soares, F., Leao, C., Machado, J., Carvalho, V.: An overview of industrial communication networks. In: Flores, P., Viadero, F. (eds.) New Trends in Mechanism and Machine Science. Mechanisms and Machine Science, vol. 24, pp. 933–940. Springer International Publishing, Switzerland (2015)
2. Stój, J.: Real-time communication network concept based on frequency division multiplexing. In: Kwiecień, A., Gaj, P., Stera, P. (eds.) CN 2012. CCIS, vol. 291, pp. 247–260. Springer, Heidelberg (2012)
3. Gaj, P., Jasperneite, J., Felser, M.: Computer communication within industrial distributed environment - a survey. IEEE Trans. Ind. Inf. 9(1), 182–189 (2013)
4. Kwiecień, A., Sidzina, M., Maćkowski, M.: The concept of using multi-protocol nodes in real-time distributed systems for increasing communication reliability. In: Kwiecień, A., Gaj, P., Stera, P. (eds.) CN 2013. CCIS, vol. 370, pp. 177–188. Springer, Heidelberg (2013)

5. Thomesse, J.P.: Fieldbus technology in industrial automation. Proc. IEEE **93**(6), 1073–1101 (2005)
6. IEC 61158 Standard: Industrial Communication Networks - Fieldbus Specifications (2007)
7. Jestratjew, A., Kwiecień, A.: Using HTTP as field network transfer protocol. In: Kwiecień, A., Gaj, P., Stera, P. (eds.) CN 2011. CCIS, vol. 160, pp. 306–313. Springer, Heidelberg (2011)
8. Jestratjew, A., Kwiecien, A.: Performance of HTTP protocol in networked control systems. IEEE Trans. Ind. Inf. **9**(1), 271–276 (2013)
9. OMG: OMG Systems Modeling Language, V1.3 (2012)
10. OMG: OMG Unified Modeling Language, Infrastructure, V2.4.1 (2011)
11. Jamro, M., Rzońca, D., Trybus, B.: Communication performance tests in distributed control systems. In: Kwiecień, A., Gaj, P., Stera, P. (eds.) CN 2013. CCIS, vol. 370, pp. 200–209. Springer, Heidelberg (2013)
12. Jamro, M., Rzonca, D.: Measuring, monitoring, and analysis of communication transactions performance in distributed control system. In: Kwiecień, A., Gaj, P., Stera, P. (eds.) CN 2014. CCIS, vol. 431, pp. 147–156. Springer, Heidelberg (2014)
13. Jamro, M., Trybus, B.: An approach to SysML modeling of IEC 61131-3 control software. In: 2013 18th International Conference on Methods and Models in Automation and Robotics (MMAR), pp. 217–222 (2013)
14. Jamro, M.: Automatic generation of implementation in SysML-based model-driven development for IEC 61131-3 control software. In: 2014 19th International Conference on Methods and Models in Automation and Robotics (MMAR), pp. 468–473 (2014)
15. Jamro, M.: SysML modeling of POU-oriented unit tests for IEC 61131-3 control software. In: 2014 19th International Conference on Methods and Models in Automation and Robotics (MMAR), pp. 82–87 (2014)
16. Jamro, M.: Development and execution of POU-oiented performance tests for IEC 61131-3 control software. In: Szewczyk, R., Zieliński, C., Kaliczyńska, M. (eds.) Recent Advances in Automation, Robotics and Measuring Techniques. AISC, vol. 267, pp. 91–102. Springer, Heidelberg (2014)
17. Jamro, M., Rzonca, D., Sadolewski, J., Stec, A., Swider, Z., Trybus, B., Trybus, L.: CPDev engineering environment for modeling, implementation, testing, and visualization of control software. In: Szewczyk, R., Zieliński, C., Kaliczyńska, M. (eds.) Recent Advances in Automation, Robotics and Measuring Techniques. AISC, vol. 267, pp. 81–90. Springer, Heidelberg (2014)
18. Hastbacka, D., Vepsalainen, T., Kuikka, S.: Model-driven development of industrial process control applications. J. Syst. Softw. **84**(7), 1100–1113 (2011)
19. OMG: MDA Specifications. OMG (2014)
20. Azmoodeh, M., Georgalas, N., Fisher, S.: Model-driven systems development and integration environment. BT Technol. J. **23**(3), 96–110 (2005)
21. Marcos, M., Estevez, E., Iriondo, N., Orive, D.: Analysis and validation of IEC 61131-3 applications using a MDE approach. In: 2010 IEEE Conference on Emerging Technologies and Factory Automation (ETFA), pp. 1–8 (2010)
22. Saifan, A., Dingel, J.: A survey of using model-based testing to improve quality attributes in distributed systems. In: Elleithy, K. (ed.) Advanced Techniques in Computing Sciences and Software Engineering, pp. 283–288. Springer, Netherlands (2010)
23. Gniewek, L.: Sequential control algorithm in the form of Fuzzy interpreted petri net. IEEE Trans. Syst. Man Cybern.: Syst. **43**(2), 451–459 (2013)

24. Olejnik, R.: Modelling of half-duplex radio access for hopemesh experimental WMN using petri nets. In: Kwiecień, A., Gaj, P., Stera, P. (eds.) CN 2014. CCIS, vol. 431, pp. 108–117. Springer, Heidelberg (2014)

25. Wenger, M., Zoitl, A., Froschauer, R., Rooker, M., Ebenhofer, G., Strasser, T.: Model-driven engineering of networked industrial automation systems. In: 8th IEEE International Conference on Industrial Informatics (INDIN) 2010, pp. 902–907 (2010)

26. Schimmel, A., Zoitl, A., Froschauer, R., Rooker, M., Ebenhofer, G.: Model-driven communication routing in industrial automation and control systems. In: 2010 8th IEEE International Conference on Industrial Informatics (INDIN), pp. 896–901 (2010)

27. Thramboulidis, K., Perdikis, D., Kantas, S.: Model driven development of distributed control applications. Int. J. Adv. Manuf. Technol. 33(3–4), 233–242 (2007)

28. Zaeh, M., Poernbacher, C.: Model-driven development of PLC software for machine tools. Prod. Eng. 2(1), 39–46 (2008)

29. Zhou, Z., Tang, B., Xu, C.: Design of distributed industrial monitoring system based on virtual token ring. In: 2nd IEEE Conference on Industrial Electronics and Applications, 2007, ICIEA 2007, pp. 598–603, May 2007

30. Tovar, E., Vasques, F.: Real-time fieldbus communications using Profibus networks. IEEE Trans. Ind. Electron. 46(6), 1241–1251 (1999)

Automatic Scenario Selection of Cyclic Exchanges in Transmission via Two Buses

Andrzej Kwiecień, Błażej Kwiecień, and Michał Maćkowski[⊠]

Institute of Computer Science, Silesian University of Technology, Gliwice, Poland
{akwiecien,blazej.kwiecien,michal.mackowski}@polsl.pl

Abstract. The paper presents the theoretical and empirical tests which examine the algorithms of automatic scenario selection of cyclic exchange in transmission via two buses. The implemented method is the base for using the second (redundant) bus for data transmission. The main aim of this method is to build two-stream data transmission so that each data stream has similar, or even the same transmission time. The flexibility of the proposed solution results from the fact that it is useful not only for two-bus transmission, but also when the transmission is via only one bus. This method allows estimating the time of the network cycle and also indicates the values of the times for acyclic transmissions. The presented method is only the introduction to designing, developing and controlling the two-stream transmission.

Keywords: Distributed control system · Redundancy · Dual bus communication · Master-slave · Industrial networks · Distributed real-time systems · Medium access

1 Introduction

The current, commonly known purpose for using a redundant transmission bus is to reach an appropriate level of reliability in data delivering in all kind of systems, not just for industrial purposes [1–3].

Considering the reliability of a system (which can be increased thanks to the redundancy) or network overload (permanent or temporary) a problem may appear regarding how to increase the flexibility of network connections in order to improve the time of data exchange. The mentioned redundancy [3–5] can also be used as a way of improving the transmission time parameters [4] and, in addition, it may increase the bandwidth [2,6].

The main goal of redundancy is to ensure a correctly operating system. In case of a system failure, appropriate prevention must be implemented to protect the system against the loss of important information (e.g. losing the object state in a real-time system). The principle aim for using a redundant system is to increase the reliability of a communication system [7].

Therefore, the idea of transmitting data via the unused bus (different than via the primary bus) obtained by dividing tasks between the redundant buses,

© Springer International Publishing Switzerland 2015
P. Gaj et al. (Eds.): CN 2015, CCIS 522, pp. 150–161, 2015.
DOI: 10.1007/978-3-319-19419-6_14

seems to be an interesting issue. It is obvious that if using an additional bus in the same conditions as the primary one (when the system is working properly), then it could increase the mentioned useful bandwidth.

Thus, when using a redundant system for parallel data transmission via two independent ways, it should be remembered that it is possible only if both buses are working properly [8]. Requirements for such redundant systems are presented in details in standard [9]. In case of a failure of one of the buses, the parallel transmission must be stopped, and instead of two independent data streams only one should be transmitted. To reach that point it is necessary to develop algorithms which allow for continuous diagnosis of the communication status, detection and classification of a failure, and switch the data transmission onto one bus. Hence, two problems arouse:

- How to organize the data transmission via two buses having only the scenario of all cyclic exchanges in the system?
- How to detect the bus failure and then instead of two channels link organize one channel link?

It should be added that these two issues must respect the rules of the real time system [10–12]. Therefore, allowing transmission via two buses in current systems should definitely save time, but only if there are some methods for failure detection in real time and algorithms for data flow control.

Faced with many issues regarding the matter, the main purpose of the paper is only to develop algorithms for creating the scenarios of exchanges in case of double bus communication. Various methods and their results are presented in this paper.

2   Master/Slave Model in Industrial Solutions

The Master/Slave Model is very popular in industrial applications. It can occur as an individual solution or can be combined with other models (e.g. ProfiNet). Its structures are often described in literature [13,14]. For the needs of this paper it would be good to mention that in case of transmission in distributed structures the execution is based on two terms:

- Cyclic transmission.
- Acyclic (triggered) transmission.

While cyclic transmissions are triggered individually by the communication coprocessor of the master station and they are not changed during data exchanges, acyclic transmissions on the other hand are executed dynamically and are triggered by the main software of the Master station. It refers to both: their type (e.g. write, read) and their numbers during one network cycle.

Figure 1 presents the scenario of exchanges in transmission via a single bus. Particular symbols in Fig. 1 stands for:

$Z1 - Z6$ – symbolic sign of a single exchange (e.g. read/write from/to any subscriber). The heights of particular squares that represent the exchanges correspond to their duration; here: all exchanges have the same durations,

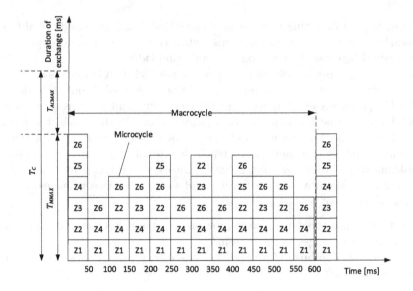

Fig. 1. Model of cyclic transmission (scenario of exchanges)

T_C – cycle duration of an exchange in the network,
T_{AC} – the remaining time for executing acyclic exchanges,
T_{MMAX} – maximum time of microcycle execution within a macrocycle,
T_{ACMAX} – maximum acceptable time for acyclic exchange.

In the abscissa in Fig. 1, the periodicity of exchanges is marked. The Fig. 1 shows that the minimal periodicity is 50 ms. Hence, the following microcycles are executed in every 50 ms.

If the requirements of real time are respected and the dependency (1) is fulfilled, then there is no ad hoc need to make any optimization of transmission – it is a very essential information at the designing phase.

$$T_C > T_{MMAX} + T_{ACMAX}. \tag{1}$$

The situation is completely different if the condition (1) is not fulfilled. In such case, the transmission via both buses can be the only way to realize the system requirements. Usually, for exchanges executing via one bus the scenario can be slightly modified so that all the microcycles can be loaded in a similar way. In case of transmission via two buses it is not essential.

3 Analysis of Redundant Master-Slave Communication System

Section three presents typical cases of analysis of monomaster systems with single and redundant buses.

3.1 The Duration of the Network Cycle in Case of a Single Bus Communication

To write the structure of a macrocycle in a formal manner, it is necessary to define the form of the macrocycle and to determine its maximum value of transmission time and the other parameters.

$\mathbf{C}_{TE}$ means the vector including values of transmission time variables according to the exchange scenario:

$$\mathbf{C}_{TE} = [T_{E1}, T_{E2}, T_{E3}, \ldots, T_{En}] \tag{2}$$

where: T_{Ei} stands for transmission time of the i exchange.

$\mathbf{P}$ stands for the vector values of all periodicities resulting from the scenario of exchanges:

$$\mathbf{P} = [p_1, p_2, p_3, \ldots, p_n] \tag{3}$$

where: p_i is the periodicity of the i exchange.

Time for executing a single i request-response transaction is:

$$T_{Ei} = T_{AM} + 2 * (T_{FP} + T_{FD} + T_{FA}) + T_{AS} + T_{FTMi} + T_{FTSi} \tag{4}$$

where:

T_{Ei} – time for executing the i request-response transmission,

T_{AM} – time measured from the moment of requesting transmission information during the automatic machine cycle of the Master till the start of the communication phase,

T_{FP} – time of frame preparation,

T_{FTMi} – time of the i frame transmission from the Master station,

T_{FTSi} – response transmission time of the i frame from the Slave station,

T_{FD} – time of frame detection,

T_{FA} – time of frame analysis,

T_{AS} – time measured from the moment of the transmission request during the automatic machine cycle of the Slave till the start of the communication phase.

Times T_{AM} and T_{AS} refer to the waiting time for the signal that finishes the cycle in the Master and Slave stations, whereas in an extreme case they mean the maximum cycle duration of the controllers. The time of frame preparation, detection and analysis are constant values for the Master and Slave stations. The transmission time of a single frame T_{FT} is represented in the below formula (5).

$$T_{FT} = \frac{L_{BT}}{V_t} \tag{5}$$

where:

T_{FT} – time of frame transmission,

L_{BT} – number of transmitted bits,

V_t – transmission bitrate.

Taking into account that in the scenario of information transmission there are n request-response transmissions, the total transmission time of the entire information in the scenario is as follows:

$$T_{TT} = \sum_{i=1}^{n} T_{Ei} \tag{6}$$

and the time for executing the i microcycle:

$$T_{MICR} = \sum_{i=1}^{l} T_{Ei} \tag{7}$$

where:

T_{TT} – the total transmission time of the entire information in the scenario,
T_{Ei} – the time of the i request-response transmission,
n – the number of request-response transmission in the scenario of the information transmission,
l – the number of exchanges in the i microcycle.

Moreover, GCD indicates the greatest common divisor of whole values of P vector, and LCM stands for the least common multiple of values of P vector. Therefore, the value L_M referring to the number of microcycles within the macrocycle can be written as:

$$L_M = \frac{LCM}{GCD}. \tag{8}$$

L stands for matrix in which the number of rows is L_M (number of microcycles), and the first column of matrix defines the number of exchanges in each microcycles. Each row of matrix describes one microcycle. Only one row of the matrix will be completed totally with the numbers of transactions. This is because whole transactions of exchanges are not executed in every microcycle. In microcycles in which only part of the transactions is executed the sequence T_{TN_i} is completed with the value 0 till the end.

$$L = \begin{bmatrix} L_{T_1} & T_{TN_i} & \cdots & T_{TN_n} \\ L_{T_2} & T_{TN_j} & \cdots & T_{TN_m} \\ \vdots & \vdots & \vdots & \vdots \\ L_{T_{LM}} & T_{TN_{LM}} & \cdots & T_{TN_z} \end{bmatrix}$$

where:

L_{T_i} – the number of transactions in the i microcycle,
seque.$\{T_{TN_i}\}$ – the sequence of numbers of transactions within the i microcycle.

The example of matrix L based on the realization of exchanges from Fig. 1 is shown in Fig. 2.

Creating matrix L we have two data for the algorithm series. Besides matrix L there is also vector T_{MICR} which defines the duration of each microcycle within the macrocycle. Having matrices and vectors: C_{TE}, P, L and T_{MICR} we have a full description of the transmission process.

$$L = \begin{bmatrix} 6 & Z1 & Z2 & Z3 & Z4 & Z5 & Z6 \\ 3 & Z1 & Z4 & Z6 & 0 & 0 & 0 \\ 4 & Z1 & Z4 & Z6 & Z2 & 0 & 0 \\ 4 & Z1 & Z4 & Z6 & Z3 & 0 & 0 \\ 5 & Z1 & Z4 & Z6 & Z2 & Z5 & 0 \\ 3 & Z1 & Z4 & Z6 & 0 & 0 & 0 \\ 5 & Z1 & Z4 & Z6 & Z3 & Z2 & 0 \\ 3 & Z1 & Z4 & Z6 & 0 & 0 & 0 \\ 5 & Z1 & Z4 & Z6 & Z5 & Z2 & 0 \\ 4 & Z1 & Z4 & Z6 & Z3 & 0 & 0 \\ 4 & Z1 & Z4 & Z6 & Z2 & 0 & 0 \\ 3 & Z1 & Z4 & Z6 & 0 & 0 & 0 \end{bmatrix}$$

Fig. 2. The form of L matrix for exchanges from Fig. 1

3.2 Defining the Value of Matrix L and Vector T_{MICR}

The algorithm for defining the values of matrix L and vector T_{MICR} is presented in the Fig. 3, where:

$$Per_{\min} = \min_{i=1...k} \{P[p_1, ..., p_k]\} \tag{9}$$

$$T_{\mathrm{MACR}} = \sum_{i=1}^{L_{\mathrm{M}}} T_{\mathrm{MICR}i} \tag{10}$$

$$T_{\mathrm{MICR}i} = \sum_{i=1}^{L_{\mathrm{T}i}} C_{\mathrm{TE}i} \tag{11}$$

$$NM \in \langle 1 ... L_{\mathrm{M}} \rangle \tag{12}$$

k – Number of periodicity values which is equal to the number of exchanges in the scenario,

T_{MACR} – Duration of the microcycle, being the sum of durations of all microcycles in the scenario of exchanges,

NM – Auxiliary variable that defines the number of the microcycle,

P, i, l – Other auxiliary variables,

Per – Current value of periodicity,

Rem – The remainder of the current value of periodicity divided by the periodicity of the next transaction,

$C_{\mathrm{TE}i}$ – Duration of particular transactions of exchanges in the i microcycle,

$T_{\mathrm{MICR}i}$ – Duration of the i microcycle.

Thanks to developing matrix L and vectors P, C_{TE} and T_{MICR}, we have a complete analytic description of the planned transmission process. Then, it helps to easily determine the maximum value of vector T_{MICR} for the planning transmission process, which in consequence enables determining the range of value T_{C} (for which the ability to realize the exchanges is available according to the scenario).

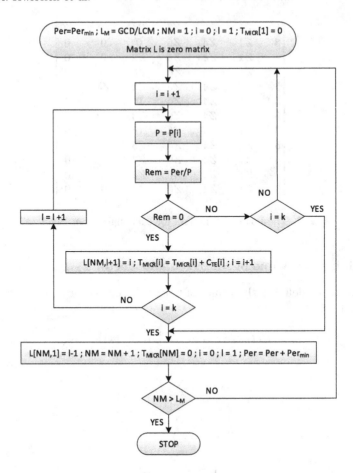

Fig. 3. Algorithm for defining the value of matrix L and vector T_{MICR}

It is hard at this moment to overestimate the importance of the analytic process as well as IT tools developed for the designing process in industrial real-time distributed systems. The designing process of both the scenario of exchanges and the entire transmission cycle is fundamental in modern industrial systems not only for maintaining the requested parameters of network work (bandwidth, useful efficiency) but also for proper execution of the transmission process in general.

3.3 The Duration of the Network Cycle in Case of a Double Bus Communication

The important issue of the method of transmission via two buses is to divide the matrix L introduced in previous section into two matrices A and B having the same structure.

$$
A = \begin{bmatrix} L'_{T_1} & T'_{TN_i} & \cdots & T'_{TN_n} \\ L'_{T_2} & T'_{TN_j} & \cdots & T'_{TN_m} \\ \vdots & \vdots & \vdots & \vdots \\ L'_{T_{LM}} & T'_{TN_{LM}} & \cdots & T'_{TN_z} \end{bmatrix} \qquad B = \begin{bmatrix} L''_{T_1} & T''_{TN_i} & \cdots & T''_{TN_n} \\ L''_{T_2} & T''_{TN_j} & \cdots & T''_{TN_m} \\ \vdots & \vdots & \vdots & \vdots \\ L''_{T_{LM}} & T''_{TN_{LM}} & \cdots & T''_{TN_z} \end{bmatrix}.
$$

where, as previously:

- sequences $\{T'_{TN_i}\}$ and $\{T''_{TN_i}\}$ include the exchanges numbers assigned to data of microcycles,
- L'_{T_i} and L''_{T_i} are the numbers of exchanges in a particular microcycle.

4   Algorithms for Searching the Form of Row of Matrices A and B

This section discusses three algorithms developed for searching the new form of row and columns of matrices A and B. The aim of the algorithm is:

- Determining the value of the first columns for matrices A and B,

$$
L' = \begin{bmatrix} L'_{T_1} \\ L'_{T_2} \\ \vdots \\ L'_{T_{LM}} \end{bmatrix} \qquad L'' = \begin{bmatrix} L''_{T_1} \\ L''_{T_2} \\ \vdots \\ L''_{T_{LM}} \end{bmatrix}.
$$

- Determining for each matrix the form of sequences representing the numbers of transactions in the i microcycle, so that:

$$
T_{MICR_i} = T'_{MICR_i} + T''_{MICR_i} \tag{13}
$$

$$
T'_{MICR_i} \cong T''_{MICR_i}. \tag{14}
$$

Algorithm 1. This algorithm is based on generating all combinations of subsequences so that the total transmission time of particular subsequences within one microcycle would be similar. The process of generating the subsequence should be repeated 2^n times, where n is the number of exchanges in the scenario of exchanges:

$$
\binom{n}{0} + \binom{n}{1} + \binom{n}{2} + \cdots + \binom{n}{n} = 2^n.
$$

where the Newton symbol $\binom{n}{k}$ means the number of subsequences in which the partial sums consist of sums where there are k components.

Moreover, after finishing each iteration, apart from the sum, the transmission structure namely subsequence T_{TN_L}, where L stands for subsequence including sums $\binom{n}{L}$ should also be remembered.

It is essential to notice that the presented algorithm has a very serious limitation. With the length of subsequence at about 100 components (which means 100 transactions of exchanges – and this is not a great value for average complex industrial systems) the number of combinations and partial sums is 2^{100}. In order to reach this result in a reasonable time, a high performance computer is necessary. It usually happens that on the industrial area, after making some changes in the scenario of exchanges there are only few minutes left to get the system working.

Algorithm 2. This algorithm is a modified algorithm of a balanced partition. Here, there is also the matrix L and its execution times (durations) C_{TEi} of particular exchanges T_{TN_i}:

$$
L = \begin{bmatrix}
L_{T_1} & T_{TN_i} & \cdots & T_{TN_n} \\
L_{T_2} & T_{TN_j} & \cdots & T_{TN_m} \\
\vdots & \vdots & \vdots & \vdots \\
L_{T_{LM}} & T_{TN_{LM}} & \cdots & T_{TN_z}
\end{bmatrix}
$$

where, as previously T_{TN_i} is a sequence including the exchanges numbers within the i microcycle, and k is the maximum time for the realization microcycle among all being executed within the macrocycle.

$$
k = \max_{i=1\dots n} \{C_{TEi}\}. \tag{15}
$$

Based on the above data after implementing the algorithm, matrices A and B were developed to describe the scenario of exchanges on particular buses. The computational complexity of this algorithm is $0(n^2 k)$.

Algorithm 3. While executing algorithm three, all transactions of exchange are assigned to one bus, and during each following iteration in these exchanges a value that can reduce significantly the difference between times of transactions execution on a particular bus is searched for. This algorithm is a very fast one (linear) and as a result it gives the matrices A and B that describe the scenario of exchanges on particular buses. However, partial sums of transaction execution on buses A and B are not always similar. This is because in some cases the difference between an element of one of the subsets and the subset of elements of a second subset is less than the least element of each subset.

In order to define the time needed for executing Algorithms 1, 2 and 3, special programmes were written. Table 1 presents the durations of algorithms execution (division of the exchanges executed on one bus into two buses) and differences of duration of the microcycle on buses A and B. Algorithms were tested for various numbers of exchanges. Table 1 presents the results for 20 exchanges.

The research results indicate that Algorithm 1 is very precise, whereas its execution time is unfortunately too long for the number of exchanges in the scenario, which is more than 25. Algorithm 2 is definitely faster than Algorithm 1

Table 1. The durations of algorithms execution and differences of duration of the microcycle on buses A and B

	Algorithm 1	Algorithm 2	Algorithm 3
Number of exchange L_{Ti} and duration of microcycle on bus A [time units] $$T'_{MICR} = \sum_{i=1}^{L_{Ti}} C'_{TEi}$$	$L_{Ti} = 7$ $T'_{MICR} = 5126$	$L_{Ti} = 8$ $T'_{MICR} = 5257$	$L_{Ti} = 14$ $T'_{MICR} = 4732$
Number of exchange L_{Tk} and duration of microcycle on bus B [time units] $$T''_{MICR} = \sum_{k=1}^{L_{Tk}} C''_{TEk}$$	$L_{Tk} = 13$ $T''_{MICR} = 5126$	$L_{Tk} = 12$ $T''_{MICR} = 5257$	$L_{Tk} = 6$ $T''_{MICR} = 4770$
$\|T'_{MICR} - T''_{MICR}\|$ [time units]	0	0	38
Duration of algorithm execution [s]	1.346	0.021	0.0001

and it is also very precise. Its execution time is influenced by input data (time for transmission execution), because the algorithm checks the conditions informing whether partial sums of transmission times have already reached the values equal to the half-sum of all transmission times. The last algorithm is certainly the fastest one. However, it generates the greatest differences of partial sums of executing transactions on buses A and B.

Based on the research results, among the three presented algorithms only Algorithms 2 and 3 should be taken into account:

- Algorithm 2 – can be used for creating matrices A and B during normal (failure-free) system work or during planned changes in the scenario of exchanges due to the fact that it is very precise, and its execution time even in case of a large number of exchanges is acceptable.
- Algorithm 3 – can be used in case of a system failure, when there is a need for fast, dynamic rebuilding of the scenario of exchanges. For instance, in the case of a subscriber interface failure the scenario should be rebuilt so that all exchanges to this subscriber would be transmitted via the functional interface connected to a particular bus. For this purpose, Algorithm 3 is suitable because of the shortest time of execution.

5   Conclusion

In the previous section the authors presented a method for automatic building of data exchange transactions according to the scenario of exchanges, including periodical exchanges. The main goal of the paper is to build two-stream data transmission so that each of the data streams has similar or even the same transmission time. The presented method divides the entire transmission process in an analytic manner, allowing at the same time the development of macrocycles and microcycles. Thanks to this, on each designing phase of the communication system it is possible to estimate a few fundamental values. They state not only completing the determined parameters of communication system, such as useful bandwidth, but also they make it possible to find the answer whether the parameters describing the transmission are easy to fulfil. This, in consequence, defines the ability to realize the entire communication process. The presented method consists of two phases:

- Defining the matrix L that includes a coded form of one-stream transmission of cyclic exchanges.
- Compiling two matrices A and B that include a new form of two streams of data transmission so that the times of their execution would be the same or very similar.

Within the presented method, the authors indicate not only the methodology but also present and test the algorithms:

- The algorithm for the coding transmission process.
- Three algorithms for segmentation of numerical sequence into two subsequences of the same values of their sum expressions.

The flexibility of the proposed method is mainly the fact that it is useful not only when we want to transmit via two buses, but also when the transmission is via only one bus. This method allows estimating the time of the network cycle and moreover, it indicates the values of times for acyclic transmissions, even in each microcycle. The presented method is only the introduction to designing, developing and controlling the two-stream transmission. However, it is not a sufficient tool (method). To make the two-stream transmission possible, another method should be developed that allow controlling communication network failures. This issue is a further objective of the next study.

Acknowledgments. This work was supported by the European Union from the European Social Fund (grant agreement number: UDA-POKL.04.01.01-00-106/09).

References

1. Kirrmann, H., Hansson, M., Muri, P.: Bumpless recovery for highly available, hard real-time industrial networks. In: IEEE Conference on Emerging Technologies and Factory Automation, 2007, ETFA, pp. 1396–1399 (2007)
2. Kirrmann, H., Weber, K., Kleineberg, O. et al.: Seamless and low-cost redundancy for substation automation systems (high availability seamless redundancy, HSR). In: 2011 IEEE Power and Energy Society General Meeting, pp. 1–7 (2011)

3. Neves, F.G.R., Saotome, O.: Comparison between redundancy techniques for real time applications. In: Information Technology: New Generations, 2008, ITNG 2008, pp. 1299–1300 (2008)
4. Kwiecień, A., Sidzina, M.: Dual bus as a method for data interchange transaction acceleration in distributed real time systems. In: Kwiecień, A., Gaj, P., Stera, P. (eds.) CN 2009. CCIS, vol. 39, pp. 252–263. Springer, Heidelberg (2009)
5. Wei, L., Xiao, Q., Xian-Chun, T., et al.: Exploiting redundancies to enhance schedulability in fault-tolerant and real-time distributed systems. Syst. Man Cybern. Part A Syst. Hum. IEEE Trans. 39(3), 626–639 (2009)
6. Sidzina, M., Kwiecień, B.: The algorithms of transmission failure detection in master-slave networks. In: Kwiecień, A., Gaj, P., Stera, P. (eds.) CN 2012. CCIS, vol. 291, pp. 289–298. Springer, Heidelberg (2012)
7. Kwiecień, A., Sidzina, M., Maćkowski, M.: The concept of using multi-protocol nodes in real-time distributed systems for increasing communication reliability. In: Kwiecień, A., Gaj, P., Stera, P. (eds.) CN 2013. CCIS, vol. 370, pp. 177–188. Springer, Heidelberg (2013)
8. Kwiecień, A., Stój, J.: The cost of redundancy in distributed real-time systems in steady state. In: Kwiecień, A., Gaj, P., Stera, P. (eds.) CN 2010. CCIS, vol. 79, pp. 106–120. Springer, Heidelberg (2010)
9. IEC 62439: Industrial communication networks: high availability automation networks. Chap. 6, entitled Parallel Redundancy Protocol (2007)
10. Gaj, P., Jasperneite, J., Felser, M.: Computer communication within industrial distributed environment-a survey. Industr. Inf. IEEE Trans. 9(1), 182–189 (2013)
11. Kwiecień, A., Maćkowski, M., Stój, J., Sidzina, M.: Influence of electromagnetic disturbances on multi-network interface node. In: Kwiecień, A., Gaj, P., Stera, P. (eds.) CN 2014. CCIS, vol. 431, pp. 298–307. Springer, Heidelberg (2014)
12. Pereira, C.E., Neumann, P.: Industrial communication protocols. In: Nof, S.Y. (ed.) Springer Handbook of Automation, pp. 981–999. Springer, Heidelberg (2009)
13. Miorandi, D., Vitturi, S.: Analysis of master-slave protocols for real-time-industrial communications over IEEE802.11 WLANs. In: 2004 2nd IEEE International Conference Industrial Informatics, INDIN 2004, pp. 143–148 (2004)
14. Xuehua, S., Min, L., Hesheng, W., Hong, W., Fei, L.: The solution of hybrid electric vehicle information system by modbus protocol. In: 2011 International Conference on Electric Information and Control Engineering (ICEICE), pp. 891–894 (2011)

Ontology-Based Integrated Monitoring of Hadoop Clusters in Industrial Environments with OPC UA and RESTful Web Services

Kamil Folkert[1] and Marcin Fojcik[2]([⊠])

[1] Institute of Informatics, Silesian University of Technology, Gliwice, Poland
kamil.folkert@polsl.pl
[2] Sogn Og Fjordane University College, Førde, Norway
marcin.fojcik@hisf.no

Abstract. Contemporary industrial and production systems produce huge amounts of data in various models, used for process monitoring, predictive maintenance of the machines, historical analysis and statistics, and more. Apache Hadoop brings a cost-effective opportunity for Big Data analysis, including the data generated in various industries. Integrating Hadoop into industrial environments creates new possibilities, as well as many challenges. The authors of this paper are involved into commercial and scientific projects utilizing Hadoop for industry as predictive analytics platform. In such initiatives the lack of standardization of monitoring of the industrial process in terms of Hadoop cluster utilization is especially perplexing. In this paper, authors propose the methodology of monitoring Hadoop in industrial environments, based on dedicated ontology and widely adopted standards: OPC Unified Architecture and RESTful Web Services.

Keywords: OPC UA · Big data · Hadoop · Process monitoring · Ontology · REST

1 Introduction

The massive amounts of data generated by contemporary computer systems caused a strong market demand for determining new ways of storing and processing of the data. This demand was an inspiration for new technologies, nowadays recognized as Big Data. According to the definition by Gartner, Big Data is high-volume, high-velocity and high-variety information assets that demand cost-effective, innovative forms of information processing for enhanced insight and decision making [1].

The trend of adoption of Big Data technologies was especially visible in the area of marketing, advertising, financial control and others in various sectors, e.g. banking, retail, social media. It also influences the industrial automation, enhancing the production process with a new tool of assessment, models generation and optimization, as well as historical data analysis platform. More and

© Springer International Publishing Switzerland 2015
P. Gaj et al. (Eds.): CN 2015, CCIS 522, pp. 162–171, 2015.
DOI: 10.1007/978-3-319-19419-6_15

more applications based on MapReduce [2] paradigm and HDFS (Hadoop Distributed File System) [3], which are the core foundation elements of Hadoop ecosystem [4,5], are applied in industrial computer systems to provide an insight into process and machine-generated data [6,7]. Apache Hadoop, a shared-nothing architecture framework for distributed data storing and analysis on cost-efficient cluster built on commodity hardware, gained a lot of attention last years. Providing support for both structured and unstructured data, Hadoop is widely adopted in many industries, including production systems based on industrial automation. In these systems Hadoop can be used not only as a high-level data analysis engine, but can also augment the low-level process execution.

Being the important part of the process, Hadoop cluster requires monitoring and maintenance to guarantee its high availability. Hadoop itself provides possibilities for monitoring of the services, however they are not based on any of the industrial standards. It is especially troublesome in projects adopting Hadoop for predictive analytics in industrial environments. The authors are involved into such projects in automotive, mining and gas&oil industries. Due to participation in these initiatives, the authors see strong demand on standardization of the monitoring of the Hadoop clusters.

In the following sections of this paper authors describe proposed standardization of Hadoop cluster monitoring in industrial environments. Section 2 provides general recapitulation of the structure of automation systems. Challenges related to the industrial integration are described in Sect. 3. Section 5 contains a description of the proposed solution for integrated Hadoop cluster monitoring. Moreover, sample implementation verifying the concept is described. In Sect. 6 the summary and conclusions are presented, as well as possible future research directions.

2 Architecture of Contemporary Industrial Automation Systems

The industrial automation systems consist of sensors, actuators, PLCs (Programmable Logic Controllers), PCs, servers and other elements. These components are interconnected with networks, both industrial and general-purpose, depending on the character of data exchange. This environment is logically divided into several layers: from field level, through control and SCADA (Supervisory Control and Data Acquisition), to MES (Manufacturing Execution System) and ERP (Enterprise Resource Planning) levels, according to ISA95 standard [8].

Modern industrial environments provide ability to persist process state (in terms of machine operations and maintenance), historical data acquisition and processing. Moreover, additional requirements for the systems may often cover also manufacturing execution, optimization of the technological process and especially collecting and processing of process data on higher levels.

These requirements are included *inter alia* in the scopes of the European Union projects aiming at optimization of the energy consumption of the factories, e.g. EMC2-Factory [9] and Factories of the Future [10]. Optimization of the

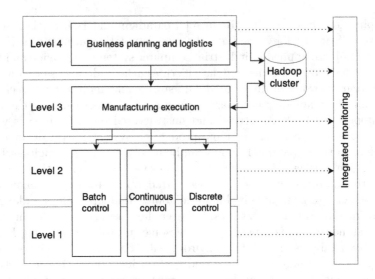

Fig. 1. Apache Hadoop cluster in ISA95 structure

production process can be realized using predictive statistical models created with data mining methods, e.g. k-means clustering [11]. Data sets used for such modeling must provide as much information about the process variables (as well as about the correlations between them) as possible to increase the quality of the model. For that reason there is a strong demand for safe and reliable data storing and processing. Hence, as mentioned in Sect. 1, the Apache Hadoop ecosystem gained wide acceptance in various business sectors as well as attention in the industrial area.

As presented in Fig. 1, Apache Hadoop can interchange data with both MES and ERP levels. Integration with manufacturing execution provides opportunity to augment the production process with insights from aggregated data processed on Hadoop cluster. Integration of ERP and MES with Hadoop creates plenty of new opportunities to correlate data from both levels to provide better understanding of the production, both from business and operational angles.

3   Challenges of Integration in Industrial Computer Systems

The requirements for the modern industrial computer systems, described in Sect. 2 above, are hard to be met without a highly heterogeneous environment. For that reason the key role of multi-level integration in design and implementation of such systems cannot be underestimated. Multi-level integration can be defined as logical or physical fit between the system components on hardware, software or data model level. Hardware and software levels of integration are relatively standardized in modern automation, e.g. with unified hardware interfaces to interconnect sensors and actuators with PLCs or software standards dedicated for vertical data exchange. However, on the data model level the integration is very often implemented ad hoc, without using a semantic approach. There are some specific ontologies, used in particular industry sectors, e.g. PRODML and WITSML for gas&oil [12], ISO 15926 [13] for asset life management or ontologies designed for smart grid solutions [14]. Unfortunately the ontologies are often defined using different formats, e.g. XML, JSON, OWL, RDF, etc. To use them in a heterogeneous environment it is required to convert them into an unified format. This conversion might lead to meta-information loss (e.g. converting XML to RDF) [15] and additionally increases utilization of the resources.

This problem was addressed during establishing of the unified standard of vertical data exchange in industrial environment. OPC Unified Architecture (OPC UA) provides a object-oriented address space, enabling implementation of various data models based on virtually any ontology. For that reason many software implementations of data models based on the ontologies mentioned above are nowadays applied to OPC UA [16–18].

One of the main reasons for successful adoption of OPC UA was its ability to provide the information not only about the process value, but also about the quality of the data. The quality can be affected by the source of the information, as well as all processing points during the transmission of the data to the destination. Due to embedding the quality marker into the variable, OPC UA can be used not only for process data exchange, but also for monitoring of the nodes of the system. However, also in that area it is required to choose and implement the right ontology for data model description.

This challenge is especially visible during integration of new components into the process. A good example is Hadoop cluster, used in the process for top-level data storing and analysis. Since there are no standards for describing Hadoop cluster for monitoring in an industrial environment the authors decided to propose an ontology and describe the data structures which can be modeled using OPC UA address space to unify the monitoring of Hadoop with other components of the system.

4   Proposition of Integrated Monitoring of Hadoop Cluster in Industrial Environment

Hadoop consists of many services, some of them may be crucial for the correct process execution and should be monitored with comprehensive and holistic approach.

The multi-level integration of Hadoop monitoring in industrial environments requires:

- on hardware level – making Hadoop accessible for the monitoring node in terms of network access on the application layer of ISO/OSI model,
- on software level – using a network protocol and data exchange mechanism allowing to read the operational parameters of the cluster,

– on data model level – defining and implementing an ontology providing semantical description of the Hadoop services.

Integration on the hardware level can be realized using a standard network setup with direct access over a switch or with routing to the subnet dedicated for Hadoop cluster (which is a common real-life scenario).

As mentioned in Sect. 3, the OPC UA protocol gained the status of de facto standard in vertical communication in industrial environments and for that reason it is commonly used also for monitoring purposes. Thus, it is proposed to use OPC UA for comprehensive Hadoop cluster monitoring. The OPC UA address space should contain parameters of specific Hadoop services and expose OPC UA variables for asynchronous read as well as for subscribing. For each session established by monitoring node as OPC UA client, the subscription should be created. The subscription should contain the monitored items created for every service which require monitoring.

The access to the OPC UA server can be realized using OPC Binary protocol or with Web Services. From the perspective of the monitoring node the technology chosen for implementation of the OPC UA server should be irrelevant. However, taking into account the nature of the Hadoop as open-source project developed and maintained in the Apache Foundation, it is recommended to choose a technology platform that can be run on the same servers on which Hadoop Data Node is running. These servers usually are based on Linux operating system, which would imply open technologies like Java, C/C++, Python or node.js implementation of the OPC UA server. On the other hand, some distributions of Hadoop can be also deployed on Microsoft Windows platform.

Data level integration of Hadoop should provide a standardized definition of the parameters of the services of the cluster. To simplify the browsing of the OPC UA address space, the services can be divided into several groups, basing on their role in the cluster [19]:

– governance (Falcon, WebHDFS, Flume, Sqoop),
– data access (Pig, Hive, HCatalog, Hbase, Storm, Spark, Solr),
– data management (Tez, Slider, Yarn, HDFS),
– security (Knox, Ranger),
– operations (Zookeeper, Oozie, Ganglia, Nagios).

Each of these services should be accessible as a separate OPC tag. Additional clusters should be modeled as additional namespaces in OPC UA address space. All of the services should provide list of parameters in one, canonical model. The ontology presented below provides semantical description of single service parameters. Using Description Logic (DL) notation, it can be described as follows:

$$Service \sqsubseteq (= 1hasServiceInfo.ServiceComponentInfo)$$
$$Service \sqsubseteq (= 1hasAlerts) \sqcap hasAlerts.Summary \sqcap (0hasAlerts.Detail)$$
$$Service \sqsubseteq (= 1hasServiceInfo.ServiceComponentInfo)$$

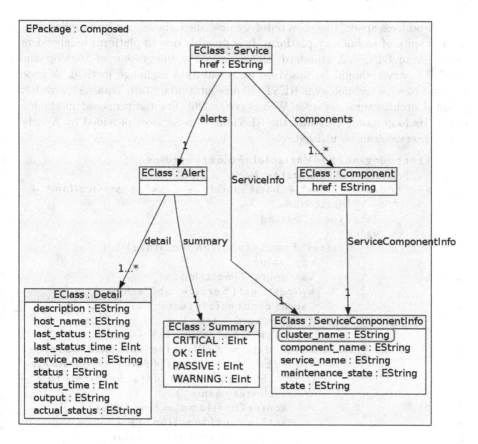

Fig. 2. Canonical domain model of Hadoop service

The notation provided above describe three facts about single Hadoop service on the class level. First, the service has one ServiceInfo of ServiceComponentInfo class. Second, the service has one Alert object with single Summary (which provides aggregated information regarding the alerts) and multiple Detail objects (representing each single occurrence of particular alert). Finally, if the service have ServiceComponents, they are described using notation equivalent to ServiceInfo.

Presented ontology describes the Hadoop services on the class level. Specific fields of each class are illustrated as JSON class model presented in Fig. 2.

5   Sample Architecture and Implementation

The cardinal feature of the proposed standard for Hadoop cluster is to separate operational data from monitoring on logical level. The separation can be implemented using data provided with Apache Ambari service. Despite Ambari is not included in every distribution of Hadoop, it can be installed and configured as external service as well.

The address space, based on ontology described above, can be supplied with data in spite of technology platform. It is possible due to platform-independent character of OPC UA standard. Hence, also the integration of Hadoop and OPC UA server should be based on universal data exchange method. A good example of such technology is REST (Representational State Transfer), a widely adopted architectural style for Web Services [20]. For the proposed methodology of Hadoop monitoring also the RESTful Web Services provided by Apache Ambari service can be utilized.

```
1    server.engine.addVariableInFolder(Service, {
2        browseName: "Cluster Name",
3          nodeId: "ns=" + nameSpaceId + ";s=" + serviceName +
             "_ClusterName",
4          dataType: "String",
5          value: {
6                refreshFunc: function (callback) {
7                    var value;
8                    var sourceTimestamp;
9                    unirest.get(Service.apiPath)
10                   .auth(credentials.user, credentials.
                        password, true)
11                   .end(function(response) {
12                       var data = JSON.parse(response.body);
13                       value = new Variant({dataType: DataType
                           .String, value: data.ServiceInfo.
                           cluster_name});
14                       sourceTimestamp = new Date();
15                       setTimeout(function () {
16                           callback(null, value,
                               sourceTimestamp);
17                       }, refreshInterval);
18                   });
19               }
20           }
21       });
```

Listing 1.1. Practical example of adding Hadoop cluster parameter to OPC UA address space

Example invocation of Ambari Web Service, followed by data deserialization and OPC UA address space variable creation is presented in Listing 1.1 The presented implementation illustrates the handling of the single parameter of the ServiceInfo class, tagged with oval marker in Fig. 2. The tag is added to appropriate OPC UA object, so the address space contains all of the parameters structured in the manner illustrated on the Fig. 2.

To verify the proposed technique an OPC UA server was implemented. Due to fast prototyping possibilities, node.js was used as the development platform. Therefore, the code presented in Listing 1.1 was developed in JavaScript and follows specific practices of that programming language, e.g. callback functions. The OPC UA server was implemented on a standard development environment

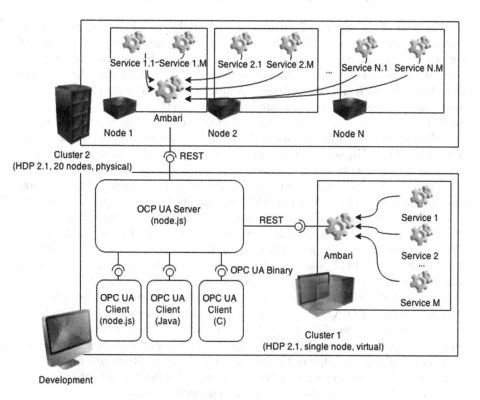

Fig. 3. The architecture of the test environment

based on Mac OS X Yosemite 10.10.2 (14C109), running on 2.4 GHz Intel Core i5 processor and 16 GB of RAM. The node.js runtime was 0.10.35.The Hadoop cluster used as data source of monitored tags was local, single-node virtual cluster and a 20-node physical cluster. Both clusters were based on Hortonworks Data Platform distribution (HDP 2.1). To verify the interoperability OPC UA clients connecting to the OPC UA server were implemented in node.js, Java and C++. The environment is presented in Fig. 3.

6 Summary and Conclusions

In this paper the authors described a standardized method of monitoring Hadoop cluster, based on OPC UA and Web Services. The proposed technique is based on the assumption that the monitored parameters should be gathered in one, unified, ontology-based model (in terms of format and contents) in one centralized location (OPC UA server). The proposed method utilizes the well-known existing standards on both hardware and software level. It can be applied to any distribution of Hadoop and implemented on a server inside or outside the cluster.

The goal of the presented attitude was to enable rapid development of the monitoring systems, as well as reduce margin for human error due to use of verified and reliable technologies as the foundation of the system. To verify the concept, an implementation basing on node.js OPC UA stack and Hortonworks Data Platform with embedded Apache Ambari was tested. Assumed advantages were achieved and implemented into ongoing projects in the area of predictive analytics based on Hadoop in industrial environments. Another advantage of providing unified monitoring data is the possibility to preprocess and aggregate data in the OPC UA server. This could be applied not only to Apache Hadoop monitoring, but also to other components of automation system.

A good example of such enhancement can be the acquisition and aggregation of alarm data from the whole industrial environment in one group of tags in OPC UA address space. This would enable clients to subscribe just aggregated system status tag instead of multiple variables, located in many objects and groups in address space. It may reduce both network traffic and server resources utilization, because the OPC UA server would not have to maintain multiple monitored items. As shown in previous research in that area, number of monitored items is directly related with utilization of the server resources and radically influences the performance of OPC UA data exchange [21,22]. Implementation of such feature requires definition of a canonical data model for alarms and warnings, based on definitions of most common identificators and numerical values, systemizing information about the type of the alarm. These considerations are not directly related with Apache Hadoop monitoring and will be a matter of further research in the area of unified monitoring of industrial environments.

Acknowledgement. This work was supported by the European Union from the European Social Fund (grant agreement number: UDA-POKL.04.01.01-00-106/09).

References

1. Gartner. Big Data definition. Gartner IT glossary. http://www.gartner.com/it-glossary/big-data/. Accessed 28 Jan 2015
2. Dean, J., Ghemawat, S.: MapReduce: simplified data processing on large clusters. Commun. ACM **51**(1), 107–113 (2008)
3. Shvachko, K., Kuang, H., Radia, S., Chansler, R.: The hadoop distributed file system. In: 2010 IEEE 26th Symposium on Mass Storage Systems and Technologies (MSST). IEEE (2010)
4. Bakshi, K.: Considerations for big data: architecture and approach. In: Aerospace Conference, 2012. IEEE (2012)
5. White, T.: Hadoop: The Definitive Guide. O'Reilly Media Inc, Sebastopol (2012)
6. Bahga, A., Madisetti, V.K.: Analyzing massive machine maintenance data in a computing cloud. IEEE Trans. Parallel Distrib. Syst. **23**(10), 1831–1843 (2012)
7. Kiss, I., Genge, B., Haller, P., Sebestyen, G.: Data clustering-based anomaly detection in industrial control systems. In: 2014 IEEE International Conference on Intelligent Computer Communication and Processing (ICCP). IEEE (2014)
8. Scholten, B.: The Road to Integration: A Guide to Applying the ISA-95 Standard in Manufacturing. Isa, USA (2007)

9. EMC2-Factory. http://www.emc2-factory.eu. Accessed 28 Jan 2015
10. Factories of the Future. http://ec.europa.eu/research/industrial_technologies/factories-of-the-future_en.html. Accessed 28 Jan 2015
11. Cupek, R., Drewniak, M., Zonenberg, D.: Online energy efficiency assessment in serial production-statistical and data mining approaches. In: 2014 IEEE 23rd International Symposium on Industrial Electronics (ISIE). IEEE (2014)
12. Cupek, R., Fojcik, M., Sande, O.: Object oriented vertical communication in distributed industrial systems. In: Kwiecień, A., Gaj, P., Stera, P. (eds.) CN 2009. CCIS, vol. 39, pp. 72–78. Springer, Heidelberg (2009)
13. Haaland Thorsen, K.A., Rong, C.: Towards dataintegration from WITSML to ISO 15926. In: Sandnes, F.E., Zhang, Y., Rong, C., Yang, L.T., Ma, J. (eds.) UIC 2008. LNCS, vol. 5061, pp. 626–635. Springer, Heidelberg (2008)
14. King, R.L.: Information services for smart grids. In: Power and Energy Society General Meeting-Conversion and Delivery of Electrical Energy in the 21st Century, pp. 1–5. IEEE (2008)
15. Van Deursen, D., Poppe, C., Martens, G., Mannens, E., Walle, R.: XML to RDF conversion: a generic approach. In: 2008 International Conference on Automated Solutions for Cross Media Content and Multi-channel Distribution, AXMEDIS 2008. IEEE (2008)
16. Stopper, M., Katalinic, B.: Service-oriented architecture design aspects of OPC UA for industrial applications. In: Proceedings of the International Multi-Conference of Engineers and Computer Scientists (2009)
17. Rohjans, S., Fensel, D., Fensel, A.: OPC UA goes semantics: Integrated communications in smart grids. In: 2011 IEEE 16th Conference on Emerging Technologies and Factory Automation (ETFA), pp. 1–4 (2011)
18. Thorsen, K.A.H., Torbjørnse, O.F., Rong, C.: Automatic web service detection in oil and gas. In: Ślęzak, D., Kim, T., Chang, A.C.-C., Vasilakos, T., Li, M.C., Sakurai, K. (eds.) FGCN/ACN 2009. CCIS, vol. 56, pp. 193–200. Springer, Heidelberg (2009)
19. Hortonworks Data Platform. http://hortonworks.com/hdp. Accessed 29 Jan 2015
20. Richardson, L., Ruby, S.: RESTful Web Services. O'Reilly Media Inc, USA (2008)
21. Clavel, M., Durán, F., Eker, S., Lincoln, P., Martí-Oliet, N., Meseguer, J., Talcott, C.: Introduction to OPC UA performance. In: Clavel, M., Durán, F., Eker, S., Lincoln, P., Martí-Oliet, N., Meseguer, J., Talcott, C. (eds.) CN 2012, CCIS 291. LNCS, vol. 4350, pp. 1–28. Springer, Heidelberg (2007)
22. Folkert, K., Fojcik, M., Cupek, R.: Efficiency of OPC UA communication in java-based implementations. In: Kwiecień, A., Gaj, P., Stera, P. (eds.) CN 2011. CCIS, vol. 160, pp. 348–357. Springer, Heidelberg (2011)

Speech Quality Measurement in IP Telephony Networks by Using the Modular Probes

Filip Rezac[(⊠)], Jan Rozhon, Jiri Slachta, and Miroslav Voznak

CESNET, z.s.p.o., Zikova 4, 160 00 Prague 6, Czech Republic
{filip,rozhon,slachta}@cesnet.cz, voznak@ieee.org
http://www.cesnet.cz/?lang=en

Abstract. The paper presents a system for monitoring the speech quality in the IP telephony infrastructures using modular probes that are placed at key nodes in the network. The system is developed mainly within research of the Czech educational and scientific network, but the tool can generally be used in any IP telephone traffic. The system is capable of using an objective intrusive methods to measure speech quality dynamically and the results are collected on a central server. Information about the speech quality are displayed in the form of automatically generated maps and tables. The article is logically divided into a description of the technology and algorithms, as well as the implementation procedure is presented and measured results were validated in real traffic. Contribution of the work consists of a new system design utilizing modular probes for the measurement of speech quality in IP telephony networks and the result of applied research enables administrators to respond and optimize network traffic efficiently.

Keywords: IP telephony · Modular probes · Speech quality · PESQ · Network monitoring

1 Introduction

Nowadays, the technology for Internet telephony – VoIP is widely used both in the corporate sector as well as in SOHO (Small Office, Home Office) environments. This is because that the audio or video packet data communications bring not only economic benefit, but also the consolidation of transmission networks. It is common for most IP telephony infrastructures that in order to reduce the payload, the stream is separated from the rest of the network using virtual LANs (Local Area Networks), even though, it is desirable to monitor the quality of calls on each node.

The system, that is presented in this article, offers an elegant way using network probes to measure and monitor speech quality on the route. The measured data are then sent to a central server in the form of Zabbix monitoring server that allows the visualization of the results in the form of network maps. Probes can be installed in the network as a virtual machine image, as well as tarball

© Springer International Publishing Switzerland 2015
P. Gaj et al. (Eds.): CN 2015, CCIS 522, pp. 172–181, 2015.
DOI: 10.1007/978-3-319-19419-6_16

packages with source code designated for compilation. The ISO image, when the user installs the whole operation system based on Linux distribution is also available, or system can be part of the BESIP (Bright Embedded Solution for IP Telephony) project [1,2].

The next section will compare the proposed system with existing tools and we will describe in details the algorithm itself and system implementation.

2   State of the Art and Speech Quality Measuring

Monitoring the quality of speech is important but often overlooked part not only of IP telephony. Before implementation of the proposed multi-agent system, an analysis of existing solutions was performed, but no complete measuring system was found. There is no system that would meet all our requirements such as multi-way and multi-node measuring at a time and the ability to collect and visualize the results. In particular, the requirement to monitor the quality of speech in many places at the same time brings many restrictions in terms of selecting an appropriate model or algorithm for speech quality determination [1].

This automatically eliminates the possibility of using subjective methods that are based on adding a human perspective in the communication chain. In the mentioned case, it is necessary that after call hangup, user connected to the SIP server subjectively evaluates each call. It is clear that such an option is not possible to implement because of several aspects. Originally, it was necessary to have a large number of users who assess the speech quality subjectively. Another important aspect is the time required to determine the final score that is computed as the mean value of all listeners. Thus, the objective methods remain the only possibility, but all of the objective methods are not suitable for the implementation. The objective evaluation of speech quality is divided into two categories non-intrusive and intrusive by the availability of the "original" voice sample. In the case of using intrusive methods, it is necessary to possess the voice sample before transferring through the communication chain and then of course the same content after it was transferred through the communication system. Otherwise, non-intrusive method is based on computing model called E-model. In this case, it is not necessary to have the original and degraded sample since the quality value is calculated from the parameters of the call.

After considering possible methods of the speech quality assessment, we decided to apply an objective intrusive approach represented by PESQ (Perceptual Evaluation of Speech Quality) model. It is standardized as ITU-T recommendation P.862. Today, PESQ is a worldwide applied industry standard for objective speech quality testing used by phone manufacturers, network equipment vendors and telecom operators. If we consider the results of subjective methods as reference [2], then PESQ achieves better results than using non-intrusive methods represented by E-model.

To evaluate speech quality, MOS (Mean Opinion Score) scale as defined by the ITU-T recommendation P.800 is applied [3]. The basic scale as prescribed by the recommendation is depicted in Fig. 1. In order to avoid misunderstanding

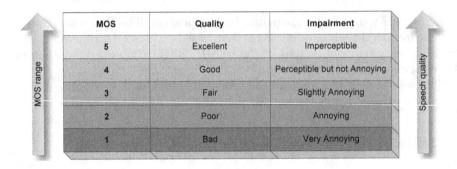

MOS	Quality	Impairment
5	Excellent	Imperceptible
4	Good	Perceptible but not Annoying
3	Fair	Slightly Annoying
2	Poor	Annoying
1	Bad	Very Annoying

Fig. 1. MOS scale

and incorrect interpretation of MOS values, ITU-T published ITU-T recommendation P.862.1 in 2004 [4]. This recommendation defines a mapping function to MOS-LQO that is used by PESQ model.

3   Design and Technology of the System

3.1   Design

Before starting the system design, the research was conducted on the topic of what objective speech quality measurement model is ideal for our purpose and what management model should be used in the implementation. Based on the results that have been published [5,6], we have made a decision that a centralized model is the most appropriate.

It works on the principle of autonomous probes which are placed at key nodes in the network and periodically carry out test calls to adjacent probes. Once the test call is made, degraded samples are sent to the server, where are compared with the original signal using PESQ model.

Server side is based on the Zabbix [7] monitoring tool, which is enhanced by our algorithms for data collection, processing and evaluation. Server also allows to monitor individual probes and respond to potential failure. An example of the system design is shown in Fig. 2.

Once the data is processed on the server, the visualization is presented in the form of a map where peaks represent the probes and the measurement paths are shown in the form of evaluated edges.

The user can easily monitor the quality of speech on different call routes and on the basis of this information he/she can optimize network flow, or edit voice codecs for VoIP traffic.

3.2   Technology

From the perspective of the technologies that were used in the implementation, it is again necessary to divide a whole system into two parts.

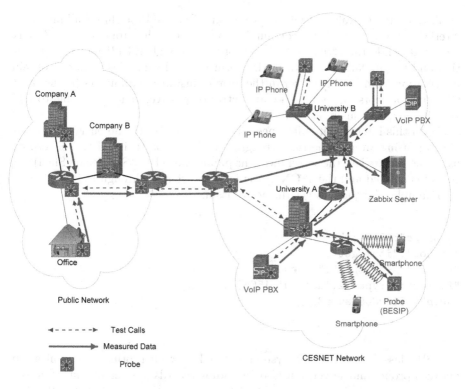

Fig. 2. Proposal deployment of a monitoring system

Server uses a virtualized platform running on Zabbix monitoring system and it was necessary to implement a communication interface using the Python programming language. Detailed description of the interface itself and ongoing communication is then given in the following section.

Probe side is developed on a specially modified OpenWRT [8] Linux distribution, which is suitable for low-cost and energy-saving devices. However, the most important parts of the probe are classes for communication with the server, applications for control periodic processes and a test call generator itself. In addition, the probe includes classes for evaluation of speech quality and communication interface with other probes. As was already mentioned, the probes may be deployed in a network in many ways, allowing rapid expansion of the system and also gain a more accurate measurement.

4   Implementation of the Algorithm

4.1   Classes

Within the implementation of the monitoring system, it was necessary to develop some custom classes, or edit the existing ones.

JSON-RPC (JavaScript Object Notation – Remote Procedure Call protocol) interface [9] was selected for communication between the probe and the Zabbix server. It sends the string as a JSON object through HTTP request. Operation and manipulation of data in this form is simple and this class can obtain information from Zabbix server in the form of instance parameters (items) that the server accepts. Whether they are active or passive, or it is possible to add your items using this interface.

With this interface, it is first necessary to authenticate and encrypt communication using an asymmetric cryptography, the probe is signed and encrypted using the private key, the server owns the public one. Then SSH (Secure Shell) [10] tunnel is created and the JSON objects are transferred through it.

Example of request to get the active item:

```
{
"jsonrpc":"2.0",
"method":"item.get",
"params": {"output":"extend",
"filter":{'type':2,'hostid':1565}},
"auth":'51awd651aw5d3a5gs',
"id":2,
}
```

SSH class is used in the system not only for encrypting communication between probes and server, but also for adding records to a remote queue on the server when calculating speech quality. Class uses a python library Paramiko [11].

It was also necessary to modify the class allows to configure Cron table – crontab. It is used for adding and removing tables commands to be executed and evaluated periodically. Such commands include both periodic updating and distribution of the probes in the network and the periodic realization of test calls between the probes for measuring purposes. This class was created in 2008, but is freely available under the GNU license and it can be used for free.

After the first start of the probe and setting up the periodic cycles via Cron, the Listener class is executed. Then it waits for commands from the server so the list of the probes can be updated in case of changes.

Once the probe receives the list of the neighboring probes from the server through JSON interface, the AsteriskCallList class creates and operates a call list, which is then used for generating test calls. These are actually controlled by AstConnection class that communicates with AMI (Asterisk Manager Interface) interface of SIP PBX Asterisk [12] which is the part of each probe.

The server side also uses scripts for database management and classes designed to run PESQ algorithm that compares the degraded sample with a reference one to determine the speech quality.

Last mentioned class, which makes the interface between the running system and classes containing particular functions is called Controller. It initializes various objects and allows the interaction and interconnection.

4.2   Communication

As in the case of technology, communication description should be divided into two parts as well. The part performed on probes and the server side.

Before deploying probes into the network, it is necessary to generate a key pair (private/public). The probe uses a private key to encrypt communication with the server. Once the installation of the probe is in progress, the IP address or domain of Zabbix server is required. Then, during the first start of the probe, it is checked whether the probe is added on the Zabbix server or not. If not, probe is added using the JSON interface. Subsequently, the list of all probes is obtained from the server from which the sip.conf file is generated and is used for test calls.

The last part of the initialization is presented by adding an item into the crontab for generating test calls from the probe to all other probes in regular intervals. Listener class is started simultaneously to tracks and updates the changes from the server. The described procedure is shown by the flow diagram in Fig. 3.

The initiation of calls is made by AstConnection class, which first started the test calls through the AMI interface of Asterisk using the reference sample and then it stores a degraded samples in a .wav format. Degraded files are stored in the form: $MAC\_caller\_probe-MAC\_called\_probe-timestamp.wav$ (e.g., $000c29b28f9a$-$000c298f419d$-$20140116092039.wav$), so they can be easily founded. These samples are then backed up and transferred using Rsync [13] to a directory on Zabbix server. The list of files on the probe that has been transmitted is gradually crawled and is added to queue on the server using SSH for further analysis. This queue is stored in a MySQL database and script comparing the reference voice sample using PESQ algorithm is periodically triggered. Returned MOS values are then inserted into Zabbix application and are further visualized. The values are stored for each probe using the following string: MOS-$MAC\_address\_of\_the\_target\_probe$ (eg. 38-$000c298f419d$). To avoid the re-analysis at the next script startup, the file is then removed from the directory. Sequence diagram of communication between the probes and the server is shown in Fig. 4.

Based on the algorithm above, test calls between all neighboring probes (also called full-mesh) should be realized. The question is, which probe starts with testing first and how to perform all the test calls without duplications. For this purpose, the method has been developed and is described below.

First, the average number of connections per node is determined. In addition, the method creates an average number of connections for each probe. The first member of the list is then moved to the end. The method is shown below, where n is the number of probes:

```
n = len(self.inp)
no_of_conns = n * (n - 1) / 2
conns_per_node = no_of_conns // n
rem_conns = no_of_conns % n
```

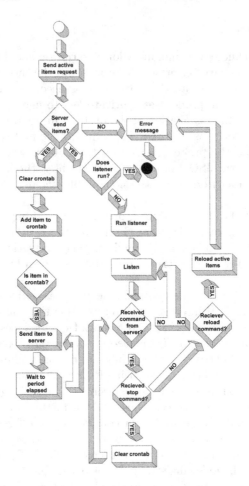

Fig. 3. Function of the crontab and listener

If we have, for example, four probes, according to full-mesh topology, the six connections should exist (complete undirected graph). After the application of the method, we obtain the average number of connections per node $no\_of\_conns = 6$. Next, integer division will create an average number of connections for each $probeconns\_per\_node = 1$. In the last step, we get the remainder after division by using modulo operation. This remainder means that two probes will need one extra connection.

The method serves to distribute test calls equally between the probes to avoid excessive load.

4.3   Visualization

The visualization script was also implemented and it allows to generate a graphical map of probes from the database of results. The map is implemented using

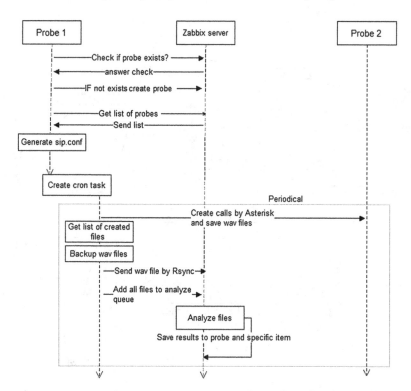

Fig. 4. Communication between probes and Zabbix server

a graph where the nodes represent individual probes and the edges are connections between them. Edge weight shows the current MOS value measured between the respective probes in both directions (if both numbers are available). The graph is generated using the Zabbix server interface and is, therefore, available directly through the web application.

In the first phase, the attributes to create a new map are defined, i.e., the height and the width in pixels of clean canvas and the name under which the map will be available in the Zabbix application. Then, a temporary graph is created using the Python package named Networkx [14], in which each probe is inserted (identifiable by their MAC addresses).

There is also a spatial distribution of probes in such a manner to avoid mutual overlap. Script then gets a list of all active probes from Zabbix server and each probe obtains an active list of items (measured speech quality), which represents the edges between the probes. Since the items are named: *MOS-MAC_address_of_the_target_probe*, it can be specified, at which edge of the graph this value has to be added. As between the two probes is always only one edge and communication is tested in both directions, both values are entered (if available) to the label of the edge. These labels are stored in 2-dimensional array, where the keys are always made from MAC address of the probe from which the test

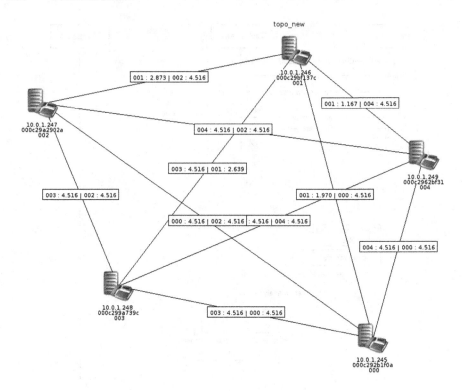

Fig. 5. Visualization of speech quality between the probes in the network

call was performed and the second key is the MAC address of the probe that was called. The algorithm thus first paired values to be displayed together on one edge and then are inserted into the graph.

Finally, it is verified that the map with the following name exists on the Zabbix server, and if so, it is deleted and recreated. Then, the graph obtained a list of all the nodes together with their coordinates is inserted into the newly created Zabbix map and descriptions are set. At the same time, the previously prepared 2-dimensional array is scanned for the labels for each edge, and if there is an entry in the array for a given MAC address, the edge is created between nodes. The result of the visualization can be seen in Fig. 5.

5   Conclusion and Future Work

Based on requests from users of Czech universities associations [15], a system for monitoring the speech quality in networks has been developed. Nowadays, the application is actively tested and individual parts are optimized. The system is capable of using network probes and objective intrusive methods to dynamically measure speech quality of VoIP infrastructures without burdening the measured network, or generate unwanted traffic. Measurement results are clearly displayed so users can monitor and adjust the network parameters.

In the future work, we would like to focus mainly on simple, multi-platform and user-friendly distribution of the probes, which will need to create installation scripts and intuitive guides. We would like to also create a Web application over the Zabbix server, where users will be able to track individual fragments of the monitored network and network probes will be shown on the map according to their geographic location based on the DNS lookup.

Acknowledgments. This work has been supported by the Ministry of Education of the Czech Republic within the project LM2010005.

References

1. Voznak, M., Macura, L., Slachta, J.: Embedded cross-platform communication server. In: 17th WSEAS International Conference on Computers, Rhodes Island, pp. 197–202 (2013)
2. Tomala, K., Macura, L., Voznak, M., Vychodil, J.: Monitoring the quality of speech in the communication system BESIP. In: 35th International Conference on Telecommunication and Signal Processing, Prague, pp. 255–258 (2012)
3. ITU-T, Recommendation P.800: Methods for objective and subjective assessment of quality. International Telecommunication Union (1996)
4. ITU-T, Recommendation P.862.1: Mapping function for transforming P.862 raw result scores to MOS-LQO. International Telecommunication Union (2004)
5. Hecht, J.: All smart, no phone. IEEE Spectr. **51**(10), 36–41 (2014). New York
6. Frnda, J., Voznak, M., Rozhon, J., Mehic, M.: Prediction model of QoS for triple play service. In: 21st Telecommunications Forum TELFOR, Belgrade, pp. 733–736 (2012)
7. Vacche, A.D., Lee, S.K.: Mastering Zabbix. Packt Publishing, Birmingham (2013)
8. OpenWRT - Wireless Freedom. https://openwrt.org/
9. Summerfield, M.: Python in Practice: Create Better Programs Using Concurrency, Libraries, and Patterns, 1st edn. Addison-Wesley Professional, Indiana (2013)
10. Barret, D.J., Silverman, R.E., Byrnes, R.G.: SSH, The Secure Shell: The Definitive Guide, 2nd edn. O'Reilly Media, Sebastopol (2005)
11. Paramiko - SSH2 Protocol for Python. http://www.lag.net/paramiko/
12. Bryant, R., Madsen, L., Meggelen, J.: Asterisk: The Definitive Guide, 2nd edn. O'Reilly Media, Sebastopol (2013)
13. Rsync. http://rsync.samba.org/
14. Networkx - High-productivity software for complex networks. http://networkx.github.io
15. CESNET z.s.p.o, CESNET2 network. http://www.cesnet.cz/services/ip-connectivity-ip/cesnet2-network/?lang=en

Simulation-Based Analysis of a Platform as a Service Infrastructure Performance from a User Perspective

Wojciech Rząsa[✉]

Rzeszow University of Technology,
Powstańców Warszawy 12, 35-959 Rzeszów, Poland
wrzasa@prz-rzeszow.pl
http://www.prz.edu.pl

Abstract. This paper describes analysis of a real-world case where growing up Platform as a Service (PaaS) provider faced and solved problems with scaling cloud-based infrastructure. Scientific, Petri net-based method was used to asses decisions taken by Heroku – the PaaS manager – and legitimacy of claims of the PaaS clients caused by not satisfying efficiency of the new solutions. Exhaustive information provided in the Internet by both parties of the conflict were used in order to create model of the application and infrastructure corresponding to the real case. The model was then used to perform reliable simulations and show that while the client's claims were well founded, but growth of the PaaS infrastructure forced and justified changes in the management algorithms.

Keywords: PaaS · Performance · Model · Simulation · Petri nets · TCPN

1 Introduction

Platform as a Service (PaaS) is a complete infrastructure that allows a client to run specific type of applications without a need to care for IT resources. Consequently, PaaS clients do not need employees trained to manage IT resources and do not have to invest in these resources. Moreover, economies of scale allow the PaaS providers manage large infrastructures efficiently, especially as the specialization in one service allows them to hire experts. However, success of a PaaS provider may result in significant challenges, performance being one of them. With the growth of the infrastructure IT solutions employed before may render inefficient. Moreover, effective scaling of the resource pool may be in conflict with performance observed by the PaaS clients. Such situation occurred in 2013 with Heroku[1] – popular platform designed to host web applications.

In this paper the problem and proposed solutions are analyzed on the basis of detailed information documenting this case in the Internet. A model is created

---

[1] http://heroku.com.

© Springer International Publishing Switzerland 2015
P. Gaj et al. (Eds.): CN 2015, CCIS 522, pp. 182–192, 2015.
DOI: 10.1007/978-3-319-19419-6_17

and simulations are performed to asses credibility of claims of the PaaS clients and accuracy of solutions deployed by Heroku. We present results that confirm that modifications of algorithms managing the PaaS infrastructure caused significant performance degradation for clients. Concurrently, we show that previous solutions could not be preserved with the growth of the infrastructure. The reason being that it would not only limit scalability of the PaaS but would also render diminished user application performance, comparable to the one observed for the new solution.

2 Related Work

There are challenging problems connected with large and dynamic infrastructures as the ones that are used to form basis for Platform as a Service. Two important and mutually connected are performance and scalability. These problems were already subject of research concerning the other dynamic and decentralized infrastructure. For years the concept of the grid system introduced by I. Foster and C. Kesselman [1] was analyzed and scientific research concerning such system formed the basis for nowadays pervasive cloud computing.

The fact that large-scale dynamic IT infrastructure must face challenging performance and scalability problems was noticed in the grid research from the beginning. There is significant number of projects that were designed to monitor, analyze, model or simulate the infrastructures and draw conclusions about correctness of applied solutions. The most popular simulator designed for the grid infrastructures is *GridSim* [2] that is designed to enable analysis of algorithms managing the infrastructure. The job scheduling algorithms can be evaluated using *Bricks* system [3]. Large data sets management solutions can be simulated using *ChicagoSim* [4] developed on the University of Chicago or *OptorSim* that emerged as a part of European DataGrid project. More complete list of abundance of the projects meant to assess infrastructure management solutions, with appropriate taxonomy can be found in the following publications [5–7].

Apart from the grid-related publications, there is a number of papers describing research concerning specifically performance problem of Platform as a Service solutions, e.g. [8,9]. Growing number of PaaS solutions resulted in research meant to enable reliable ranking of the providers [10,11].

These solutions, however, are not frequently used by companies that eagerly adopted other results of grid-related research. Consequently, managers of quickly growing-up infrastructures have to face the problems that could be avoided early if only appropriate research was carried out using the grid-related experiences.

The research presented in this paper was carried out using concepts that emerged while the research related to efficiency of the grid infrastructure and applications [12]. In short, a High-level model of application and infrastructure is created, then the model is automatically transformed to the executable model in Timed Colored Petri Nets [13] (TCPN) to perform reliable simulation. Prototype implementation of this concept was used before to analyze different aspects of multi-layer web application [14]. This paper does not describe the method concepts that can be found in [12,15], but focuses on the benefits that can be gained from exploiting simulation methods for analysis of distributed, dynamic infrastructures.

3   The PaaS Efficiency and Scaling Problems

Heroku – manager of the analyzed PaaS – founded in 2007 is one of the first cloud-based platforms created as Platform as a Service (PaaS) for web applications. At the beginning it supported only Ruby on Rails[2] (RoR) applications that were run in one or more private containers.

Single-threaded nature of then available RoR application servers had important implications in the PaaS architecture (Fig. 1). Each container in the first Heroku stack provided one single-threaded application server and serving more requests concurrently required more containers. In this architecture, important strength of Heroku was its *intelligent routing* ensuring that HTTP requests were forwarded only to idle containers. If no container was available, the requests were queued by the Heroku HTTP *router*.

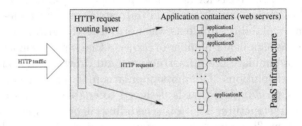

Fig. 1. Simplified architecture of the PaaS infrastructure

At the beginning of 2013 James Somers of Genius Engineering Team – one of Heroku clients – published information showing that the PaaS silently given up using the efficient for its users *intelligent routing* [16]. Currently, Heroku documentation clearly states that they are using *random routing* i.e. the PaaS routes HTTP request to a randomly chosen container running specific application [17]. J. Sommers described his observations and presented results of methodically created and carried out simulation of previous and new routing solutions. He noticed that behavior of the PaaS infrastructure became significantly inefficient from the client's point of view, especially for significantly loaded applications utilizing considerable pool of containers.

As it is shown in this work, observations of J. Sommers and his team were certainly valid. However, the other point of view should also be considered. As the infrastructure of Heroku grown, the PaaS had to preserve means to scale efficiently and that required scalable routing infrastructure. Preserving the *intelligent routing* required either single instance of the HTTP router or at least

---

[2] http://rubyonrails.org.

a single repository of information about idle and working containers. The necessary singleton in the architecture of the PaaS certainly limited its ability to scale. With *random routing* the HTTP routers can be multiplied creating what Heroku now calls *routing mesh.*

4   The Goal

The two points of view presented in the previous section – the clients and manager's of the infrastructure – justify two definitions of efficient infrastructure. Convinicing simulations carried out by J. Sommers present the client's point of view and only one aspect of the infrastructure. The other important aspect is scalability of the PaaS infrastructure.

The basic solution that could be deployed by PaaS to scale was to run more HTTP routers, independently proceeding according to the *intelligent routing* mechanism. Two such routers from the client's point of view would certainly be worse then one, they should however be significantly more efficient then *random routing*. However, large number of *intelligent routers* would finally lead to degradation of HTTP routing quality to this provided by *random routing*. Certainly, the PaaS infrastructure that currently serves more then 4 million applications and 5 billion requests per day needs significant number of HTTP routers.

The aim of this paper is to respond to two crucial questions: how many *intelligent routers* cause degradation of routing quality to *random routing* and shouldn't Heroku run a mesh of *intelligent routers* instead of the mesh of *random routers* to improve performance of client's applications?

To answer these questions a model of the infrastructure with two routing algorithms and web application using multiple containers was created. Using simulation performance of the application for different HTTP routing algorithms and different loads of the web application were compared. It is shown that modeling and simulation can provide valuable information, that can be used by companies to design transitional as well as final architectures of their systems.

5   Model of the PaaS and Application

The web application was modeled using Ruby[3] based Domain Specific Language as a set of **processes** that were capable of receiving and responding to requests. Each request caused specific CPU load and the response was delayed accordingly. Definition of the **program** modeling behavior of single container serving HTTP requests is presented on Listing 1.1. When event :**data_received** occurs, the request is processed and a response is sent using **send_data** statement.

HTTP requests are forwarded to application by the HTTP **router** that was modeled in two versions: *intelligent* and *random*. The latter simply forwards every reqest to a randomly chosen web server (Listing 1.2).

---

[3] http://rubylang.org.

```
1   program :webserver do |opts|
2     on_event :data_received do |data|
3       cpu do |cpu|
4         tm = opts[:request_times][data.content[:content][:length]]
5         tm / cpu.performance
6       end
7       send_data to: data.src, size: data.size * 10,
8                 type: :response, content: data.content
9     end
10  end
```

Listing 1.1. Model of web server

```
1   program :random_router do |servers|
2     on_event :data_received do |data|
3       if data.type == :request
4         server = servers.sample # get random app server
5         send_data to: server, size: data.size, type: :request,
6                   content: { from: data.src, content: data.content }
7       else # :response
8         send_data to: data.content[:from], size: data.size,
9                   type: :response, content: data.content[:content]
10      end
11    end
12  end
```

Listing 1.2. Model of *random router*

In the *intelligent router* model presented on Listing 1.3 the requests from clients are enqueued by the router after they are received. The router tries to serve this queue immediately after a new request is received and after a response is sent to a client. Serving the queue consists on sending request from the queue to an idle web server and removing the server from the list of available ones. This continues until no web server is available or there is no request in the queue. Responses are forwarded to appropriate clients, and the servers that processed them are added to the list of the idle ones.

The web client defined on Listing 1.4 sends two types of HTTP requests: :short and :long with defined frequency. Type of request indicates expected time required to serve it. The probability of sending long request as well as time required to serve particular type of request is defined as the program parameter.

For each program defined in the model a number of processes are started on separate nodes to model activities of analyzed configuration. The nodes are connected with network segments that for this simulation cause negligible delay.

6   Service Quality Assessment

To asses application performance from the perspective of its user, service time for subsequent requests is usually measured and presented as histograms of request count for subsequent time slots. Such presentation can be useful to visualize application performance, or to understand better or worse experience of application users. It is, however, hard to reliably compare histograms and thus they are not convenient to draw conclusions about changes in application performance.

```
1  program : router do | servers |
2    request_queue = []
3    on_event : data_received do | data |
4      if data.type == : request
5        request_queue << data # enqueue request
6        register_event : process_request # try to serve request
7      else data.type == : response
8        servers << data.src # the web server is idle
9        send_data to: data.content [: from], size: data.size,
10                  type: : response, content: data.content [: content]
11       register_event : process_request # try to serve a request
12     end
13   end
14   on_event : process_request do
15     unless servers.empty? or request_queue.empty?
16       data = request_queue.shift # first request from queue
17       server = servers.shift # first idle server
18       send_data to: server, size: data.size, type: : request,
19                 content: { from: data.src, content: data.content }
20       register_event : process_request unless request_queue.empty?
21     end
22   end
23 end
```

Listing 1.3. Model of *intelligent router*

```
1  program : wget do | opts |
2    sent = 0
3    on_event : send do
4      length = rand < opts [: long_prob] ? : long : : short
5      content = { number: sent, length: length }
6      target = opts [: targets][rand opts [: targets].length]
7      send_data to: target, size: 1024.bytes, type: : request,
8                content: content
9      sent += 1
10     register_event : send, delay: opts [: delay] if sent < opts [: count]
11   end
12   on_event : data_received do | data |
13     # stats collecting code here
14   end
15   register_event : send
16 end
```

Listing 1.4. Model of client program

Therefore, in order to reliably asses efficiency Application Performance Index (Apdex) [18] was used. It is an *open standard* developed to provide credible value that represents performance of application from the user's perspective and designed to better reflect user experience. It allows to credibly define Service Level Agreement between application provider and its users or between infrastructure provider and application provider.

The concept of $Apdex_t$ is to divide request service times to three groups: satisfying users (served in time not longer then t), being tolerated by users (served in time longer then t but not longer then $4t$) and frustrating users (served in time longer then $4t$). Having a sample of request service times one can compute value of Apdex for this sample using the following equation:

$$Apdex_t = \frac{S_t + \frac{T_t}{2}}{S_t + T_t + F_t} \tag{1}$$

where S_t is number of requests that were classified as satisfying users, T_t is number of requests classified as being tolerated by users and F_t is the number of requests that frustrate users.

The Apdex is used by a growing number of companies that offer monitoring of web application performance. What is important for this work, it is also used by NewRelic[4] that is Heroku partner providing extension for the PaaS users to monitor their applications [19].

7   Results of the Analysis

The model described in Sect. 5 was used to perform simulation and observe performance of the application. The actual configuration used in the analysis consisted of 10 web servers and a client sending requests with required frequency. The request service times were set to 50 ms for :short requests and to 5000 ms for :long requests. Probability of sending long request was set to 1 %. These values are obviously simplified in comparison to real applications, but they correspond to the general trend and more importantly are consistent with description of the problem published by Heroku [20].

With these parameters settled, the simulations were performed depending on routing solutions and load of application. Simulations were carried out for single random router and for different number of intelligent routers, starting from 1 up to 100. Application load ranged from about 30 % to about 100 %. Each experiment consisted on sending 10^4 requests. Performance of application in subsequent configurations is assessed using $Apdex_t$ with t parameter set to 0.5 s – the default value used to asses web application server response time.

The first experiment confirmed results obtained by J. Sommers and his team and correspond to their observations of real system behavior (Fig. 2). Performance of application is significantly reduced when PaaS uses *random routing* – $Apdex_{0.5\,s}$ value fell from 99.9 % to less then 65 % even for about 30 % load. Value of the index perfectly matches conclusions that can be drawn from histogram of :short request service times for this simulation presented in the Fig. 3. One can observe that even if application is not significantly loaded, *random routing* causes short request service times to exceed 3 s instead of being efficiently served in less then 100 ms. It should also be noticed that *intelligent routing* ensures stable performance of application even for significant loads while *random router* causes constant degradation of request service time. While *intelligent routing* allows application to ensure satisfaction of most clients up to load of the order of 90 %, with *random router* when application is loaded in 50 %, half of requests is served with unacceptable time. With *random routing* application loaded in 50 % behaves similarly to application loaded in almost 100 % when *intelligent routing* is in use. This clearly shows that claims of J. Sommers and his team to Heroku had their reasons.

The other part of the research was meant to verify scalability of *intelligent routing* without central repository of idle and working web servers. Therefore,

---

[4] http://newrelic.com/.

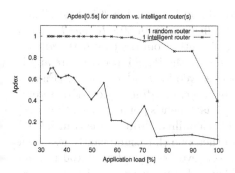

Fig. 2. Apdex comparison for 1 random router and 1 intelligent router

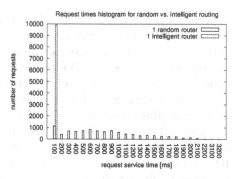

Fig. 3. Histogram comparing request times for 1 random and 1 intelligent router

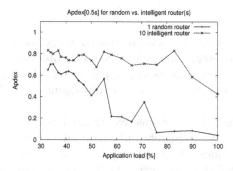

Fig. 4. Comparison of random routing and 10 intelligent routers

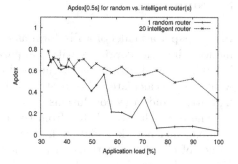

Fig. 5. Comparison of random routing and 20 intelligent routers

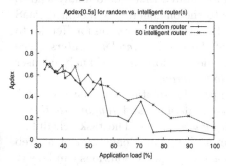

Fig. 6. Comparison of random routing and 50 intelligent routers

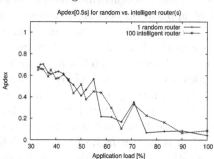

Fig. 7. Comparison of random routing and 100 intelligent routers

the number of independently working intelligent routers was increased and performance of application was compared to the one obtained for *random routing*.

Application performance for infrastructure with 10 *intelligent routers* instead of one is noticeably worse but it is still significantly better then for *random router* (Fig. 4). Increasing number of *intelligent routers* to 20 further degrades application performance (Fig. 5), but it would still be more beneficial for clients then *random routing*. However, simulations indicate that for 50 *intelligent routers* application performance does not differ significantly from the case with *random routing* (Fig. 6) while for 100 *intelligent routers* in the PaaS infrastructure application performance corresponds to the one observed for *random routing* (Fig. 7).

The results of the simulation show that *intelligent routing* deployed by Heroku at the beginning is perfect solution from the PaaS client's point of view. However, it is impossible to preserve it for the sake of scalability. Probably as transitional solution, it could be beneficial to use more then one (but significantly less then 100) independent *intelligent routers*. However, even the number of 100 routers would certianly not solve scalability problems of infrastructure with 4 million applications and 5 billion requests per day.

8   Summary

In this paper a model of a Platform as a Service was used to analyze its performance using real world case with real parameters. Simulation allowed to observe behavior of analyzed infrastructure and collect wide variety of parameters that were used to calculate industry standard performance index for subsequent configurations and draw conclusions.

The research showed that from the PaaS client perspective, the claims concerning efficiency degradation had strong basis. Significantly reduced performance of client applications on modified PaaS infrastructure can be observed. On the other hand it can also be concluded that the initial PaaS configuration was impossible to preserve for the scalability reasons. It has been shown that transitional solutions that could be helpful for limited time would finally lead to client application performance comparable to the one offered by modified PaaS solutions. Consequently, we must conclude that there is no reason for Heroku to run a mesh of *intelligent routers* instead of the mesh of *random routers*, since a mesh of only 50 *intelligent routers* degraded performance of analysed user application to the level offered by the *random routing*.

There probably should be found a compromise solution between the *intelligent routing* – perfect for clients, but not scalable – and *random routing* – perfectly scalable, but not efficient for the PaaS clients. The task scheduling problem is known to be hard, especially in such dynamic environment as web applications. Using simulation methods to assess proposed solutions can be significantly beneficial, as one can easily implement HTTP routing algorithms and test them without employing large computing infrastructures.

Acknowledgment. The equipment used for this research was purchased within the project no RPPK.01.03.00-18-003/10, which was co-financed by the European Union from the European Regional Development Fund within Regional Operational Program for the Podkarpackie Region for 2007–2013.

Author expresses his gratitude to dr Dariusz Rzońca for his valuable remarks that helped to improve this paper.

References

1. Foster, I., Kesselman, C., Tuecke, S.: The anatomy of the grid: enabling scalable virtual organizations. Int. J. Supercomputer Appl. **15**(3), 200–222 (2001)
2. Sulistio, A., Cibej, U., Venugopal, S., Robic, B., Buyya, R.: A toolkit for modelling and simulating data grids: an extension to gridsim. Concurrency Comput. Pract. Experience (CCPE) **20**(13), 1591–1609 (2008). Wiley Press, New York, USA
3. Takefusa, A., Tatebe, O., Matsuoka, S., Morita, Y.: Performance analysis of scheduling and replication algorithms on grid datafarm architecture for high-energy physics applications. In: Proceedings of the 12th IEEE International Symposium on High Performance Distributed Computing (HPDC-2012), pp. 34–43 (2003)
4. Ranganathan, K., Foster, I.: Decoupling computation and data scheduling in distributed data-intensive applications. In: Proceedings of 11th IEEE International Symposium on High Performance Distributed Computing (HPDC-2011), Edinburgh, July 2002
5. Dobre, C., Pop, F., Cristea, V.: New trends in large scale distributed systems simulation. In: Internatioal Conference on Parallel Processing Workshops ICPPW, pp. 182–189 (2009)
6. Sulistio, A., Yeo, C.S., Buyya, R.: A taxonomy of computer-based simulations and its mapping to parallel and distributed systems simulation tools. Int. J. Softw. Pract. Experience **34**(7), 653–673 (2004). Wiley Press, USA
7. Sulistio, A., Yeo, C.S., Buyya, R.: Simulation of Parallel and Distributed Systems: A Taxonomy and Survey of Tools. http://www.cs.mu.oz.au/~raj/papers/simtools.pdf
8. Zhou, J., Zhou, B., Li, S.: Automated model-based performance testing for PaaS cloud services. In: Computer Software and Applications Conference Workshops (COMPSACW), pp. 644–649, July 2014
9. Zhang, W., Huang, X., Chen, N., Wang, W., Zhong, H.: PaaS-oriented performance modeling for cloud computing. In: Computer Software and Applications Conference (COMPSAC), pp. 395–404, July 2012
10. Atas, G., Gungor, V.C.: Performance evaluation of cloud computing platforms using statistical methods. Comput. Electr. Eng. **40**(5), 1636–1649 (2014)
11. Garg, S.K., Versteeg, S., Buyya, R.: A framework for ranking of cloud computing services. FGCS **29**(4), 1012–1023 (2013)
12. Rząsa, W.: Timed colored petri net based estimation of efficiency of the grid applications. Ph.D. AGH University of Science and Technology, Kraków, Poland (2011)
13. Jensen, K., Kristensen, L.: Coloured Petri Nets. Modeling and Validation of Concurrent Systems. Springer, Heidelberg (2009)
14. Dec, G., Rząsa, W.: Modeling multilayer distributed web application with TCPN (Modelowanie wielowarstwowej rozproszonej aplikacji www z zastosowaniem TCPN). In: Trybus, L., Samolej, S. (eds.) Projektowanie, analiza i implementacja systemów czasu rzeczywistego, pp. 137–148. WKŁ, Warszawa (2011). (in Polish)
15. Rząsa, W.: Synchronization algorithm for timed colored petri nets and ns-2 simulators. In: Kwiecień, A., Gaj, P., Stera, P. (eds.) CN 2013. CCIS, vol. 370, pp. 1–10. Springer, Heidelberg (2013)
16. Somers, J.: Heroku's Ugly Secret. http://genius.com/James-somers-herokus-ugly-secret-annotated
17. Heroku HTTP Routing. https://devcenter.heroku.com/articles/http-routing

18. Apdex specification. http://apdex.org/index.php/category/specification/
19. Apdex: Measuring user satisfaction. https://docs.newrelic.com/docs/apm/new-relic-apm/apdex/apdex-measuring-user-satisfaction
20. Routing Performance Update. https://blog.heroku.com/archives/2013/2/16/routing_performance_update

Applying Software-Defined Networking Paradigm to Tenant-Perspective Optimization of Cloud Services Utilization

Dominique Jullier[1], Marek Konieczny[2]([✉]), and Sławomir Zieliński[2]

[1] Open Systems AG, Räffelstrasse 29, 8045 Zürich, Switzerland
dominique@jullier.ch
[2] Department of Computer Science, AGH University of Science
and Technology, Kraków, Poland
{marekko,slawek}@agh.edu.pl

Abstract. The article describes results of research in the area of software-defined networking applied to multi-cloud infrastructures. We present an innovative load balancer designed in accordance with a unique set of requirements, stemming from taking the cloud tenant's perspective on cloud service usage effectiveness. We developed a proof of concept implementation of the load balancer and drawn the conclusions regarding the system itself and its applicability to production environments. Our solution is more comprehensive than others available in contemporary SDN environments, and offers more functionality than just service chaining realized through flow reroutes.

Keywords: Software-defined networking · Cloud environemnt · Load balancing

1 Introduction

The development in the field of network technologies was nearly standing still for a long time in comparison with the progress in other fields of IT, such as software engineering. Nevertheless, in recent years, this started to change. The development and adoption of software-defined networking paradigm opened a variety of possibilities both to the application developers and providers of networking services. The networks are becoming programmable, and therefore more flexible and easier to adjust to their users' expectations. Programmable networks, combined with virtualization technologies, form a powerful toolset that can be used by application vendors, who are frequently the clients of processing power, storage and network vendors.

Network Function Virtualization (NFV) augments the Software-Defined Networks (SDNs) in terms of cost reduction. It allows to implement all network functions as virtual appliances. Load balancing is one of the network functions which can be virtualized. Noteworthy, the functionality can be of value both to the cloud providers and clients (tenants). A virtualized proxy can define service

© Springer International Publishing Switzerland 2015
P. Gaj et al. (Eds.): CN 2015, CCIS 522, pp. 193–202, 2015.
DOI: 10.1007/978-3-319-19419-6_18

utilization characteristics in a software-defined network, by receiving all requests and distributing them to backend servers. Such proxies can be based on various criteria or algorithms [1].

In this paper we address the issue of providing a virtualized load balancing service by using software-defined networking paradigm. In the chosen approach, an SDN controller instructs switches which operations to perform on packets and where to forward them. Note that in the decision making process the controller may use criteria and data not available to typical load balancers (such as dynamic information of individual servers load, various context information [2] or current monetary costs of resources usage). By utilizing many sources of information, multi-criteria load balancing functionality can be effectively delivered. The software presented in the article is designed for tenants of cloud services, offering their applications to the end users. The tenants wish to optimize their applications in terms of cost effectiveness, as well as other, performance-related metrics.

The paper organization is as follows. Section 2 surveys related technologies and paradigms. Then, based on the conclusions drawn upon the survey, Sect. 3 specifies requirements to be fulfilled by the load balancer. Section 4 discusses the architecture of the solution, while Sect. 5 presents some concentrates on the prototype implementation and its functional testing. The article ends with conclusions in Sect. 6 and acknowledgments.

2   Related Work

This section surveys the contemporary load balancing techniques and identifies the areas in which they lack support for clients of multi-site virtualized environments. Then, it analyzes the emerging software-defined networks technologies in context of their applicability to fill in the identified gaps.

Network services are often replicated on multiple servers in order to increase processing capacity and improve reliability. Typically, they have a similar architecture – a single data center or an enterprise has a front-end load balancer [3,4], which assigns client requests to a chosen server instance. A dedicated load balancer using consistent hashing is a popular solution today. A good example of such approach is Steel App implemented by Riverbed [5]. The solution can be installed in a resilient and scalable manner. It implements many load balancing algorithms and can be deployed as software, virtual appliance or a dedicated hardware appliance. The other approach assumes that load balancing is performed by software in the integration layer of the service e.g. Adaptive ESB is a goal-driven framework for adaptive execution of SOA based services [6]. In such a case, communication between services is rerouted to the different instances of service components. The drawbacks include costs and limited customizability.

Inside data centers, most of load balancing solutions are based on the local infrastructure monitoring only. Multi-cloud environments are typically not supported. The leading cloud service provider – Amazon – uses the Elastic Load Balancing [7]. The system balances the traffic incoming through Amazon EC2

instances in their respective availability zones. It can also detect whether the
load is unbalanced and distribute it evenly among healthy instances. Current
research is focused more on supporting multi-tenant single-site than on multi-
site environments [8].

Inter-data center load balancing is not a trivial task, mainly due to the lack
of standardized metrics that could be used for assessing the particular locations.
Inter-cloud load balancing is even more complex, because of differences in services offered by distinct providers. A methodology that can be used to evaluate
cloud services was proposed in [9]. The methodology defines tests, metrics and
interpretation guidelines for testing process outcomes. Therefore, it can useful
the development of cloud service scoring and load balancing solutions.

One of the key promises of SDNs is their flexibility and scalability. Therefore, research work on load balancing is also conducted in the software-defined
networking domain. For example, the Plug-n-Serve [10] solution balances load
across unstructured network using the OpenFlow [11] protocol. The proposed
solution allows to monitor the overall network congestion and network topology.
Based on that data the controller is able to balance traffic based on response
times of HTTP requests. The solution installs flows in a reactive manner (a controller decides upon forwarding direction after being sent a packet that was not
matched in a switch flow table). Research based on updating flows in a proactive
way is also conducted. In the environment described in [12], decisions regarding
the balancing, based on wildcard matching are taken by specialized algorithms,
which generate wildcard rules. Unfortunately the solution supports only static
weights, which can be assigned to server instances (e.g. server instance should
receive 50 % of all requests).

While the traditional balancers are designed for services specific to ISO/OSI
layers (e.g. L4 or L2), the SDN-based ones are open for balancing on different services [13]. In addition, SDN controller can take advantage of multiple matching
schemas. Also when a new network device is added, it can be dynamically discovered and handled by controller. These features, accompanied with the ease
of starting a new controller instance, and offer a great degree of flexibility at
low monetary cost to their user. To summarize, none of the reviewed, representative solutions addresses tenant-perspective multi criteria load balancing in
multi-cloud architectures. Our motivation was, therefore, to try and implement
a unique solution for that specific functionality.

3   Load Balancing Service Functionality

This section overviews the functionality of the presented load balancing system
and leverages the considerations presented in Sect. 2 to list the key functional
and non-functional requirements. The list of requirements is then reflected in
the load balancing parameters and application descriptors, discussed later in the
article.

Requirement 1: Load Balancing Based on Servers' State. Each server
    instance reports the values of key load-related metrics, such as CPU load,

memory usage and network load. The load balancer distributes the incoming requests based on these reports, ensuring optimal overall utilization of server instances and avoiding overloads.

Requirement 2: Load Balancing Based on the Costs of Resources' Utilization. The load balancer supports a way to evaluate the costs of using particular server instances. Based on the costs, the load balancer is able to use cheaper servers whenever possible. More expensive servers will be still used in case the cheaper servers reach their performance limits.

Requirement 3: Load Balancing Based on Location Preferences. In order to reduce the request processing times and traffic between the data centers, location preference mechanism is implemented. In this way, the user is able to express his preference for server instances located close to the computation client.

Requirement 4: High-Availability. The load balancer implementation includes techniques ensuring its high-availability. In particular, the following events do not cause a service outage:
- server instance failure,
- SDN controller unreachability,
- inter-site connection breakdown.

Requirement 5: Open Standards-Based Implementation. The load balancer is implemented as a NFV module. It is based on open technologies, in order to reduce the costs of its installation, usage, and further development. Moreover, it uses standardized protocols and APIs wherever possible. Therefore, although developed and tested in an environment consisting of virtual switches, nothing prohibits it from managing hardware.

4   System Architecture

Figure 1 presents the architecture of the load balancing system. The whole solution consists of three major parts: the load balancer and SDN controller, SDN-enabled switches, and instrumented server (virtual machine) instances. The server instances, which host tenant's applications, are instrumented by monitoring agents, which collect relevant metrics regarding the server state, such as CPU load, memory usage, network load, and the costs of resources. The values of the metrics are exposed by an API. The load balancer's architecture is modular, and the functions of particular modules reflect the required key functionalities. The monitoring module periodically gathers the information from the monitoring agents and collects it in case it is requested by the decision taking module. The Policy module stores applications specific descriptors including weights that reflect the need for CPU power, network bandwidth, memory, location (e.g. preferring local sites), and the cost sensitivity. The data collected by the monitoring and policy modules are used by the scoring module each time it is required to compute the list of servers suitable for processing a new task. Such event is triggered by a new application user request coming to any site and forwarded by an appropriate switch to the controller and, in turn to the decision taking module. The decision taking module uses the list of feasible task allocation options,

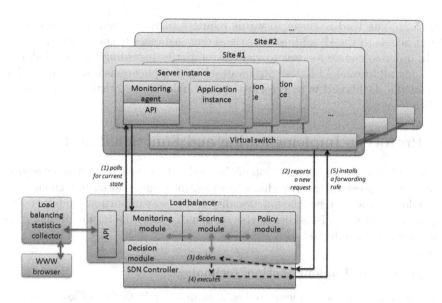

Fig. 1. System architecture

produced by the scoring module, and decides upon the flows to be installed in the network of virtual switches. The flows, having been installed in the virtual switches by an SDN controller, will effectively redirect the request to the application instance optimal from the tenant's point of view. The switches used in this scenario are not required to reason upon the network topology. All they need to maintain is a flow table and an interface that allows for manipulating the flows. In case there is a new, unmatched flow, they simply signal the event to the controller.

We assume that every virtual switch instance handles three distinct network domains (alternatively, one can use three switch instances per site), in order to isolate networks of three different functions from each other. The networks are as follows:

– Internet access (*brInternet* domain): This network provides the server instances with Internet access. Administrative tasks like updating the application instances are possible over this network.
– Monitoring and management network (*brAdmin* domain): The network provides access to the monitoring agents and virtual switches. It is used to gather performance-related data and install user traffic handling rules (flows).
– Application-specific overlay network (configured on *brLoadbalancer* switch): The network is used to forward client requests to appropriate server instances.

Load balancing service is transparent to the application users. User requests are directed to entry points located on sites. The entry points use virtual IP addresses specified by the application provider in application descriptors (see

Sect. 5.2). The load balancer forwards requests to the chosen server instances. Direct access to the IP addresses of the server instances is not possible.

In order to enable its user to view the statistics regarding the load balancer decisions, additional API, designed for dashboard applications is introduced. The task of the dashboard is to present the load balancing statistics in a readable and concise way.

5   Prototype Implementation and Functional Tests

A proof-of-concept implementation of the load balancing system was developed. This section presents the key characteristics, including chosen implementation technologies, and sketches the functionality of the most important modules building up the prototype. In addition, this section discusses the results of functional tests that were conducted.

5.1   Choice of Implementation Technologies

The load balancer was implemented as a module for POX OpenFlow controller [14]. It uses POX capabilities for managing flow tables in OpenFlow-compatible switches. The controller functionality, including the flow table manipulation is exposed by a convenient, Python-based API. Therefore, a logical decision was to implement the decision taking, scoring, monitoring, and policy modules also in Python.

Regarding the APIs, used both for monitoring the server instances and reporting load balancing statistics, the choice was to use the Representational State Transfer (REST) guidelines. Additionally, although the load balancer is not switch-dependent, switches differ in terms of OpenFlow support and available extensions. Therefore, for the proof of concept development and testing, we decided to work with Open vSwitch [15], the most popular open source virtual switch.

5.2   Policy Module

The following metrics representing the system state were taken into consideration: CPU load, memory usage, network load and cost of the resource. In order for the load balancer to figure out which metrics are more important than the others, each application comes with a descriptor. Among other things, the descriptor denotes whether CPU, memory or network is critical for the application. Also a fourth metric, called location factor, was used to mark the preference for the nearest server location. By definition, the maximum weight for CPU, memory, network, and proximity is 10. The weights work as multipliers for the metrics delivered by the agents. The calculated score can then be used to decide which system should be delivered the user request (see Sect. 5.3). The application descriptor defines virtual ip address of each site, the mapping beetween networks' addresses and the sites topology. However, these settings are irrelevant for scoring module.

5.3 Scoring Module

The scoring module creates a list of probabilities for the weighted random pick function. At first, the algorithm queries the policy module for application-specific weights of performance-related metrics. Then, it queries the monitoring module for performance-related metrics. If the utilization of any of the key resources is over 90 % the assigned score is 1, and therefore the probability of choosing the instance is minimal.

Next, a sum of products of metrics values and their respective weights in square is computed. Weights in square have a bigger impact on the score than linear ones. The solution was chosen because a linear calculation (i.e., simply multiplying the state with the weight) showed too little impact during the examined test cases. Next, the proximity-related weight is added (for servers located in the particular site) or subtracted (for remote servers). Finally, the result is normalized, so it comes to a value between 0 and 99. In the last step, out of the calculated result, an individual score is calculated, by subtracting the result from 100. Due to the previous normalization, a server can get a score ranging from 1 (inappropriate) to 100 (very suitable).

5.4 Decision Module

The decision taking module consults the scoring, policy and monitoring and decides upon flow propagation (see Sect. 4). It performs a weighted random pick from the list of server choice probabilities produced by the scoring module, and chooses a server instance to handle a particular request. In the next step, the module generates appropriate flow table entries and installs them in the switches. Noteworthy, after installing the flow, the switches become capable of handling the new connection even in the case of controller's failure. Figure 2 depicts the most important conversation between a switch and the load balancer. At the begining, a new request enters an Open vSwitch (nw_src: client IP, nw_dst: entry point IP). Because it is the first request of that specific client to the service, there is no appropriate flow table entry. Therefore, the switch forwards the packet to the controller. With the help of the load balancer, the controller decides what to do

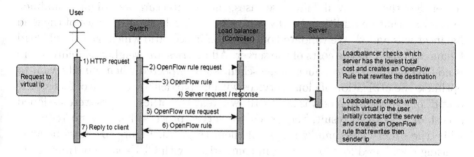

Fig. 2. Sequence diagram of the load balancer actions

with the packet, and installs a new flow table entry in the switch. The switch now forwards the packet to the chosen server instance (nw_src: client IP, nw_dst: server instance IP). The server processes the request as if it came straight from the client and returns a reply (nw_src: server instance IP, nw_dst: client IP). The switch intercepts the reply and queries the controller (and, in turn, the load balancer) for entry point (virtual) IP that was initially used. The controller installs another flow table entry in the switch. The switch is now able to forward the reply to the client (nw_src: entry point IP, nw_dst: client IP).

There are at least two more things to note here. First, it may well happen that the chosen server instance is located at a remote site and the packet will be forwarded through an inter-site tunnel. Such case is similar from the controller point of view; the only significant difference is that multiple switches are involved and more flows are installed. Second, although this has nothing to do with switch-controller conversation, it is worth to note that the load balancer's decisions can be based on information coming from OSI layers 2–7, or even from outside the particular network. Therefore, the decision taking module can hardly be implemented on site.

5.5 Sample Setup

The proof-of-concept implementation was tested in a two-site scenario. This section presents a very simple configuration built from three PCs (one acting as a client, two – simulating cloud service providers) connected to the Internet. It shows that even so simple installation can benefit from using the load balancer.

The operating system used on all hosts was Ubuntu 14.04. Site 1 runs a HP 6530b notebook, powered by an Intel Core 2 Duo Processor T9600 with 2.80 GHz and 4 GB of memory. Site 2 uses a Lenovo T420 Notebook, powered by an Intel Core i7-2640M CPU with 2.80 GHz and 8 GB of memory. The server instances are KVM guests, on which Ubuntu 13.10, monitoring agents and test applications are installed.

The topology used for testing consisted of four servers located in two different sites (see Fig. 3). Both sites had their own, external entry point (a public IP used to access it). The goal was to realize different load balancing strategies between these servers based on the current state, location and cost of each individual server. For the high-availability use cases, the architecture was slightly modified in the way that an additional POX controller with a backup-instance of the load balancer was added. We validated our solution in scenario which uses simple load balancing based on the costs of the server. All metrics are weighted equally, there is no location preference. Such usage schemes are expected when tenants want to optimize the utilization of cloud services with no particular resource of focus. The cost of one of the servers was set to 10, whereas the other servers were assigned a cost of 1. As the result, the expensive server processed nearly no requests, while the other three handled most of them (Fig. 4a). Figure 4b shows benefits of using cost-based load balancing in comparison with CPU-based optimization. A tenant will benefit in case of normal or low service usage, while in the case

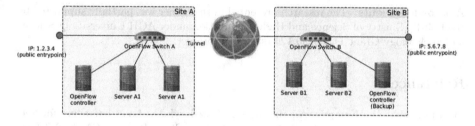

Fig. 3. High-Availability topology with fallback controller

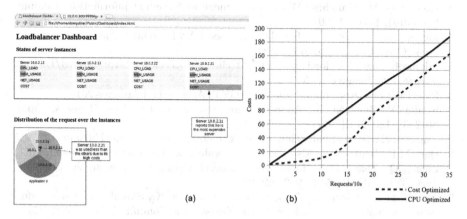

Fig. 4. Prototype implementation: (a) dashboard view, (b) benefits of using cost opitimization policy

of heavy load of cheaper server instances, the cost will increase. As mentioned earlier, such load balancing is not possible using contemporary solutions (2).

6   Conclusions

Classic, hardware-based load balancing solutions suffer from the lack of standardization. Solutions based on open standards, e.g., OpenFlow, allow the customer to avoid the vendor lock-in. That advantage should persuade customers towards choosing solutions such as the one presented in the article and opens a chance for SDN-based solutions to be adopted widely.

In the article we presented our research on tenant-perspective optimization of cloud services utilization. As a part of our work we developed a virtualized multi-criteria load balancing system based on software-defined networking paradigm. Later, we presented the prototype solution architecture and a proof of concept implementation as well as an evaluation scenario in which cost effectiveness was the main load balancing criteria. The presented results are promising and the prototype forms a good basis for development of more complex algorithms in the future.

Acknowledgments. The research presented in this paper was partially supported by the Polish Ministry of Science and Higher Education under AGH University of Science and Technology Grant 11.11.230.124 (statutory project).

References

1. Piórkowski, A., Kempny, A., Hajduk, A., Strzelczyk, J.: Load balancing for heterogeneous web servers. In: Kwiecień, A., Gaj, P., Stera, P. (eds.) CN 2010. CCIS, vol. 79, pp. 189–198. Springer, Heidelberg (2010)
2. Konieczny, M.: Enriching WSN environment with context information. Comput. Sci. **13**(4), 101–114 (2012)
3. McQuerry, S., Jansen, D., Hucaby, D.: Cisco LAN Switching Configuration Handbook. Cisco Press, London (2009)
4. Microsoft: network load balancing technical overview. http://goo.gl/D6Spr3. Accessed 30 January 2015
5. RiverBed: SteelApp. http://goo.gl/eAklYT. Accessed 30 January 2015
6. Szydło, T., Zieliński, K.: Adaptive enterprise service bus. New Gener. Comput. **30**(2–3), 189–214 (2012)
7. Amazon: elastic load balancing. http://goo.gl/M69X9g. Accessed 30 January 2015
8. Bari, M.F., Boutaba, R., Esteves, R., Granville, L.Z., Podlesny, M., Rabbani, M.G., Zhang, Q., Zhani, M.F.: Data center network virtualization: a survey. Commun. Surv. Tutor. IEEE **15**(2), 909–928 (2013)
9. Jarząb, M., Zieliński, K., Zieliński, S., Grzegorczyk, K., Piascik, M.: PaaS performance evaluation methodology. Int. J. Next-Gener. Comput. **1**(1), 1–24 (2015)
10. Handigol, N., Seetharaman, S., Flajslik, M., McKeown, N., Johari, R.: Plug-n-serve: load-balancing web traffic using OpenFlow. In: ACM SIGCOMM Demo (2009)
11. McKeown, N., Anderson, T., Balakrishnan, H., Parulkar, G., Peterson, L., Rexford, J., Shenker, S., Turner, J.: OpenFlow: enabling innovation in campus networks. ACM SIGCOMM Comput. Commun. Rev. **38**(2), 69–74 (2008)
12. Wang, R., Butnariu, D., Rexford, J.: OpenFlow-based server load balancing gone wild. In: Proceedings of the 11th USENIX Conference on Hot Topics in Management of Internet, Cloud, and Enterprise Networks and Services, Hot-ICE 2011, Berkeley, p. 12. USENIX Association (2011)
13. Koerner, M., Kao, O.: Multiple service load-balancing with OpenFlow. In: 2012 IEEE 13th International Conference on High Performance Switching and Routing (HPSR), pp. 210–214. IEEE (2012)
14. NOX: POX OpenFlow controller. http://goo.gl/Mzkoss. Accessed 30 January 2015
15. OVS: Open VSwitch. http://openvswitch.org/. Accessed 30 January 2015

Color-Aware Transmission of SVC Video over DiffServ Domain

Slawomir Przylucki[(⊠)] and Dariusz Czerwinski

Lublin University of Technology, 38A Nadbystrzycka Str, 20-618 Lublin, Poland
{s.przylucki,d.czerwinski}@pollub.pl
http://www.pollub.pl

Abstract. For the scalability extension (SVC) of the H.264/AVC coding algorithm, video data belonging to different layers have different priority depending on their importance to the quality of the video and the decoding process. This creates new challenges but also new opportunities for video streaming systems, especially those that use the Internet infrastructure. Therefore, mechanisms of the traffic engineering used in the IP network should as far as possible take into account internal distribution of priorities inside video streams. The paper proposes the Weighted Priority Pre-marking (WPP) algorithm for color-aware SVC video streaming over a DiffServ network. This solution takes into account relative importance of Network Abstraction Layer Units (NALU) and does not require any changes in the DiffServ marker algorithm. Presented results of simulations show that the color-aware transmission system, based on the WPP packet pre-marking, can provide better perceived video quality than the standard (best effort) streaming of multi-layered SVC video as well as single-layered AVC video.

Keywords: DiffServ · SVC · QoS · Packet marking

1 Introduction

Video streaming over IP is not a trivial task since IP was developed as a transmission environment providing no build-in Quality of Service (QoS). Therefore, the IETF recommended the Differentiated Service (DiffServ) architecture [1,2] as one of the methods to guarantee QoS parameters. The idea of differentiated services is based on a simple model, which classifies IP packets to the appropriate aggregates and therefore offers scalability and manageability. Packet classification and marking can take place at the data source (pre-marking) or at the edge of the DiffServ network. The first solution, pre-marking, is especially promising because it allows the source of a video stream to indicate priority of the individual packet. This opens the possibility to take into account the specific characteristics of video coding, and prioritization of these packages, which are of great importance to maintain acceptable level of the perceived video quality [3]. Taking into account the diversity of end-user expectations and problems posed

© Springer International Publishing Switzerland 2015
P. Gaj et al. (Eds.): CN 2015, CCIS 522, pp. 203–212, 2015.
DOI: 10.1007/978-3-319-19419-6_19

by the characteristics of modern video transmission systems, the Scalable Video Coding (SVC) is a highly attractive solution for video streaming. It enables the scalability in spatial, temporal, and quality (SNR), while keeping compression at high efficiency [4]. Scalability leads to a hierarchical stream of the SVC video. In this hierarchical structure, the priority of the data depends on the position within the individual layers as well as on the inter-layer relationships. For this reason, methods used for H.264/AVC may not be in a simple and direct way applied to the case of the system for the streaming of the layered video. In this paper, we develop a new Weighted Priority Pre-marking (WPP) algorithm, which takes into account relative importance of data within SVC video stream and does not require any changes in the DiffServ marker algorithm. It allows to obtain a better perceived video quality than for standard SVC video transmission without pre-marking. The streaming system employing the WPP algorithm also showed superiority (in terms of the video quality) over the system for the H.264/AVC video transmission with the TypeMapping pre-marking [5,6].

The remainder of this paper is organized as follows. The principles of the DiffSev architecture and the process of the packet marking for video transmission systems is presented in Sect. 2. Section 3 describes the coding rules for the scalability extension (SVC). The priority and bit stream ordering is also discussed. The developed WPP pre-marking algorithm is presented in Sect. 4 along with a discussion of other proposed methods of packets pre-marking. Simulation and experimental results are in Sect. 5. Finally, a brief conclusion is made in Sect. 6.

2   Service Differentiation

The principles of DiffServ have been described in the RFC2475 [1]. When classifying and marking packets according to DiffServ rules, the Differentiated Services (DS) field is used. It is located in the IP packet header. The first three bits of the DS field are used to determine the traffic class, while the next three define the packets rejection probability. Last two bits are left unused [7]. All these 6 bits create so called Differentiated Services Code Point (DSCP). Traffic management, according to DiffServ model, is carried out only in the boundary nodes of the DiffServ domain. Packets forwarding inside DiffServ network is performed in core nodes in accordance with established principles. These principles are called Per Hop Behavior (PHB). PHB can be defined on the basis of resources allocated to them or the priorities in relation to other PHB, which have common characteristics such as buffer management mechanisms, rejection policies and packets scheduling algorithms. The packet classification and marking on the edge of the DiffServ network leads to distinguish four basic service classes [7,8].

According to the IETF [9,10], a service class designed for video streaming is the Assured Forwarding (AF) class. The AF PHB offers different levels of forwarding assurances for IP packets, while accomplishing a target throughput for each network aggregate. This PHB defines IP packet forwarding for four AF subclasses with three dropping precedences. Packet drop priorities are usually identified by colors: green for the lowest drop precedences, yellow for the middle and red for the highest one. In order to carry out the task of marking packets belonging to class AF, IETF recommended Two Rate Three Color Marker (TRTCM) [11].

Despite the above-mentioned standards and recommendations, the question remains, how to properly use that AF class to fulfill the requirements of video transmission. The IETF in RFC4594 [7] recommends AF3x PHB for services that require near-real-time packet forwarding of VBR traffic and are not delay sensitive. These characteristics are consistent with the requirements of video streaming applications. Further, the IP sources should pre-mark their packets with DSCP values or the router topologically closest to a video source should perform Multifield Classification and mark all packets as AF3x. However only pre-marking scenario can potentially apply the hierarchical structure of the video stream and the relative importance of the different frames transmitted over the IP network.

3   SVC Coding

The H.264/SVC standard was created as an extension of the H.264 codec. The idea of a scalable video coding is to enable the removal of parts of the coded video data by rejecting certain layers. The resulting stream is clearly, lower quality but it will continue to be correctly decoded by the receiver. In order to achieve this in practice, layered coding schema has been implemented by generation a multi-layer bit stream consisting of one base layer (BL) and several enhancement layers (EL). The terminal devices with different technical parameters can choose to receive partial bit stream, e.g., base layer in case of mobile devices, and all layers for HD screens [4,12]. The hierarchical structure of SVC bit stream is presented in Fig. 1.

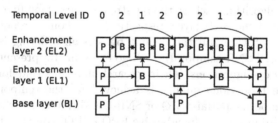

Fig. 1. Hierarchical SVC bit stream

According to the H.264/SVC, each spatial dependency layer requires dedicated prediction module to perform both motion-compensated prediction and intra prediction within the layer. Additionally, SVC coding algorithm introduces new modules closely related to the video quality. First one, a SNR refinement module, provides the mechanisms for quality scalability within each layer and a second one, the inter-layer prediction module, is responsible for the dependency management between subsequent spatial layers. As the end result, different temporal, spatial and SNR levels are simultaneously integrated into a single scalable bit stream.

Functionally, the H.264/SVC is divided into two parts, the Video Coding Layer (VCL) and the Network Abstraction Layer (NAL) [13]. The VCL produces the coded representation of the source video and the NAL formats these video data by means of the header information [4]. A NAL unit consists of a header and a payload part. The three fields inside the header are relevant to the presented issues: DID (dependency id) which indicates the inter-layer coding dependency level of a layer representation, QID (quality id) which indicates the quality level of an SNR layer representation and TID (temporal id) which indicates the temporal level of a layer representation.

At this point, the question arises how to properly and most efficiently, in the context of the perceived video quality, carry out the mapping the content information (priority) inside NALU to the DSCP field used in the DiffServ architecture. The proposed answer is presented in the next section.

4   Priority-Aware Packet Pre-marking

The video packets need to be classified into different priorities according to their relative importance before any color marking algorithm can be applied. This issue contains a few sub-problems. The first one is the method of determining the priority of the single NALU. The second issue is the choice of the method for associating the priority with classes in the DiffServ AF PHB. Finally, the third sub-problem is the selection of appropriate algorithm of the DiffServ color marker.

The first sub-problem is directly related to the methods of the Fix Priority Ordering (FPO). Taking into account the specific characteristics of the SVC encoding, obvious solutions are: temporal-based (or frame-based), spatial-based and SNR-based algorithms, where the bit stream is arranged first by the temporal, the spatial and the SNR layer, respectively. It should be noted that in the JSVM reference software [14] the default bit stream ordering is spatial-based (layer order: spatial-SNR-temporal). S. Xiao et al. in [15] presented test results for various FPO configurations. Their research indicates that for a wide range of bit rates, the best video quality is achieved for the temporal-based FPO (layers order: temporal-spatial-SNR) or SNR-based (layer order: SNR-temporal-spatial). They also proposed the adaptive FPO (APO) configuration. The APO method arranges the H.264/SVC bit stream according to contribution of different layers to the whole performance within a given Group of Pictures (GoP). Unfortunately, since each GoP may have different characteristics, the optimal bit stream may vary from one GoP to the other in the same video sequence. Therefore, three FPO configurations (JSVM default and best two from [15]) were selected for further testing.

With respect to the second and the third of the above-mentioned sub-problems, short review of related work is presented in the next subsection followed by description of our proposal of the WPP algorithm.

4.1 Related Work

The review of the literature leads to the conclusion that the vast majority of the proposed solutions require the changing of the packet marking algorithm. Chih-Heng Ke et al. presented an Enhanced Token Bucket Three Color Marker (ETBTCM) with consideration of both the network condition and the relative importance of video packets [16]. Another proposal has been published in [17]. An Improved Two Rate Three Color Marker (ITRTCM) marks video packets according to the current vacancy degrees of the token buckets and the relative importance of the packets. Both markers use a number of thresholds for dividing the relative significance into a few grades. Unfortunately, the more divisions of the relative significance of the video packets, the higher possibility of the mistake. That problem has been solved in [18]. Authors describe there a new marker called Priority-Aware Two Rate Three Color Marker (PATRTCM), which allows to minimize inaccuracy when the token count is close to the thresholds. Further development of the use of ITRTCM has been proposed in [19]. The solution consists of source marking scheme based on NALU priorities and ITRTCM as edge router marker.

Another group of solutions is focused on the QoE-based traffic management techniques. Study in [20] demonstrates the QoE-aware traffic management for scalable mobile video delivery within the MEDIEVAL architecture. The neuro-fuzzy scheme, described in [21], regulates output rate using a buffer and ensures that video streams from host to client conform to desired traffic conditions. The latest developments in this area are also using artificial intelligence algorithms and based on Software Defined Network principles [22]. Finally, we should also mention a very promising approach bonding marking policies and analysis of traffic patterns [23].

4.2 The Weighted Priority Pre-marking

Let us assume that the relative importance of the NALU is represented by (S_i, SNR_j, T_k) where S_i is the i-th spatial layer, SNR_j is the j-th quality layer and T_k is the k-th temporal layer. Let $P(S_i, SNR_j, T_k)$ be the priority of the NALU. We assign the weights to the individual layers in such a way that layers has weight of value: 1, 2 and 3 for the most important layer, less important layer and the least important layer, respectively. In such case, the priority for given NALU can be expressed by Formula (1)

$$P(S_i, SNR_j, T_k) = W_S S_i + W_{SNR} SNR_j + W_T T_k \qquad (1)$$

where W_S, W_{SNR}, W_T are weights for spatial, SNR and temporal layers, respectively.

In order to use the DiffServ marking algorithm TRTCM, the mapping of priorities to three packet colors (DSCP codes) is necessary. Let H be the highest value of priority calculated according to (1). The priority after mapping can be obtained using the formula as follows.

$$P_{\mathrm{mapped}}(S_i, SNR_j, T_k) = \left\lceil \frac{3}{H} P(S_i, SNR_j, T_k) \right\rceil . \qquad (2)$$

The last step is to apply the following rules of marking.

- if $P_{\text{mapped}}(S_i, SNR_j, T_k) \leq 1$ then use pre-marking as green,
- if $P_{\text{mapped}}(S_i, SNR_j, T_k) = 2$ then use pre-marking as yellow,
- if $P_{\text{mapped}}(S_i, SNR_j, T_k) = 3$ then use pre-marking as red.

That algorithm we named Weighted Priority Pre-marking (WPP) and the assessment of its usefulness has been presented in the next section.

5   Test Scenarios

The test scenarios has been divided into two phases. In the first phase, for the purposes of comparison, we selected three solutions that use different SVC FPO bit stream ordering. Therefore, the quality was analyzed for following schemas:

- S-SNR-T: pre-marking order S-green, SNR-yellow, T-red (JSVM default),
- T-S-SNR: pre-marking order T-green, S-yellow, SNR-red,
- SNR-T-S: pre-marking order SNR-green, T-yellow, S-red.

The conclusion of this stage we used as a starting point for the assessment of proposed WPP algorithm. The simulation of this algorithm is the content of the second phase of our test. The obtained results are compared with two other scenarios of the video transmission in the DiffServ domain. The first one is the transmission of the pre-marked stream of single layer H.264/AVC video [5,6] and the second one is the transmission of the SVC stream without pre-marking. All compared solutions were tested at the comparable network configuration and used the same video source.

5.1   Testbed Configuration

The coding part of the testbed system is built up based on the JSVM [14,24]. The network part consists of two elements: the software framework SVEF [25] and the network simulator GNS3. The first one, EVEF, allows to obtain the desired order in the SVC bit streams and was responsible for the proper generation and processing of the SVC traces. This package also has been used to estimate the quality of the transmitted video. The DiffServ domain was implemented in GNS3. The TRTCM marker (in color-aware mode) was configured on the ingress router. The structure of developed testbed is presented in Fig. 2.

The test video was a Foreman sequence, which has 2000 frames with a GoP size of 8. The structure of the stream consist of the following layers:

- spatial (SL0 – QCIF, SL1 – CIF, SL2 – CIF),
- temporal for SL0 and SL1 (TL0 – 3.75 Hz, TL1 – 7.5 Hz, TL2 – 15.0 Hz),
- temporal for SL2 (TL0 – 3.75 Hz, TL1 – 7.5 Hz, TL2 – 15.0 Hz, TL3 – 30 Hz),
- SNR (3 SNRL layers: BL and two EL for each spatial-temporal layer).

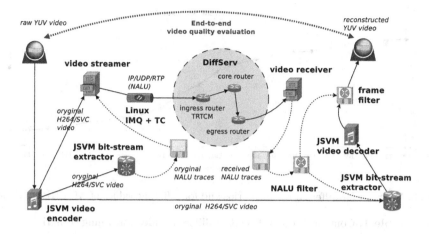

Fig. 2. The testbed structure

Above video bit stream has competed with one ON-OFF background traffic flow, which had an exponential distribution with mean packet size of 1000 bytes, burst time 200 ms, idle time of 50 ms, and rate of 500 kbps. Also the test network trasmitted one FTP traffic flow of 640 kbps. The DiffServ routers implemented the Weighted Random Early Detection (WRED) mechanism for active queue management. The WRED parameters include a minimum threshold, a maximum threshold, and a maximum drop probability. In our simulations, these parameters were specified respectively as 2, 4, 0.1 for red packets, 4, 6, 0.05 for yellow packets, and 6, 8, 0.025 for green packets. The final assessment of video quality based on PSNR metrics.

5.2 Simulation Results

Phase I. At this phase, all three selected pre-marking strategies have been simulated for the AF PHB overload ranging from 1.0 to 1.2. Each test was repeated 20 times and average values of received PSNR are presented in Fig. 3.

Analyzing the results shown in the Fig. 3, it is difficult to identify a clear winner. From the perspective of the typical IP network behavior, it seems to be reasonable to concentrate on the area of small and medium-size congestions. With respect to the performed simulation, it is the range of congestion from 1.05 to 1.15. In this area the best quality guarantee methods of protecting data associated mainly with the SNR and the spatial layers. For this reason, in the second phase, the highest level of protection is given for these two layers.

Phase II. Based on results from the phase I, the weights in Eq. (1) are assigned values, 1 for the SNR layer, 2 for the spatial layer and 3 for the temporal layer. Next, the priority for any given NALU were calculated according to the Eq. (1). The H coefficient has value 15 for used video sequences

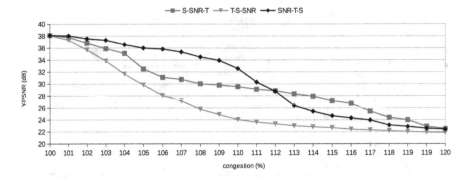

Fig. 3. Y-PSNR for different test scenarios and for different values of AF PHB overload

Table 1. Comparison of Y-PSNR in different video streaming scenarios

Scenario	Link overload		
	105 %	110 %	115 %
AVC with frame-type based pre-marking	32.4 dB	27.2 dB	25.5 dB
SVC without pre-marking	32.5 dB	29.6 dB	27.1 dB
SVC with WPP pre-marking	36.1 dB	31.2 dB	27.8 dB

(the least important triple in the Foreman sequence is (2,2,3) therefore $H = 1 * 2 + 2 * 2 + 3 * 3$). The last step was to apply the rules of marking presented in Sect. 4.

To justify our algorithm, the simulation results for the video transmission of SVC stream with the WPP pre-marking have been compared to the transmission of the SVC sequence without pre-marking (TRTCM was configured in blind mode) and the video coded by H.264/AVC coder with pre-marking based on simple frame type mapping (I,P,B) [5,6]. In all cases the Foreman video sequence has been used. The simulation results are shown in Table 1.

In the case of the use of the WPP pre-marking, the video quality improvement is observed for relatively small values of congestion (especially in the range from 1.0 to 1.1). This is due to better protection of spatial and SNR layers. Without pre-marking, mechanism inside the transmission system cannot protect the low and the lowest layers very well and at the same time losses of higher spatial and SNR layers are relatively high so end-user has no or little benefit from the SVC coding. The transmission of the H.264/AVC video with random losses of P and B macroblocks (pre-marking algorithm preferred I frames) causes numerous errors in the mechanisms of motion vectors reconstruction and inter-frame prediction. These phenomena very quickly (for relatively small values of overload) manifest themselves as important video quality degradation. For the higher values of congestion, advantage of SVC over AVC coding slowly disappears. The same can be said about the relationship between transmission with and without pre-marking. Even WPP algorithm cannot prevent loss of a substantial part of video data.

6   Conclusion

In this paper the relationship between the relative importance of NALUs and the packet pre-marking for the H.264/SVC video has been studied. We proposed The Weighted Priority Pre-marking algorithm for color-aware SVC video streaming over DiffServ network. In contrast to other proposed solutions, our approach is consistent with the DiffServ model and does not require changing the marking schema at the edge of DiffServ domain. Thus, the proposed algorithm can be applied to any IP network using the principles of the services differentiation.

By comparing the simulation results with standard streaming solution based on single layer H.264/AVC and best-effort H.264/SVC transmission, a simple conclusion can be drawn that the proposed pre-marking algorithm can well reflect the relative importance inside SVC bit stream and allows users to take advantage of the scalability extension of H.264/AVC.

References

1. Blake, S., Black, D., Carlson, M., Davies, E., Wang, Z., Weiss, W.: An architecture for differentiated services. IETF RFC2475 (1998)
2. Zhang, F., Macnicol, J., Pickering, M.R., Frater, M.R., Arnold, J.F.: Efficient streaming packet video over differentiated services networks. IEEE Trans. Multimed. **8**(5), 1005–1009 (2006)
3. P.10/G.100 Amendment 1: New Appendix I - Definition of Quality of Experience (QoE). ITU-T (2007)
4. Schwarz, H., Marpe, D., Wiegand, T.: Overview of the scalable video coding extension of the H.264/AVC standard. IEEE Trans. Circuits Syst. Video Technol. **17**(9), 1103–1120 (2007)
5. Ziviani, A., Wolfinger, B.E., De Rezende, J.F., Duarte, O., Fdida, S.: Joint adoption of QoS schemes for MPEG streams. Multimed. Tools Appl. **26**(1), 59–80 (2005)
6. Przylucki, S.: Efficiency of IP Packets pre-marking for H264 video quality guarantees in streaming applications. In: Kwiecień, A., Gaj, P., Stera, P. (eds.) CN 2012. CCIS, vol. 291, pp. 120–129. Springer, Heidelberg (2012)
7. Babiarz, J., Chan, K., Baker, F.: Configuration guidelines for DiffServ service classes. IETF RFC4594 (2006)
8. Baker, F., Polk, J., Dolly, M.: A Differentiated Services Code Point (DSCP) capacity-admitted traffic. IETF RFC5865 (2010)
9. Heinanen, J., Baker, F., Weiss, W., Wroclawski, J.: Assured forwarding PHB group. IETF RFC2597 (1999)
10. Grossman, D.: New terminology and clarifications for DiffServ. IETF RFC3260 (2002)
11. Heinanen, J., Guerin, R.A.: A two rate three color marker. IETF RFC2698 (1999)
12. Unanue, I., Urteaga, I., Husemann, R., DelSer, J., Roesler, V., Rodriguez, A., Sanchez, P.: A tutorial on H.264/SVC scalable video conding and its tradeoof between quality, coding efficiency and performance. InTech (2011)
13. Wiegand, T., Sullivan, G.J., Bjontegaard, G., Luthra, A.: Overview of the H.264/AVC video coding standard. IEEE Trans. Circuits Syst. Video Technol. **13**(7), 560–576 (2003)

14. Reichel, J., Schwarz, H., Wien, M., Vieron, J.: Joint Scalable Video Model 9 of ISO/IEC 14496–10:2005/AMD3 Scalable Video Coding. Joint Video Team (JVT), Documents JVT-X202 (2007)
15. Xiao, S., Wu, C., Li, Y., Du, J., Kuo, C.C.: Priority ordering algorithm for scalable video coding transmission over heterogeneous network. In: Proceedings IEEE the 22th International Conference Advanced Information Networking and Applications (AINA 2008), pp. 896–903 (2008)
16. Ke, C.-H., Shieh, C.-K., Hwang, W.-S., Ziviani, A.: A two markers system for improved MPEG video delivery in a DiffServ network. IEEE Commun. Lett. **9**(4), 381–383 (2005)
17. Chen, L., Liu, G., Zhao, F.: An improved marking mechanism for real-time video over DiffServ networks. In: Ip, H.H.-S., Au, O.C., Leung, H., Sun, M.-T., Ma,W.-Y., Hu, S.-M. (eds.) PCM 2007. LNCS, vol. 4810, pp. 510–519. Springer, Heidelberg (2007)
18. Wang, H., Liu, G., Chen, L., Wang, Q.: A novel marking mechanism for packet video delivery over DiffServ Networks. In: 2011 IEEE International Conference on Multimedia and Expo (ICME), pp. 1–5 (2011)
19. Chen, L., Liu, G.: A delivery system for scalable video streaming using the scalability extension of H.264/AVC over DiffServ networks. In: IIHMSP 2008 International Conference on Intelligent Information Hiding and Multimedia Signal Processing, pp. 665–669 (2008)
20. Amram, N., Fu, B., Kunzmanny, G., Melia, T., Munaretto, D., Randriamasy, S., Sayadi, B., Widmer, J., Zorzi, M.: QoE-based transport optimization for video delivery over next generation cellular networks. In: IEEE Symposium on Computers and Communications (ISCC), pp. 19–24 (2011)
21. Mushtaq, M.S., Augustin, B., Mellouk, A.: Empirical study based on machine learning approach to assess the QoS/QoE correlation. In: 17th European Conference on Networks and Optical Communications (NOC), pp. 1–7 (2012)
22. Alreshoodi, M., Woods, J., Musa, I.K.: QoE-enabled transport optimisation scheme for real-time SVC video delivery. In: 9th International Symposium on Communication Systems, Networks and Digital Signal Processing (CSNDSP), pp. 865–868 (2014)
23. Sierszen, A.: Reduction of reference set for network data analyzing using the bubble algorithm. Image Process. Commun. **313**, 319–328 (2014)
24. Joint scalable video model - reference software. http://www.hhi.fraunhofer.de/en/fields-of-competence/image-processing/research-groups/image-video-coding/svc-extension-of-h264avc/jsvm-reference-software.html?NL=0
25. SVEF: scalable video-streaming evaluation framework. http://svef.netgroup.uniroma2.it/#download

Method of Visual Information Processing Based on Grid-Calculations

Olga Dolinina[✉], Alexander Ermakov, and Alexandr Shvarts

Yury Gagarin State Technical University of Saratov, Saratov, Russia
{olga,ermakov,shvarts}@sstu.ru

Abstract. Processing video information is a complex problem requiring a variety of methods to solve. Local computer resources are generally used to process home video. Local computer networks resources are more suitable with larger size video files. Such an approach is expedient in processing an average volume of information and fulfills the medium requirements of processing time. This method has several disadvantages which can be eliminated with use of GRID-systems. However, GRID-calculations also demand the increasing of the efficiency by finding the number of tasks in a job and parameters of processing environment. An outline is given in this paper comparing methods of video processing based upon Grid calculations and cloud computing.

Keywords: GRID-calculations · Cloud computing · Video processing · Tasks scheduling · Efficiency of visual information processing

1 Introduction

Existing methods of video information processing have been developed to provide the most compact and efficient way of storing video files. Such a requirement is fulfilled with intraframe compression, in which discrete cosine transform is used, and interframe compression, in which prediction of macroblock offset is used. Videofile structure is suitable for three algorithmic approaches to its processing: parallel intraframe compression processing; parallel or distributed frame processing; distributed processing of independent images groups.

From a computational point of view, given methods can be implemented by using one of the computational frameworks. Among the existing approaches there are local computers, special processing hardware, local area networks (LAN), computing clusters.

The main drawback of the existing hardware processing methods is that they perform very wide range of task. Methods based on software video information processing have much more flexible structure. Therefore, from the point of view of cost-time ratio, using LAN resources for processing is more suitable for many tasks. However, this method is limited by the environment scalability and there is also a problem of processing efficiency.

Efficiency here means achieving of minimal processing time by utilizing only those computing nodes which can decrease the time of processing.

© Springer International Publishing Switzerland 2015
P. Gaj et al. (Eds.): CN 2015, CCIS 522, pp. 213–221, 2015.
DOI: 10.1007/978-3-319-19419-6_20

2 Using of Grid-Computing for Video Information Processing

The GRID technology provides support for different classes of applications offering them access to virtually unlimited resources [1]. With GRID-computing it is possible to implement video processing with LAN resources and avoids disadvantages because of the number of nodes and using modern processing reliability and security systems [2]. However, existing methods with GRID-computing do not have tools that provide efficiency of video processing. In order to solve this problem a new method was introduced. It is based on calculation of the time of video processing in the GRID-environment with given amount of available resources [3]. Generally, video processing time t_J can be presented as

$$t_J = T_{calc} + T_{transfer}, \tag{1}$$

where

T_{calc} – time of processing all data,
$T_{transfer}$ – time of transferring all data.

In order to define time of processing all data the term of grid performance (GP) is introduced. It represents the number of tasks executed by one node in one time unit. For a node, which is responsible for resource detection, performance test is run and time t_{test} spent on processing it is measured. This time is considered to be one performance unit. It means that this node executes one task in one time unit. For all other nodes the test is run and the performance is calculated as following

$$GP_j = \frac{t_j}{t_{test}}, \quad j = 1, \ldots GN, \tag{2}$$

where

t_j – time spent on test task execution by node j,
GN – number of computing nodes available.

Calculated values of GP_j represent the number of tasks executed by node j in the same time as by the test node and can be used as basic characteristic. This characteristic represents the speed of nodes comparing to each other. Moreover, it is possible to calculate the processing time comparing to any other node with previously calculated processing time

$$t_i = t_j \cdot \frac{GP_j}{GP_i}, \tag{3}$$

where

t_j and GP_j – processing time and performance of node j,
t_i and GP_i – processing time and performance of node i.

However, in GP calculation a number of problems arise as the performance of computing node is not a constant due to the changes in background utilization and delays in tasks processing.

It is necessary to use a system that would ignore insignificant fluctuations on a short period of time. In this work performance groups are introduced. Performance group is a set of computing nodes with the same performance. Grouping is executed according to average test results of each node. Resulting group is associated with average performance of all its nodes.

The number of the local nodes in one computing site can be presented as

$$GN = \sum_{i=1}^{N_{\text{gr}}} \sum_{j=1}^{N_i} G_{ij},$$ (4)

where

N_{gr} – number of node groups,
N_i – number of nodes in group i.

For the nodes newly added into the system the first task received is a test task and it is considered that $GP = 1$. After that the performance value is based in the processing time of the test node. The proposed principle of performance calculation means that the processing time is calculated as follows

$$T_{\text{calc}} = TN_{\text{p}} \cdot GP_{\text{test}} \cdot t_{\text{test}} + TN \cdot t_{\text{delaycalc}},$$ (5)

where

GP_{test} – performance of the test node,
TN – number of tasks,
t_{test} – processing time of the test node,
$t_{\text{delaycalc}}$ – duration delay in processing by each node,
TN_{p} – number of tasks, executed by the group of nodes with performance value of GP_{test}, which is calculated as follows

$$TN_{\text{p}} = \frac{TN}{\sum_{i=1}^{N} GN_i \cdot \frac{t_{\text{test calc}} + t_{\text{test transfer}}}{t_{i\,\text{calc}} + t_{i\,\text{transfer}}}},$$ (6)

where

N – number of node groups,
GN_i – number of nodes in groups i,
$t_{\text{test calc}}$ – processing time of one task by the test node,
$t_{i\,\text{calc}}$ – processing time of one task by node i,
$t_{\text{test transfer}}$ – transferring one task by the test node,
$t_{i\,\text{transfer}}$ – transferring one task by node i,

$$t_{\text{calc}} = GP \cdot tt,$$ (7)

$$t_{\text{test transfer}} = JV \cdot \frac{GN}{U} + TN \cdot t_{\text{delaytransfer}}, \tag{8}$$

where

U – data channel speed,
JV – video file size,
GN – number of nodes,
$t_{\text{delaytransfer}}$ – delay duration in data transfer,
GP – node performance,
tt – processing time of one task.

Data transfer time can be calculated as follows

$$T_{\text{transfer}} = 2 \cdot JV \cdot \frac{GN}{U} + TN \cdot t_{\text{delaytransfer}} + t_{\text{offset}}, \tag{9}$$

where t_{offset} – offset time of processing finish, defined by the number of delays.

Proposed model has two manageable variables. One of them is TN – number of tasks. Each of them can contain one or more groups of pictures (GOP). So the processing of one video file can be split into various numbers of tasks. Therefore there are JK frames with overall size JV. Each task is presented by TK frames and its size is TV, where $\frac{TK}{K}$ – an integer, K – number of frames in one GOP. Therefore

$$TK = \frac{JK}{TN}, \quad TV = \frac{JV}{TN}. \tag{10}$$

So, TN can be calculates as follows

$$TN = \frac{JK}{NK} \cdot K, \tag{11}$$

where NK – number of GOP in one task The second manageable variable is a composition of the computing site. It means that it is possible to use all or only some of available computing nodes (N node groups, GN_i nodes in each).

The rest of parameters are constant for one process. Thus, variation of site composition and number of tasks affect the processing time. Considering given features, new method was introduced. This method is aimed to increase efficiency of video processing. The main idea is that there is such number of tasks, that minimal time spent on data transfer is achieved by decreasing delays and maximizing utilization of the computing nodes.

So, video processing time in GRID-environment can be represented by following model

$$t_{\text{J}} = TN_{\text{p}} \cdot GP_{\text{test}} \cdot t_{\text{test}} + TN \cdot t_{\text{delaycalc}} + \frac{2 \cdot JV \cdot GN}{U} + TN \cdot t_{\text{delaytransfer}} + t_{\text{offset}}, \tag{12}$$

The resulting formula is adopted as a mathematical model that reflects video stream processing. According to this model, video stream processing consists of independent subtasks. Each of the subtasks has one or more GOP. The analysis of adequacy is required in order to make the use of this model possible. Large amount of practical experiments and simulation using formula (12) was made to obtain the experimental data. The choice of the criteria for evaluating the adequacy is based on the data distribution type. The hypothesis of normal distribution of the data is introduced. The developed model contains four values (t_{test}, $t_{\text{delaycalc}}$, $t_{\text{delaytransfer}}$, t_{offset}) with a random character, so it is necessary to assess their impact on the distribution. From empirical samples of random variables obtained during the experiments an estimate of their average values and estimates of their deviation are obtained. These values are considered to be the input data to generate distributed independent random numbers selected from the quasi-random sequence. Obtained values allowed calculation of the standard deviation for each variable and their standard deviation. To assess the deviation a "method of transferring errors" was used, according to which for small deviations it is possible to use the Taylor series with retaining of the first terms of the expansion. Computer modeling allowed to construct a histogram the shape of which corresponds to a normal distribution, with a value of kurtosis equal to 0.04, and the value of the asymmetry equal to -0.005. Thus, there is no reason to reject the null hypothesis.

Hence, the procedure for assessing the adequacy can be done through the construction of confidence intervals and evaluation of checked values covering. The hypothesis H_0 about experimental data normal distribution was introduced. To approve it, one must get the exponential average and calculate the absolute difference between the average and all values of the experiment. The result shows that the residues are white noise, so there is no reason to reject the null hypothesis. To assess the model adequacy there was used Student's t- test. The hypothesis $H_0 : \bar{o}_i = \hat{o}_i$ where $i = 1, 2$ was introduced and the alternative hypothesis $H_0 : \bar{o}_i \neq \hat{o}_i$ where $i = 1, 2$, $\hat{o}_i$ – expectation of estimation of experimental data and mathematical modeling data. Derived from the dispersion values with significance level satisfies the null hypothesis. Thus, the constructed confidence intervals for the average cover unobserved truth with a probability of 95 %. Thus, the constructed confidence intervals for the average cover unobserved truth with a probability of 95 %.

So it is proved that the proposed mathematical model is adequate to the experimental results. Based on the proposed model, the decision was made that it is possible to choose a size of the task (i.e. the number of GOP in the same job) that will allow the most efficient use of computing nodes and LAN.

Proposed method consists of 6 steps:

Step 1. Assume that the task size is 1 GOP, calculate TN. Assume that the number of groups is 1. Assume that GN is equals to the number of nodes.

Step 2. Using the partition method define such GN, that the average $f(GN)$ value of the following n points is not less than $f(GN)$.

Step 3. Select GN existing nodes from the network.

Step 4. Using the partition method define minimal $f(GN)$ for existing nodes.

Step 5. Using the partition method define minimal $f(GN)$, so that the main condition is that there is such X that the result of (13) is an integer. So

X – number of GOP in one task.

$$TN = \frac{JK}{NK} \cdot X. \tag{13}$$

Step 6. In case of improvement in $f(GN, TN)$ go to Step 4. Otherwise, the optimal value is achieved.

The result of this method is predicted processing time and values of TN and GN. Original video file is split into TN pieces, each of them will be a separate task to be processed. Obtained characteristics are passed to the scheduler of GRID-environment as target set of resources, which must be used for processing.

In order to check efficiency of proposed method experiments were held, using developed program complex, based on OurGrid system [4]. OurGrid consists of three parts: MyGrid – services of task management; Peer – services of resource management; Gum – service of task execution on computing nodes. Developed complex consists of two parts: interface and computing. Interface part implements a number of functions, required for composing and running user tasks. Input data for this part is video file and processing task. Output is processed video file. Computing part is passed to computing node with the file itself and is responsible for processing the data.

Practical experiments for approbation of the method were done. The dependence $t_j = f(TN)$, where $TN = 10 \ldots 3600$; $TK = 1$; $TV = 550.7\,\mathrm{KB}$; $N = 1$; $GN = 4$; $GP = 1$; $GF = 1$; $t_{\text{test}} = 10.5\,\mathrm{msec} \cdot 10^{-2}$. Conducted experiments produced data that can be used for comparison of proposed method and common method called Storage Affinity (SA) [5], which is used in OurGrid for large amounts of data processing. In this case it is assumed that each task contains only one GOP. In the experiment the set of input parameters was: $JK = 30 \ldots 3600\,\mathrm{s}$; $NK = 10$ frames; time of one frame processing $NKK = 30\,\mathrm{ms} \cdot 10^{-2}$; size of each frame $NKV = 0.5\,\mathrm{MB}$; $GN = 9$.

The experiment was carried on according the following algorithm:

Algorithm "experiment"
begin
    Selection of the Grid-framework, setting of the values $GN = \sum_{i=1}^{N_{\mathrm{gr}}} \sum_{j=1}^{N_i} G_{ij}$
    Selection of the parameters TN, JV, TV
    Formation of the file of the "job" $J = \{T_1, T_2, \ldots, T_{TN}\}$
    The initialization of t_j
    Cycle k=1; while k<TN; k++
    begin
      if (number of processes < GN)
      begin
        create process $P_i(T_k)$
      end
      else
      begin
        wait while (number of processes >= GN)

end
end
end of t_j calculation
Union of all video segments into video file
end

The described algorithm processes all video segments using the values of the parameters initiated by the user. Key output information of the experiment is the time of processing t_j. Each parallel process in this algorithm can be represented by the following sequence of operations:

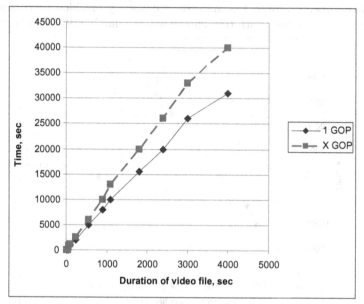

Fig. 1. Correlation between the length of video file and processing time

Algorithm "Proccess (T)"
begin
    Selection of the free node G_j from the list of the nodes where $i = 1 \ldots GN$
    Transmission of the data of the task T to the node G_j
    Processing()
    Getting a processed video segment()
end

Algorithm "processing"
begin
    input video
    VK = decoding (video)
    cycle from i=VK; while i<=VK; i++
    begin

$processing(frame_i)$
   end
   $coding(frame, VK)$
end

Figure 1 shows the correlation between the length of video file and processing time for SA method (*1 GOP*) and proposed method (*X GOP*). Both of them are linear. Therefore, *X GOP – 1 GOP* ratio is almost the same for video files of all presented volumes of files. For example, in this case the gain is around 25 %. For other experiments this gain varies from 0 % to 50 %.

The experiments proved the increase of efficiency of video processing with decreased processing time with the same or smaller number of computing nodes.

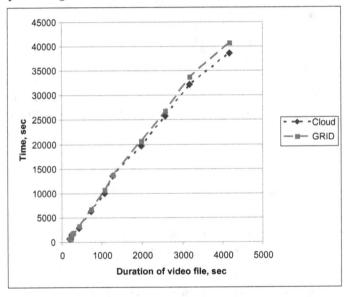

Fig. 2. Comparison of the proposed method run in GRID and in Cloud

3 Using of Cloud-Technologies for Visual Information Processing

Modern tendencies in distributed and high load computing are connected with cloud computing. The main idea of this approach is that maintenance and management of computing nodes is done in special datacenters, so that the consumers access resources through the Internet. The most important advantages of this approach are:

1. Dynamic management of amounts of computing resources required for each task.

2. Effective utilization of resources and networking infrastructure, provided by datacenters.
3. No need of buying equipment or hiring specially trained personnel.

In order to test the proposed method in cloud environment, an experiment was conducted using Windows Azure platform. One computing node was presented by virtual machine with Microsoft Windows Server R2 2008 and service pack HPC 2008 R2 SP1. This service pack is required to create computing clusters of several computers. All the computing nodes were situated inside one datacenter, so that the overheads in networking were minimized and each node was utilized as effective as possible.

Experimental results presented on Fig. 2 shows that the proposed method can be used also on cloud platform, for example – Windows Azure. In the nearest future cloud computing is likely to replace GRID-systems, because of the lower cost and flexibility in management computing resources.

4   Conclusions

In this paper a new method of video processing using distributed computing has been proposed. It has been demonstrated that this method can be effectively used to decrease overhead during video processing and for flexible resource management in a distributed environment. It has been shown that the efficiency of the described method is higher than one of the commonly used Storage Affinity method. Future research will be aimed to optimize the proposed method for a cloud platform, including not only dynamic node selection, but also node parameters changes.

References

1. Preve, N.P. (ed.): Grid Computing: Towards Global Interconnected Infrastructure. Computer Communications and Networks. Springer, London (2011)
2. Polak, M., Kranzlmüller, D.: Interactive videostreaming visualization on grids. Future Gener. Comput. Syst. **24**(1), 39–45 (2008)
3. Dolinina, O., Ermakov, A.: Video information processing using GRID-computing. In: Telematika'2010: Telecommunication, Web-Technologies, Supercomputing, pp. 197–203 (2010) [in Russian]
4. Andrade, N., Cirne, W., Brasileiro, F., Roisenberg, P.: OurGrid: an approach to easily assemble grids with equitable resource sharing. In: Feitelson, D.G., Rudolph, L., Schwiegelshohn, U. (eds.) JSSPP 2003. LNCS, vol. 2862, pp. 61–86. Springer, Heidelberg (2003). Revised Paper
5. Silva, D.P., Cirne, W., Brasileiro, F.: Trading cycles for information: using replication to schedule bag-of-tasks applications on computational grids. In: Kosch, H., Böszörményi, L., Hellwagner, H. (eds.) Euro-Par 2003. LNCS, vol. 2790, pp. 169–180. Springer, Heidelberg (2003)

An Energy Saving Solution in Integrated Access Networks

Glenda Gonzalez[1]([⊠]), Tülin Atmaca[1], and Tadeusz Czachórski[2]

[1] Laboratoire SAMOVAR, UMR 5157 IMT/Télécom SudParis, Evry, France
{glenda.gonzalez,tulin.atmaca}@telecom-sudparis.eu
[2] Polish Academy of Sciences, Warsaw, Poland
tadek@iitis.gliwice.pl

Abstract. In convergent access networks, access nodes (composed by optical and wireless components) can serve Base Stations (BSs). In the access node, the wireless receivers are one of the components with high energy consumption level. We can save energy by implementing transmission techniques to receive data in some periods of times, and switching off the wireless receivers the rest of time. However, this strategy will result in increased access delay for packets waiting in the BSs until transmission to the access node. In this paper, we aim at developing a framework for modeling the reception of data by the wireless receivers at the access nodes. We present an algorithm, a numerical formulation and an asymptotic approximation. Comparison between simulation and numerical results is presented and the analysis of results is summarized.

Keywords: Energy saving · Efficient integrated networks · Simulation

1 Introduction

The new generations of integrated access networks face a big challenge because the wireless systems evidence an elevated consumption of energy. Different studies affirm that Base Stations (BSs) are one of the network components with higher energy consumption [1,2]. Consequently, some works have been focused on designing methods for power consumption reduction in BSs. For saving energy, in [3] the authors propose a vacation model based on queuing theory to model the system as a $M/G/1$ queue with server vacation. Exhaustive and gated disciplines are also considered. In [4], the authors consider a wireless network consisting of one access point and a wireless node, they seek scheduling algorithms that conserve battery power at the wireless node. In [5], a scheme is presented to store information in mobile relay nodes to be relayed to another nodes or BSs at a later instance of time based on channel conditions. The authors present in [6], a mathematical model to determine the energy dissipation of a node as a function of its sleep period and its distance to the destination for convergecast data patterns.

In our work [7], an algorithm of bandwidth allocation and scheduling on convergent access networks was presented. In that work, the packet transmission

© Springer International Publishing Switzerland 2015
P. Gaj et al. (Eds.): CN 2015, CCIS 522, pp. 222–231, 2015.
DOI: 10.1007/978-3-319-19419-6_21

was synchronised. Some rules to turn on/off transmitters at Optical Network Unit (ONU) were implemented to reduce the energy consumption. However, the receivers at ONU were not considered to save energy. Our proposal is focused on switch active/saving algorithm for receivers in convergent access nodes.

The remainder of this paper is organized as follows. The system model is presented in Sect. 2. Section 3 introduces our proposed scheme. A performance analysis (including a numerical formulation and an asymptotic approximation) is described in Sect. 4. A set of results are reported in Sect. 5. Section 6 concludes the paper along with possible lines of future work.

2   System Model

In Fig. 1, the optical-wireless access network structure is presented. The Optical Line Terminal (OLT) is located at the central office and directly connected to the core network. Each access node, named ONU as in classical Passive Optical Network (PON), is connected to the OLT through one splitter. The ONUs are hybrid equipments, they provide wireless connectivity to one group of BSs and wired connectivity to fixed customers. One ONU serves a number n of BSs. In this example, each ONU is covering 6 cells. A tri-sector antenna is considered in the ONU. The traffic pattern considered (Fig. 2) was presented in [8]. It is based on real measurements observed on a normal weekday. When the traffic load is low, one sector of the tri-sector antenna is dedicated to the data traffic reception and the other two to the rest of traffic. Once the BSs receive data from final customers, they store the packet in one local buffer until the receiver (at ONU) is in active state. Packets at BS are inserted into a first-in first-out (FIFO) queue.

3   Energy Efficiency-Algorithm

Each BS is equipped with one buffer to store the data traffic going to the ONU until transmission is possible. Transmission of upstream data (from BSs to ONU)

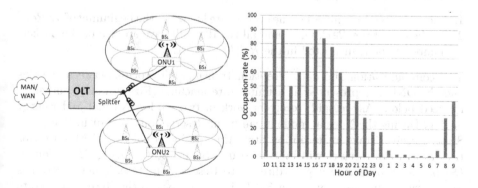

Fig. 1. Access network architecture   **Fig. 2.** Pattern traffic considered

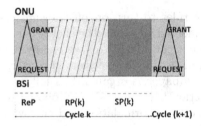

Fig. 3. Communication process between ONU and BSs

will be synchronized; this allows to wireless receivers turning into normal and energy saving states periodically. The communication between ONU and BSs (Fig. 3) uses two messages: REQUEST and GRANT. REQUEST message is used to inform to ONU the queue length of BSs. The GRANT message is used to inform to BSs the length of Reception Period (RP) and Saving Period (SP). This process is done between one ONU and all the BSs served (by the current ONU), at the same time.

Receivers work in two modes: Normal and Energy Saving. During the Normal mode it can be in Reservation state or in Reception state. Lets denote ReP the time spent in Reservation State, and let be $RP(k)$ and $SP(k)$ the time spent in Reception state and Saving state, respectively, during the cycle k. In ReP all the BSs send a REQUEST message to the ONU, the transmission window requested by each BS is based on its queue length. ONU allocates one sub-channel and transmission rate to each BS. After, it determines the next $RP(k)$ and $SP(k)$. Then, ONU sends GRANT messages to the BSs. GRANT message contains this information. ReP is constant during all the service cycles. All the BSs start data transmission after the ReP. When the $RP(k)$ ends, the wireless receiver turns into Saving state (according to $SP(k)$). During the $SP(k)$, there is no data packet reception by the wireless receiver, and the packets in the BSs are stored in one buffer to be sent in the forthcoming cycle. Some assumptions are considered for the proposed model:

1. Packets to be transmitted in the cycle k, are known at the beginning of its ReP.
2. At the end of the ReP, ONU and all the BSs will have an idea of the packets number to be transmitted in the $RP(k)$ to start.

This communication and data reception/transmission process between the BSs and the ONU is represented as a finite state machine (Figs. 4 and 5) and summarized below. As we said, receivers work in two modes: Normal and Energy Saving. During the Normal mode it can be in Reservation State or in Reception State. Receiver stays in Reception State time period equal to $RP(k)$; once this time is out, the receiver will be turned into Saving state during $SP(k)$ length ($SP(k) > 0$), otherwise it goes directly to Reservation state. When receiver is in Energy Saving state, once the $SP(k)$ is out, it goes to the Reservation state to restart the communication process and it defines the begins of a new service cycle. During Reservation state, receiver receives the REQUEST message of all

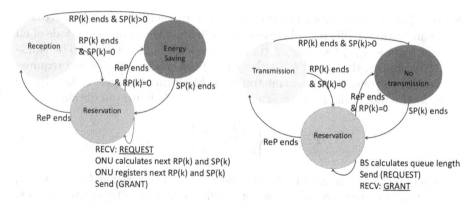

Fig. 4. State machine for receivers   **Fig. 5.** State machine for BSs

the BSs. After, ONU calculates the $RP(k)$ and $SP(k)$ for the current cycle. These values are registered by ONU and then a GRANT message is transmitted to all the BSs. Once the ReP is out, the receiver goes to the Reception state ($RP(k) > 0$) or goes to the Energy Saving state ($RP(k) = 0$ and $SP > 0$)).

BSs work also in two modes: Transmission and No Transmission. During the Transmission mode they can be in Transmission or in Reservation state. BS stays in Transmission state a time period equal to $RP(k)$; once this time is out, if the receiver goes to Saving mode ($SP(k) > 0$), then BS goes to No Transmission state; otherwise, it goes directly to Reservation state. When BS is in No Transmission state, once the $SP(k)$ is out, it goes to the Reservation state. During Reservation state, BS calculates the queue length of its buffer, it sends a REQUEST message to the ONU, and it receives a GRANT from ONU. Once the ReP is out, the BS goes to Transmission state ($RP(k) > 0$) or to No Transmission state ($RP(k) = 0$ and $SP(k) > 0$).

The algorithm uses link adaptation to change transmission parameters over a link. Specifically, data transmission rate is adapted every cycle, based on traffic arrival to BSs. ONU is in charge of obtain the link rate value during the ReP. Let be C the total channel capacity for uploading traffic from all BSs to ONU. Lets denote n as the maximum number of channels that can be served simultaneously, we must consider the following equation:

$$C = \sum_{i=1}^{n} r_i(k) \tag{1}$$

where $r_i(k)$ is the transmission rate of channel i allocated to BS_i for the next $RP(k)$. The algorithm to allocate the link rate to each channel follows the next steps:

Step 1: Each BS calculates its queue length at ReP begins.
Step 2: Every BS sends to the ONU the information obtained in Step 1 by using the REQUEST message.

Step 3: The synchronization of data reception between BSs and ONU, is reduced to minimize the difference between the individual transmission periods of all BSs. Let be $R_i(k)$ the amount traffic in the local buffer at BS_i, to be transmitted during the period k. Let be $TP_i(k)$ the transmission period required for BS_i to transmit its amount traffic $R_i(k)$, during the cycle k. Let denote $r_i(k)$ the rate of data transmission of BS_i in cycle k, we can obtain:

$$TP_i(k) = \begin{cases} \frac{R_i(k)}{r_i(k)} & \text{if } R_i(k) > 0 \\ 0 & \text{if } R_i(k) = 0 \end{cases}. \tag{2}$$

It is important to remark that $r_i(k)$ and $R_i(k)$ can be variable. Consequently, $TP_i(k)$ can be also variable from one cycle to another. The synchronization problem can be solved as:

$$\forall_i \text{ Minimize } |TP_i(k) - TP_j(k)| \tag{3}$$

with $i, j = 1, 2, 3, \ldots, n$ and $i \neq j$. In this step, ONU determines the link rate of each channel by using the next equation

$$r_i(k) = \begin{cases} \frac{R_i(k) \cdot C}{\sum_{l=1}^{n} R_l(k)} & \text{if } \sum_{l=1}^{n} R_l > 0 \\ 0 & \text{if } \sum_{l=1}^{n} R_l = 0 \end{cases}. \tag{4}$$

Determining Saving Period. For any BS, $TP_i(k)$ is obtained by using the Eq. (2). The cycle length is obtained regardless the constraints of delay considered to provide QoS. Let be Th the "time of life" of one packet buffered in one BS, assuming l packets in the current BS at the beginning of the ReP, the last packet in the queue (p_l) will experiment the maximum access delay. From the precedent subsection, we can summarize that $RP(k) = TP_i(k), \forall 1 \leq i \leq n$. Consequently, the maximum length of $SP(k)$ is conditioned to:

$$ReP + RP(k-1) + SP(k-1) + ReP + TT_{l-1} \leq Th \tag{5}$$

where TT_{l-1} is the transmission time of $(l-1)$ packets buffered in the current BS. We can obtain an approximated length of $SP(k)$ by using the next expressions:

$$SP(k) = Th - (2ReP + RP(k-1) + SP(k-1) + TT_{l-1}) \tag{6}$$

if $Th > 2ReP + RP(k-1) + SP(k-1) + TT_{l-1}$, else $SP(k) = 0$.

4 Performance Analysis

4.1 Numerical Formulation

In this section, we derive Numerical Formulation (NF) of the system. This formulation is done in terms of energy efficiency. Particularly, we focus on the percentage of energy efficiency improvement ω, compared between with and without applying our proposed algorithm. Let be $E_{\mathrm{ONU}}(k)$ the energy consumed by one ONU during cycle k, we can obtain $E_{\mathrm{ONU}}(k)$ by using:

$$E_{\mathrm{ONU}}(k) = E_{\mathrm{WR}}(k) + E_{\mathrm{ER}}(k) + E_{\mathrm{WT}}(k) + E_{\mathrm{ET}}(k) + E_{\mathrm{BF}}(k) \qquad (7)$$

where $E_{\mathrm{WR}}(k)$, $E_{\mathrm{ER}}(k)$, $E_{\mathrm{WT}}(k)$, $E_{\mathrm{ET}}(k)$, and $E_{\mathrm{BF}}(k)$ are the energy consumed by the set of wireless receivers, the set of Ethernet receivers, the set of wireless transmitters, the set of Ethernet transmitters and the energy consumed by the rest of components in the ONU, respectively, during the cycle k. $E_{\mathrm{WR}}(k)$ is given by the sum of the energy consumed by all the wireless receivers at ONU. Let be M the number of wireless receivers at ONU, we can express $E_{\mathrm{WR}}(k) = \sum_{w=1}^{M} E_w(k)$. In this sense, let be $E_{ReP}(k)$, $E_{RP}(k)$ and $E_{SP}(k)$ the energy consumed by one wireless receiver during the ReP, $RP(k)$ and $SP(k)$, respectively is given by:

$$E_w(k) = E_{ReP(k)} + E_{RP(k)} + E_{SP(k)}. \qquad (8)$$

$E_{RP(k)}$ is obtained by the energy consumed for the reception of REQUEST messages of all BSs, and the energy that the wireless receivers spend in sensing state (receiver is in active mode without receiving data transmission). Denote P_{rec}, P_{sen} and P_{s} as the power consumption of the receiver in receiving, sensing and energy saving mode, respectively. We can rewrite the Eq. (8) for one wireless receiver as:

$$E_w(k) = \sum_{i=1}^{n} (REQ * P_{\mathrm{rec}} + C_{\mathrm{T}} * P_{\mathrm{sen}}) + E[RP] * P_{\mathrm{rec}} + E[SP] * P_{\mathrm{s}}. \qquad (9)$$

REQ is assumed constant value, it is the time period from ReP begin until REQUEST message reception. C_{T} is the time period from REQUEST reception until RP begin. During C_{T} the ONU executes the algorithm to obtain $RP(k)$ and $SP(k)$. $E[RP]$ and $E[SP]$ are the expected values of $RP(k)$ and $SP(k)$, respectively, and the will be obtained regardless to QoS constraints. The energy consumption during the same length of time, assuming same number of packets in the BS buffers, if our algorithm is not applied is:

$$E_0(k) = \sum_{i=1}^{n} (REQ * P_{\mathrm{rec}} + C_{\mathrm{T}} * P_{\mathrm{sen}}) + E[RP] * P_{\mathrm{rec}} + E[SP] * P_{\mathrm{sen}}. \qquad (10)$$

Let be T_{pkt} the average transmission time of one packet, with $RP = TT_{(l-1)} + T_{\mathrm{pkt}}$. To derive $E[RP]$ and $E[SP]$, by using the Eq. (6) and taking expected value in both sides, we get the expression:

$$E[SP] = \frac{Th - (2 \cdot E[RP] + 2 \cdot ReP - T_{\mathrm{pkt}})}{2}. \qquad (11)$$

Now, denote $\widetilde{\lambda}$ as the packet arrival rate (in Mb) averaged (bits per unit of time) for the current ONU. Note that $\widetilde{\lambda}$ may be time varying and its long-term average is denoted as λ. Once the receiver is in normal state, the BSs start uploading all the buffered packets. During the reception of data packets, there may also packets arriving at the BSs. These packets, as we said before, will be buffered and then be served in the next service cycle. Therefore we have the following relationship:

$$C \cdot RP(k) = \tilde{\lambda} \cdot (ReP + SP(k-1) + RP(k-1)). \tag{12}$$

Since $\tilde{\lambda}$ is unknown to the ONU, its long-term average λ is used as an estimate, where we assume that λ is a known parameter. In this sense, the estimated uploading period RP satisfies:

$$C \cdot E[RP] = \lambda \cdot (ReP + E[RP] + E[SP]). \tag{13}$$

After taking expectation in both sides, and by using the Eq. (11), we obtain the expected value of RP as:

$$E[RP] = \frac{\lambda \cdot (Th + T_{\text{pkt}})}{2 \cdot C}. \tag{14}$$

With this result and by using the Eq. (11) we can obtain the energy consumed by the wireless receiver.

4.2 Asymptotic Approximation

We propose an Asymptotic Approximation (AA) for the RP and the SP. Without considering the ReP, and with $\tilde{\lambda}$ as the packet arrival rate (in Mb) averaged (number of packets per unit of time). Again, note that $\tilde{\lambda}$ may be time varying and its long-term average is denoted as λ. We propose a linear system of two equations with two variables. In solving the system of equations we try to find values for $E[RP]$ and $E[SP]$. Let be C a constant value denoting the transmission capacity of the channel (in number of packets). Lets denote Th the "time of life" of one packet buffered in one BS. The equation system to solve is:

$$C \cdot RP = \lambda \cdot (RP + SP)$$
$$2 \cdot RP + 2 \cdot SP = Th$$

Taking expectation in both sides of two equations and finding one solution for the equation system we get:

$$E[RP] = \frac{\lambda \cdot Th}{2 \cdot C}$$
$$E[SP] = \frac{(C - \lambda) \cdot Th}{2 \cdot C}$$

The energy efficiency can be improved by a factor:

$$\omega = \frac{E_0 - E_w}{E_0}. \tag{15}$$

5   Simulation Results

In this section, we present simulation and numerical results. We develop a Mat-Lab program to simulate the system where one ONU is serving 6 BSs. Only one packet length of 800 bits was considered. The ReP has a duration of 50 ms, the P_{rec} has a value of 787 mW, the P_{sen} of 503 mW and P_s of 44 mW. The size of one REQUEST packet used is 20 bits. The results were obtained with the 95 % confidence intervals.

We present in Fig. 6 the average $RP(k)$ in each cycle. RP is the $E[RP]$ obtained by NF and AP, and the $RP(K)$ by Simulation. The value of $Th = 1$ s and $\lambda = 5 \%$ of C (in number of packets). As we can appreciate, the NF and the AP results represent a lower bound for simulation results. The relative error between the simulation results and the NF is 0.018. The relative error between the simulation results and the AP is also 0.018. Figure 7 shows the average $SP(k)$ obtained in each cycle. SP is the $E[SP]$ obtained by NF and AP, and the $SP(K)$ obtained by Simulation. In the stationary state, NF is a lower bound for simulation results, and AA an upper bound for simulation results. In this case, the relative error between the simulation result and the NF is 0.0012, and between the AA and the simulation result is 0.11. The results obtained with the NF are most precise than the results obtained with the AA. This is due to the fact that the AA does not consider all the parameters of the system. Specifically, the ReP and the T_{pkt} were not considered in our proposition.

The energy consumed by the receiver applying our algorithm is presented in Fig. 8. We can see the results obtained by NF, AA and simulation. All the results are very similar. As expected, the energy saving reduce when the arrival rate increase because the $RP(k)$ is also increasing. The Th value used is 0.7 s. In Fig. 9 we can observe the factor of energy efficiency improved ω, during low load traffic period. The Th value is 0.7 s. As we expected, the improvement is very representative approximative to 87 % for very low traffic load, and when the traffic load is increasing ω is reducing.

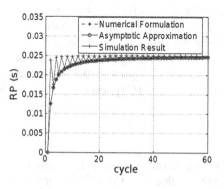

Fig. 6. RP obtained each cycle

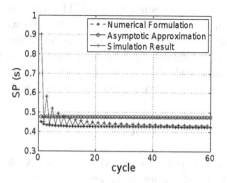

Fig. 7. SP obtained each cycle

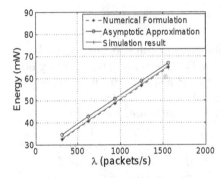

Fig. 8. Energy consumption ($Th = 0.7$ s)   **Fig. 9.** Improvement factor ($Th = 0.7$ s)

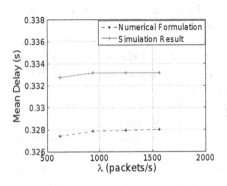

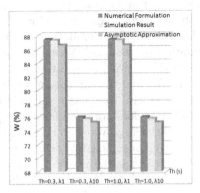

Fig. 10. Mean access delay

Fig. 11. Impact of Th on the receiver performance ($\lambda 1 = 0.01C$ and $\lambda 10 = 0.1C$)

Figure 10 shows the mean access delay of packets at BSs, under varying arrival rate. Only NF and simulation results are showed. When we increase the arrival rate, both results increase. The average of relative error is 0.015, which indicates the good performance of our NF. To study the impact of Th on the performance of receiver (Fig. 11), we ran some simulations for different values of Th. We could observe that in the period of low traffic load, the Th parameter does not impact significatively the performance of the receiver. We present the results for $Th = 0.3$ s, $Th = 1.0$ s, $\lambda = 0.01C$ (packets/s) and $\lambda = 0.1C$ (packets/s).

6   Conclusions

In this paper, we propose an energy-efficiency algorithm for traffic transmission in fixed-mobile integrated networks. Each ONU calculates the $RP(k)$ for data transmission from BSs in the end of the network to the current ONU, and it determines $SP(k)$ of its wireless receivers under packet delay constraint. Packet transmissions are synchronized between the ONUs and their associated BSs.

The adaptive link rate is used to allow this process. The performance of the proposed algorithm in terms of energy saving is evaluated by using MatLab. The results show that our algorithm can achieve significant energy level while keeping the access delay corresponding to QoS constraints. In future, we will consider the study by Markov chain and queuing theory.

References

1. Zhou, S., Gong, J., Yang, Z., Niu, Z., Yang, P.: Green mobile access network with dynamic base station energy saving. In: ACM Mobicom. Beijing (2009)
2. Oh, E., Krishnamachari, B.: Energy saving through dynamic base station switching in cellular wireless access networks. In: IEEE GLOBECOM, pp. 1–5 (2010)
3. Laun, K.M., Yue, O.: Modeling synchronous wakeup patterns in wireless sensor networks: server vacation model analysis. In: IEEE 16th International Symposium of Personal, Indoor and Mobile Radio Communications (2005)
4. Pasca, S., Srividya, V., Premkumar, K.: Energy efficient sleep/wake scheduling of stations in wireless networks. In: ICCSP, pp. 382–386 (2013)
5. Kolios, P., Friderikos, V., Papadaki, K.: Mechanical relaying in cellular networks with Soft-QoS guarantees. In: IEEE GLOBECOM, pp. 1–6 (2011)
6. Zhang, Y., Feng, C.-H., Demirkol, I., Heinzelman, W.B.: Energy-efficient duty cycle assignment for receiver-based convergecast in wireless sensor networks. In: IEEE GLOBECOM, pp. 1–5 (2010)
7. Gonzalez, G., Atmaca, T.: An integrated bandwidth allocation for energy saving in fixed-mobile networks. In: IEEE CAMAD (2013)
8. 3GPP TR 36.922 V10.1.0: 3rd generation partnership project: technical specification group radio access network; Evolved Universal Terrestrial Radio Access (E-UTRA); Potential solutions for energy saving for EUTRAN; Release 10, Technical report (2011)

Mobility Robustness in LTE Based on Automatic Neighbor Relation Table

Konrad Połys and Krzysztof Grochla[✉]

Institute of Theoretical and Applied Informatics,
Polish Academy of Science, Bałtycka 5, 44-100 Gliwice, Poland
{kpolys,kgrochla}@iitis.pl

Abstract. We evaluate the effect of automatic updates to the LTE Neighbor Relation Table (NRT) to optimize the efficiency of handovers in the network. We propose an algorithm which monitors the handovers within the LTE network and detects events when the procedure was started to early (handover to early) or too late (handover too late). Basing on the observed statistics in Operation and Maintenance server (OAM) the proposed algorithm updates the NRT to maximize the network throughput and minimize the probability of loss of connectivity. Performed simulations show that correctly forbidden handovers can reduce number of unnecessary signaling, reduce risk of fail during handover and reduce the time required to perform the handovers.

Keywords: LTE ANR · Neighbour relation table · SON · No HO · Handover to early

1 Introduction

In the last few years we observe a fast growth of mobile networks. The global traffic is more than doubled every year [1]. In response to this the operators are rapidly extending the wireless networks, to increase the number of base stations and allow for higher total network throughput by higher frequency reuse. This results in decrease of the average cell size and increased complexity of the modern wireless networks.

The wireless networks for mobile devices consist of multiple base stations and provide coverage over a large area. The structure of the network may be quite complex, with different size of the cells [2]. As the mobile clients move, the communication between them and the network is handled by different eNodeBs (base stations). If the data or voice transmission is in progress while the client moves from the range of one cell to another, the handoff procedures are initiated to provide the client with continuous transmission. Handoff is a process of transferring an ongoing call or data session from one base station to another in wireless networks.

In LTE networks the handover (HO) procedure is initiated by the network. It is crucial for the network performance and reliability to start the procedure in the

© Springer International Publishing Switzerland 2015
P. Gaj et al. (Eds.): CN 2015, CCIS 522, pp. 232–241, 2015.
DOI: 10.1007/978-3-319-19419-6_22

right moment. Each HO procedure causes a short distraction of the communication and requires network signaling to put some overhead to the transmission [3]. Starting the HO too often would generate unnecessary signaling, but starting it too late may result in increasing the probability of losing the connectivity. In some cases the HOs between two of the base stations are very often ineffective – for example when there is a small area with higher signal to second base stations within the range of the first one, caused e.g. by signal tunneling. In such situation the mobile terminal can switch between the two base stations multiple times, what is often called "ping-pong".

One of the ways to limit the effect of wrong handovers: and improve the network performance, defined within the Long Term Evolution (LTE) standard, is a possibility to block handovers between certain Base Stations (BSs). It results from existence of Neighbor Relation Table (NRT) [4]. Performed simulations show that correctly forbidden handovers can reduce number of unnecessary HOs, risk of fail in HO and also reduce the time in HOs. In this work we propose an algorithm to detect the situation when handovers are decreasing the network performance or increasing the probability of loosing a connectivity with the network and update the NRT to limit it.

The rest of the paper is organized as follows: in second section we describe the handover procedure in LTE and describe the function of Automatic Neighbor Relation (ANR) in handover management. Next we propose an algorithm to detect and block the incorrect handovers. In Sect. 4.1 we describe the simulation model used to analyse the performance of the proposed scheme. In Sect. 5 we discuss the results. We finish the paper with a short conclusion.

2   Handover Management in LTE Through Automatic Neighbour Relation

In LTE the handover procedure is started when the network software decides that it is more efficient to use another eNB for the connection. This may be triggered by multiple different reasons – in [5] the following criteria are enumerated: received signal strength, network connection time, available bandwidth, power consumption, monetary cost, security and user preferences. However in LTE typically the handoff procedure is initiated in response to change in measured signal level, reported by the UE.

2.1   Wrong Handovers

The LTE Mobility Robustness Optimization standard defines the following three types of wrongly initiated handovers. *Handover too late* occurs when radio link failure happens before the handover was initiated and the UE needs to re-establish the connectivity to the network. *Handover too early* happens when an UE needs to reconnect to another base station within very short time after finishing the previous handover, it is often called ping-pong handover when the UE reconnects back to the previous base station after a short service in another

BS. *Handover to wrong cell* occurs in the target cell after a handover has been completed, and the UE needs to re-establish its radio link to a BS which is not the source cell nor the target cell.

The signal level fluctuation, reported by the UE, may lead to initiation of several handovers between two eNBs in a short time period, when both of them are capable of serving the wireless client. As the number of handoffs increase, forced termination probability and network load also increases. Therefore, handover initiation algorithm should avoid unnecessary handoffs.

The correct handover initiation procedure should limit the number of wrong handovers of all three types described above. This may implemented through the correct triggering of the procedure using e.g. hysteresis [6]. When the signal level drops below some level which indicates the minimum acceptable signal level the hysteresis is ignored, to maintain the network connectivity. Some advanced algorithms for handover initiation have been proposed in the literature, based e.g. on fuzzy logic [7], but the simple threshold on the signal level is easy to measure and is directly related to the service quality. The LTE standard allows also to track the number of wrong handovers and propagate the information between the eNBs via X2 interface [8].

2.2 Automatic Neighbor Relation Functions

The manual management of mobile network is very time consuming and costly task, thus the network management protocols [9] and self-optimization procedures have been defined to automatically configure parameters of LTE base stations [10]. Within the LTE set of standards the Automatic Neighbor Relation (ANR) functionality has been defined [4]. The ANR algorithm gathers the list of neighbors per each base station and defines what neighbor relation should be established between them, as e.g. the X2 protocol link establishment.

The ANR function is based on the updates of a Neighbor Relation Table (NRT). The eNodesB software detects new neighbors and adds them to the NRT. There is also a function which removes outdated information from the ANR about the nodes which have not been heard for a long time. The LTE standard does not define the algorithm for the neighbor detection and removal, leaving it for the vendors. Typically it is being filled basing on the reports of signal level measurements provided by the UEs.

The NRT consist of three fields: No remove flag, No Handover flag and No X2 flag, as it is shown on Fig. 1. In this paper we concentrate on the No Handover flag (No HO), which allows to block handovers between two neighboring base stations. When it is set to true the handovers are not initiated.

3 Updates of NRT

The NRT may be configured manually, by the network operator or automatically either by a SON software implemented on the eNodeB or through the Operation and Maintenance (OAM) server in central location, as can be seen on the Fig. 1.

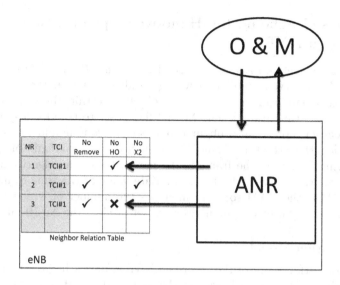

Fig. 1. Neighbor relation table

The content of the ANR table may be based on configuration defined by the network operator, or automatically calculated using the self-optimizing networks (SON) functions. The algorithm how to update the NRT has not been defined by the LTE standard and is left for vendors or network operator.

A few proposals of ANR procedures can be found in the literature. Most of the papers concentrate on the filling of the table and automatic neighbor detection. Mueller, Bakker and Ewe [11] show optimistic simulation results for LTE ANR function and propose additional mechanism of PCI report blacklisting for more robust operation. The interesting paper by Dahlen et al. [12] present the real-life pre-launch commercial LTE cluster ANR operation test results. Their observations regarding the UE reporting of non-obvious neighbor relations while not reporting the obvious ones are worth noticing. The observed phenomenon suggests that the distributed approach for neighbor cell list may be not sufficient for long-term handover optimization and some aid from a centralized OAM server could be beneficial. In [13] Cai et al. confirm this approach and propose a layered approach to ANR collection.

The problem of handover optimization through ANR is not analyzed deep enough in the literature, what limits the possible performance gains. The authors of [14] propose how to exploit knowledge of handover probability among cells derived from a handover history to reduce the amount of scanned cells, what limits the cost of scanning, but does not influence the probability of executing handover to early or to late. Similar approach is proposed in [15], with a dynamic neighbor cell list management scheme to enhance convergence and alleviate missing neighbor problems. The paper [16] proposes to use the simulated annealing approach to optimize the HO parameters for LTE, such as handover hysteresis.

4 Proposed Scheme for Handover Optimization Through ANR

In this work we evaluate a simple method for setting the No HO flag in NRT basing on the observed number of wrong handovers within the specific time frame (another handover or signal loss is observed within the time defined as the length of the time frame). The No HO flag is set to true for a specific pair of cells whenever more than configured number of wrong handovers is detected within the length of the time frame. The time frame is a configurable period of time, the length of the time frame is the parameter of the algorithm. Within our study we evaluate what is the appropriate value of the time frame length. We assume that the OAM software monitors the number of wrong handovers between the base stations in the network to set or clear the No HO flag.

4.1 Simulation Model

To verify the performance of the described NRT update schema we have implemented a simulation model based on the map and signal propagation model representing sample city, for which we were able to acquire the data about the location of the LTE base stations. The simulation model has been implemented using OMNeT++ with INET framework. It consists of buildings, base stations (eNodeB) and number of nodes (UE). Nodes moves with RandomWaypoint mobility model [17]. Scenario area is 3 x 5 km and buildings layout is taken from the center of the Hannover, Germany (Fig. 2) by parsing files from OpenStreetMap project [18]. Base stations locations are taken from [19]. In the simulation there were 20 nodes moving in 8 h time. Increase in number of nodes or simulation length is problematic because of complexity and very long running time. We modeled the signal attenuation due to the buildings using the FreeSpace [20] propagation model. In equal time interval the Signal To Interferences and Noise Ratio (SINR) for every node was calculated and the next the possible bitrate, based on SINR and Modulation and Coding Scheme (MCS) table. Described in more details in Bitrate Calculation subsection.

In this research the move of the nodes is generated by simple Random Waypoint mobility model because the goal was to simulate the events of handovers and a result the characteristic of the move can be omitted. The reproduction of the buildings in Hannover helps in modeling of the sudden signal changes. Velocity of users was set to 50 km/h, the change of this parameter can influence on the number of detected wrong handovers in 8 h of simulation but should not influence on overall handovers ratio. The simulation was simplified by not using traffic model because it is assumed that UE is a smartphone with data transmission enabled, so data session is sustained all the time [21].

4.2 Bitrate Calculation

In LTE a single user can be assigned to only one modulation and coding scheme (MCS) in each transmission time interval (TTI) or scheduling period. The possible modulation and coding schemes are presented in the Table 1. The LTE

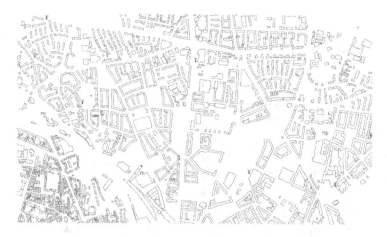

Fig. 2. Fragment of Hannover's buildings from simulation. Red lines are the walls of buildings (Color figure online).

standard does not define a method of MCS selection – a comparative study of different methods is given in [22], however for the sake of this work we use a simple method used in [23] on the basis on mapping tables [24]. The table is indexed by *Channel Quality Indicator* (CQI). The efficiency η_i of the i-th transmission mode (CQI_i) is derived using the equation:

$$\eta_i = r_i \cdot \log_2(M_i). \tag{1}$$

The throughput offered to a particular client can be calculated using the equation above as a function of the code rate r_i and the amount of radio Physical Resource Blocks (PRBs) allocated to this client. We assume that the Round Robin scheduler is used and each client is allocated with the same amount of PRBs, proportional to the channel bandwidth (see Table 2). The code rate is determined by the MCS used for particular client, which is selected basing on the CQI reported by the client. The CQI in the simulation is calculated basing on the SINR (determined using signal propagation model) and threshold given in the Table 1.

5 Results

The simulation study has been divided into two parts: the learning phase and the evaluation phase. During the first part nodes were moving for gather all handovers and detect the pairs of base stations for which large number of wrong handovers is detected. During this phase we counted the number of events when a mobile client was handed over from a source base station to the destination and back within a configured time. Six configurations with different thresholds were considered (Table 3). Every configuration differs by two parameters – time frame and number of repetitions. The pair of nodes was considered as causing

Table 1. MCS table

CQI index	MCS	Modulation	Code rate x 1024, r_i	Modulation size, M_i	SINR Threshold
1	0	QPSK	78	4	−5.45
2	2	QPSK	120	4	−3.63
3	4	QPSK	193	4	−1.81
4	6	QPSK	308	4	0
5	8	QPSK	449	4	1.81
6	10	QPSK	602	4	3.63
7	12	16QAM	378	16	5.45
8	14	16QAM	490	16	7.27
9	16	16QAM	616	16	9.09
10	18	64QAM	466	64	10.90
11	20	64QAM	567	64	12.72
12	22	64QAM	666	64	14.54
13	24	64QAM	772	64	16.36
14	26	64QAM	873	64	18.18
15	28	64QAM	948	64	20

Table 2. PRBs allocated per TTI

Bandwidth (BW) [MHz]	1.5	2.5	5	10	15	20
Numer of PRBs per TTI	12	24	50	100	150	200

Table 3. Parameters set

Time frame	5	5	5	10	10	10	20	20	20
Repetition number	3	5	10	3	5	10	3	5	10

wrong handovers when a ping-pong handover occurred within the specified time frame and the event was repeated more than defined value. For example if time frame is equal 5 s and repetition number is equal 3 it means that if between pair of BS A and BS B occur 3 ping-pongs or other wrong handovers in last 5 s, HO from A to B will be forbidden. In the second part it was assumed that the table of forbidden handovers is filled basing on the observations from the first part and simulations were repeated 10 times with different random seeds to get average result. From each simulation values such as number of HO, time spent in HO, signal losts, time with signal lost and average bitrate were compared to the case where no bans were in operation.

We started the analysis by a selecting values for time frame and threshold on number of handovers that allow to detect the wrong handovers. The Fig. 3

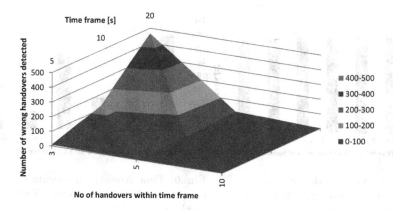

Fig. 3. No of handovers detected as wrong for different time frame and threshold values

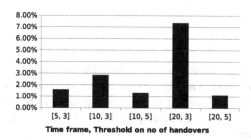

Fig. 4. Percentage of handovers detected as wrong for different thresholds and time frame sizes

shows the results of the first part of simulations, presenting how many events considered as wrong handovers were detected for different periods and for different thresholds. More than 3 quick handovers between a pair of base station happen very rarely even if we extend the time frame length is set to 20 s. If we take very long time frame (as e.g. 20 s) the number of events considered as wrong handovers is very high. Basing on the results of this test 5 configurations were selected for further analysis, for which the handovers were blocked between pairs of base stations where wrong handovers were observed. The percentage of handovers detected as wrong for those 5 configuration is shown on Fig. 4.

The influence of the blocking of the handovers was evaluated in terms of the time spend during handovers and the influence of blocking on the average network throughput. The blocking of some handovers always decreases the throughput, because when the handover is blocked the client is not connected to the base station which offer the strongest signal, but stays to more distant base station. Our evaluation shown on Fig. 5 proves that the decrease in total average throughput was very low, reaching 2 % for the configuration [20,3] which blocked largest number of HOs. The proposed method allowed to decrease the time spent in handovers by more than 50 % – see Fig. 6.

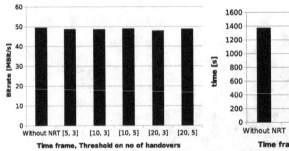
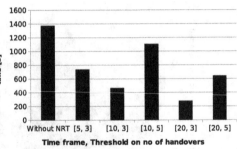

Fig. 5. Average throughput for different parameters of handover blocking algorithms

Fig. 6. Time required to execute the handovers for different parameters of handover blocking algorithms

6   Conclusion

The optimization of mobility robustness in LTE by the updates to ANR table is relativelly easy method. The simulaiton model presented shows that it allows significantly decrease the amount of unecessery handovers and limits the time spend during the handover procedure, while having almost no influence on the network performance.

Acknowledgement. The work is partially supported by NCBIR Project LIDER/10/ 194/L-3/11/.

References

1. Cisco: Cisco visual networking index: global mobile data traffic forecast update 2013–2018. Cisco Public Information (2014)
2. Domański, A., Domańska, J., Czachórski, T.: The impact of self-similarity on traffic shaping in wireless LAN. In: Balandin, S., Moltchanov, D., Koucheryavy, Y. (eds.) NEW2AN 2008. LNCS, vol. 5174, pp. 156–168. Springer, Heidelberg (2008)
3. Rostański, M., Mushynskyy, T.: Security issues of IPv6 network autoconfiguration. In: Saeed, K., Chaki, R., Cortesi, A., Wierzchoń, S. (eds.) CISIM 2013. LNCS, vol. 8104, pp. 218–228. Springer, Heidelberg (2013)
4. 3rd Generation partnership project: technical specification 25.484, v12.0.0 (2014)
5. Yan, X., Şekercioğlu, Y.A., Narayanan, S.: A survey of vertical handover decision algorithms in fourth generation heterogeneous wireless networks. Comput. Netw. **54**(11), 1848–1863 (2010)
6. Grochla, K., Połys, K.: Influence of the handoff threshold hysteresis on heterogeneous wireless network performance. In: Proceedings of 7th International Working Conference HET-NETs (2013)
7. Dhand, P., Dhillon, P.: Handoff optimization for wireless and mobile networks using fuzzy logic. Int. J. Comput. Appl. **63**(14), 31–35 (2013)

8. 3GPP: Telecommunication management; Study on enhancement of OAM aspects of distributed self-organizing network (SON) functions. TS 32.860, 3rd Generation Partnership Project (3GPP). December 2014
9. Słabicki, M., Grochla, K.: Performance evaluation of SNMP, NETCONF and CWMP management protocols in wireless network (2014)
10. Chrost, L., Grochla, K.: Conservative graph coloring: a robust method for automatic PCI assignment in LTE. In: Kwiecień, A., Gaj, P., Stera, P. (eds.) CN 2013. CCIS, vol. 370, pp. 268–276. Springer, Heidelberg (2013)
11. Mueller, C.M., Bakker, H., Ewe, L.: Evaluation of the automatic neighbor relation function in a dense urban scenario. In: Vehicular Technology Conference (VTC Spring), 2011 IEEE 73rd, pp. 1–5. IEEE (2011)
12. Dahlén, A., Johansson, A., Gunnarsson, F., Moe, J., Rimhagen, T., Kallin, H.: Evaluations of LTE automatic neighbor relations. In: Vehicular Technology Conference (VTC Spring), 2011 IEEE 73rd, pp. 1–5. IEEE (2011)
13. Cai, T., van de Beek, J., Sayrac, B., Grimoud, S., Nasreddine, J., Riihijarvi, J., Mahonen, P.: Design of layered radio environment maps for RAN optimization in heterogeneous LTE systems. In: IEEE 22nd International Symposium on Personal Indoor and Mobile Radio Communications (PIMRC), pp. 172–176. IEEE (2011)
14. Becvar, Z., Mach, P., Vondra, M.: Self-optimizing neighbor cell list with dynamic threshold for handover purposes in networks with small cells. In: Wireless Communications and Mobile Computing (2013)
15. Watanabe, Y., Matsunaga, Y., Kobayashi, K., Sugahara, H., Hamabe, K.: Dynamic neighbor cell list management for handover optimization in LTE. In: Vehicular Technology Conference (VTC Spring), 2011 IEEE 73rd, pp. 1–5. IEEE (2011)
16. Capdevielle, V., Feki, A., Fakhreddine, A.: Self-optimization of handover parameters in LTE networks. In: 11th International Symposium on Modeling and Optimization in Mobile, Ad Hoc and Wireless Networks (WiOpt), pp. 133–139. IEEE (2013)
17. Gorawski, M., Grochla, K.: Review of mobility models for performance evaluation of wireless networks. In: Gruca, A., Czachórski, T., Kozielski, S. (eds.) Man-Machine Interactions 3. AISC, vol. 242, pp. 573–584. Springer, Heidelberg (2014)
18. Openstreetmap project. http://www.openstreetmap.org
19. Rose, D.M., Jansen, T., Werthmann, T., Türke, U., Kürner, T.: The IC 1004 urban hannover scenario-3D pathloss predictions and realistic traffic and mobility patterns. In: European Cooperation in the Field of Scientific and Technical Research, COST IC1004 TD(13)8054 (2013)
20. Rappaport, T.S., et al.: Wireless Communications: Principles and Practice, vol. 2. Prentice Hall PTR, New Jersey (1996)
21. Foremski, P., Gorawski, M., Grochla, K.: Source model of TCP traffic in LTE networks. In: Czachórski, T., Gelenbe, E., Lent, R. (eds.) Information Sciences and Systems 2014, pp. 125–135. Springer, Switzerland (2014)
22. Fan, J., Yin, Q., Li, G.Y., Peng, B., Zhu, X.: MCS selection for throughput improvement in downlink LTE systems. In: Proceedings of 20th International Conference on Computer Communications and Networks (ICCCN), pp. 1–5. IEEE (2011)
23. Piro, G., Baldo, N., Miozzo, M.: An LTE module for the NS-3 network simulator. In: Proceedings of the 4th International ICST Conference on Simulation Tools and Techniques, pp. 415–422. ICST (Institute for Computer Sciences, Social-Informatics and Telecommunications Engineering) (2011)
24. Kawser, M.T., Hamid, N.I.B., Hasan, M.N., Alam, M.S., Rahman, M.M.: Downlink SNR to CQI mapping for different multiple antenna techniques in LTE. Int. J. Inf. Electron. Eng. 2(5), 757–760 (2012)

Influence of Corpus Size on Speaker Verification

Adam Dustor[✉], Piotr Kłosowski, Jacek Izydorczyk, and Rafał Kopański

Institute of Electronics, Silesian University of Technology,
Akademicka 16, 44-100 Gliwice, Poland
{adam.dustor,piotr.klosowski,jacek.izydorczyk}@polsl.pl
rafakopanski@gmail.com

Abstract. The scope of this paper is to check influence of the size of the speech corpus on the speaker recognition performance. Obtained results for TIMIT corpus are compared with results obtained for smaller database ROBOT. Additionally influence of feature dimensionality and size of the speaker model was tested. Achieved results show that the best results can be obtained for MFCC features. The lowest EER for larger TIMIT database are 4 times worse than the best result for ROBOT corpus which confirms that biometric systems should be tested on as large data sets as possible to assure that achieved error rates are statistically significant.

Keywords: Biometrics · Security · Speaker verification · Feature extraction

1 Introduction

Division of Telecommunication, a part of the Institute of Electronics and Faculty of Automatic Control, Electronics and Computer Science of the Silesian University of Technology, for many years specializes in speech and speaker recognition [1–6]. One of the results of conducted research is presented in this paper which is devoted to speaker identification and verification.

Speaker recognition is the process of automatically recognizing who is speaking by analysis speaker-specific information included in spoken utterances. This process may be divided into identification and verification. The purpose of speaker identification is to determine the identity of an individual from a sample of his or her voice. The purpose of speaker verification is to decide whether a speaker is whom he claims to be. Most of the applications in which voice is used to confirm the identity claim of a speaker are classified as speaker verification.

The paper is organized in the following way. At first fundamentals of speaker recognition like feature extraction and speaker modeling are presented. Next applied in this research TIMIT corpus is briefly described. At last achieved speaker recognition results are shown.

© Springer International Publishing Switzerland 2015
P. Gaj et al. (Eds.): CN 2015, CCIS 522, pp. 242–249, 2015.
DOI: 10.1007/978-3-319-19419-6_23

2 Speaker Verification

Typical structure of speaker verification system consists of feature extraction, similarity calculation, construction of speaker model and making an accept/reject decision. Basic structure of speaker verification system is shown in Fig. 1. Speech signal is cut into short fragments, which usually last for 20–30 ms known as speech frames. Feature extraction is responsible for extracting from each frame a set of parameters known as feature vectors. Extracted sequence of vectors is then compared to speaker model by pattern matching. The purpose of pattern matching is to measure similarity between test utterance and speaker model. In verification the similarity between input test sequence and claimed model must be good enough to accept the speaker as whom he claims to be.

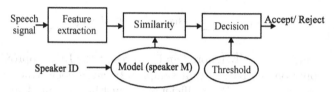

Fig. 1. Speaker verification scheme

3 Feature Parameters

Feature extraction is one of the most important procedures in speaker recognition. Applied parameters should be stable in time – physical and mental state of the speaker should have low impact on recognition performance which means low intraspeaker variability. Features should also posses high interspeaker variability in order to discriminate well between speakers. The most often applied features are parameters based on frequency spectrum of the speech like linear prediction coefficients LPC and parameters derived from cepstrum as LPCC and MFCC.

3.1 LPC Parameters

Calculation of these parameters is based on the linear model for speech production shown in Fig. 2. The excitation of the filter is either a quasi-periodic impulse sequence for voiced sounds or a random sequence for unvoiced ones. Intensity of the excitation is controlled by the gain factor G. The transfer function $V(z)$ and radiation $R(z)$ describe the vocal tract and the air pressure at the lips respectively. As a result combining these parts an all-pole transfer function is obtained

$$H(z) = G(z)V(z)R(z) = \frac{G}{A(z)} = \frac{G}{1 - \sum\limits_{k=1}^{p} a_k z^{-k}}, \tag{1}$$

where a_k are predictor coefficients and p is the prediction order. The LPC model of the speech signal specifies that a speech sample $s(n)$ can be represented as a linear sum of the p previous samples plus an excitation term

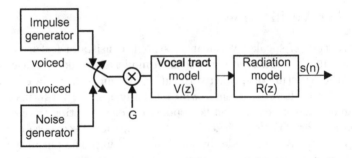

Fig. 2. The linear model of speech production

$$s(n) = \sum_{k=1}^{p} a_k s(n-k) + Gu(n). \tag{2}$$

The excitation $u(n)$ is usually ignored and as a result the LPC approximation of $s(n)$ depends only on the past output samples. Because vocal tract changes its configuration over time, the LPC coefficients a_k must be computed adaptively for short intervals (10 ms to 30 ms) known as frames during which time-invariance is assumed. The a_k are usually computed using autocorrelation method, which is based on minimizing the mean-square value E of the prediction error

$$E = \sum_{n=0}^{N-1+p} \left[s(n) - \sum_{k=1}^{p} a_k s(n-k) \right]^2. \tag{3}$$

The a_k parameters may be found after solving the linear equations resulting from

$$\frac{\partial E}{\partial a_i} = 0, \qquad i = 1, 2, \ldots p. \tag{4}$$

Defining the autocorrelation function as

$$r(\tau) = \sum_{i=0}^{N-1-\tau} s(i)s(i+\tau), \tag{5}$$

where N is the number of samples in a frame and assuming that speech samples outside the frame of interest are zero, the autocorrelation method yields the Yule-Walker equations given by [7]

$$\begin{bmatrix} r(0) & r(1) & r(2) & \cdots & r(p-1) \\ r(1) & r(0) & r(1) & \cdots & r(p-2) \\ r(2) & r(1) & r(0) & \cdots & r(p-3) \\ \vdots & \vdots & \vdots & \ddots & \vdots \\ r(p-1) & r(p-2) & r(p-3) & \cdots & r(0) \end{bmatrix} \begin{bmatrix} a_1 \\ a_2 \\ a_3 \\ \vdots \\ a_p \end{bmatrix} = \begin{bmatrix} r(1) \\ r(2) \\ r(3) \\ \vdots \\ r(p) \end{bmatrix}. \tag{6}$$

In order to solve these equations, the Levinson-Durbin recursion algorithm is applied.

3.2 LPCC and MFCC Parameters

These parameters are based on cepstrum which is defined as the inverse Fourier transform of the log of the signal spectrum. As it can separate excitation from the vocal tract it posses high discrimination capabilities and is widely used in speaker recognition applications. Cepstral parameters can be derived from LPC coefficients (known as the LPCC coefficients) or from the filter-bank spectrum (known as the mel frequency cepstral coefficients MFCC).

MFCC parameters are based on the nonlinear human perception of the frequency of sounds. They can be computed as follows: window the signal, take the FFT, take the magnitude, take the log, warp the frequency according to the mel scale and finally take the inverse FFT. Mel warping transforms the frequency scale to place less emphasis on high frequencies [8].

LPCC parameters can be calculated from the transfer function $H(z)$ in (1). However, it requires calculating poles of the $H(z)$ and as a result more computationally efficient recursion formula [7] is used

$$
c(n) = \begin{cases} a_n + \sum_{k=0}^{n-1} \frac{k}{n} a_{n-k} c(k), & 1 \le n \le p, \\ \sum_{k=n-p}^{n-1} \frac{k}{n} a_{n-k} c(k), & n > p. \end{cases} \tag{7}
$$

4 Pattern Recognition

Speaker recognition is based on similarity calculation between test utterance and the reference model. As a result, the problem of construction of a good model is crucial.

One of methods used to create voice model is based on vector quantization VQ. Speaker is represented as a set of vectors that possibly in the best way represent speaker. This set of vectors is called a codebook. In this case during recognition each test vector is compared with its nearest neighbour from the codebook and the overall distance for the whole test utterance is computed. Calculation of normalized distance D for M frames of speech is given by

$$
D = \frac{1}{M} \sum_{i=1}^{M} \min(d(x_i, c_q)) \qquad 1 \le q \le L, \tag{8}
$$

where x_i is a test vector and c_q a code vector from a codebook of size L. As it can be seen for M frames and L code vectors its necessary to calculate ML distances. The most often used measure of similarity is an Euclidean distance

$$
d(x_i, c_q) = \sum_{k=1}^{p} (x_i(k) - c_q(k))^2 \tag{9}
$$

where p is a dimension of a vector. How to find the best codebook for speaker from a lot of training data? To solve this problem a kind of clustering technique is required, which can find a small set of the best representative vectors of a speaker. One of applied algorithms is k-means procedure.

K-means algorithm is an iterative procedure and consists of four major steps. At first arbitrarily choose L vectors from the training data, next for each training vector find its nearest neighbour from the current codebook, which corresponds to partitioning vector space into L distinct regions. The third step requires updating the code vectors using the centroid of the training vectors assigned to them and the last step – repeat steps 2 and 3 until some converge criterion is satisfied. The converge criterion is usually an average quantization error expressed in the same way as in (8) with an exception that x_i is a training vector.

5 TIMIT Corpus

In order to test applied algorithms TIMIT [9] corpus was used. The DARPA TIMIT Acoustic-Phonetic Continuous Speech Corpus has been designed to provide speech data for the acquisition of acoustic-phonetic knowledge and for the development and evaluation of automatic speech recognition systems. TIMIT has resulted from the joint efforts of several sites under sponsorship from the Defense Advanced Research Projects Agency – Information Science and Technology Office (DARPA-ISTO). Text corpus design was a joint effort among the Massachusetts Institute of Technology (MIT), Stanford Research Institute (SRI), and Texas Instruments (TI). The speech was recorded at TI, transcribed at MIT, and has been maintained, verified, and prepared for CD-ROM production by the National Institute of Standards and Technology (NIST). TIMIT contains a total of 6300 sentences, 10 sentences spoken by each of 630 speakers from 8 major dialect regions of the United States.

6 Research Procedure

During training and testing of the speaker verification system the same signal processing procedure was used. Speech files, before feature extraction, were processed to remove silence. Voice activity detection was based on the energy of the signal. Next signal was preemphasized and segmented. Hamming windowing was applied. For each frame LPC analysis was applied to obtain LPC coefficients. LPC parameters were then transformed into LPCC coefficients using Eq. (7). From each frame MFCC parameters were also computed. In case of TIMIT corpus each model was trained using 5 sentences. There were 500 speaker models tested. Each speaker provided 5 test sequences. As a result there were 1 250 000 verification trials – 2500 valid trials (500*5) and 1 247 500 impostor trials (500*5*499) for each combination of dimensionality of the feature vector and the size of the speaker model.

Verification performance was characterized in terms of the two error measures, namely the false acceptance rate FAR and false rejection rate FRR. These measures correspond to the probability of acceptance an impostor as a valid user

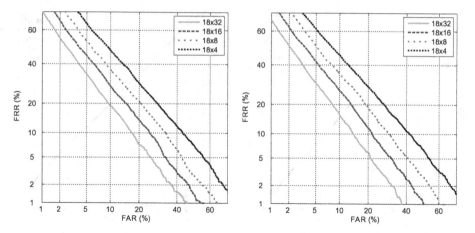

Fig. 3. Influence of model complexity on speaker verification (LPC parameters, 18 features per frame)

Fig. 4. Influence of model complexity on speaker verification (LPCC parameters, 18 features per frame)

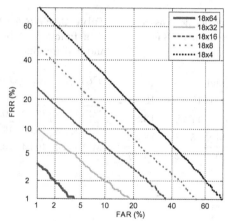

Fig. 5. Influence of model complexity on speaker verification (MFCC parameters, 18 features per frame)

and the probability of rejection of a valid user. Changing the decision level, DET curves which show dependence between FRR and FAR can be plotted. Another very useful performance measure is an equal error rate EER which corresponds to error rate achieved for the decision threshold for which $FRR = FAR$. In other words EER is just given by the intersection point of the main diagonal of DET plot with DET curves.

Speaker verification performance for LPC, LPCC and MFCC with 18 parameters per frame are shown in Figs. 3, 4 and 5 respectively. As the lowest errors were obtained for MFCC features, verification performance was also tested for 12 and 24 MFCC parameters per frame which was shown in Figs. 6 and 7. Achieved EER values are shown in Tables 1 and 2.

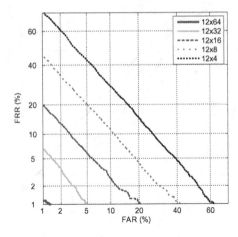

Fig. 6. Influence of model complexity on speaker verification (MFCC parameters, 12 features per frame)

Fig. 7. Influence of model complexity on speaker verification (MFCC parameters, 24 features per frame)

Table 1. *EER* for verification as a function of number of centroids (18 features per segment)

Table 2. *EER* for MFCC parameters as a function of number of centroids and feature dimensionality

EER [%]	Speaker model			
	18 x 4	18 x 8	18 x 16	18 x 32
LPC	24.4	20.4	17.08	13.89
LPCC	23.84	19.44	15.97	12.52
MFCC	18.68	12.89	7.72	4.36

EER [%]	Number of centroids				
	4	8	16	32	64
MFCC 12	17.72	10.49	5.4	2.56	1.12
MFCC 18	18.68	12.89	7.72	4.36	2
MFCC 24	19.04	13.36	8.16	4.68	2.24

7 Conclusion

Comparing obtained results, definitely the best features are MFCC. For the best combination of dimensionality and complexity of the model (12 parameters per segment and 64 codevectors per speaker model) achieved $EER = 1.12\%$ is definitely better than $EER = 12.52\%$ for LPCC and $EER = 13.89\%$ for LPC features. From Tables 1 and 2 it can be seen that number of codevectors per speaker model has great impact on verification errors and is more important than dimensionality of feature vector.

It is very interesting to check influence of the applied corpus and the number of speakers on verification errors. In paper [5] similar research was done for the Polish corpus ROBOT [10] which consists of only 30 speakers. The best achieved value of $EER = 0.27\%$ is 4 times smaller than the lowest $EER = 1.12\%$ for TIMIT with 500 speakers. These results directly show that in order to reliably test speaker recognition procedures one of the most important problems is the correct choice of corpus. It must contain as many speakers as possible.

Unfortunately good speech corpus is quite expensive and must be purchased from Linguistic Data Consortium LDC [11].

Although achieved value of $EER = 1.12\%$ is quite low, it must be further decreased in order to implement speaker verification procedures as an additional security level in a mobile phone which is the final goal of our research.

Acknowledgment. This work was supported by The National Centre for Research and Development (http://www.ncbir.gov.pl) under Grant number POIG.01.03.01-24-107/12 *(Innovative speaker recognition methodology for communications network safety)*.

References

1. Kłosowski, P.: Speech processing application based on phonetics and phonology of the polish language. In: Kwiecień, A., Gaj, P., Stera, P. (eds.) CN 2010. CCIS, vol. 79, pp. 236–244. Springer, Heidelberg (2010)
2. Dustor, A.: Voice verification based on nonlinear Ho-Kashyap classifier. In: International Conference on Computational Technologies in Electrical and Electronics Engineering SIBIRCON 2008, pp. 296–300. Novosibirsk (2008)
3. Dustor, A.: Speaker verification based on fuzzy classifier. In: Cyran, K.A., Kozielski, S., Peters, J.F., Stańczyk, U., Wakulicz-Deja, A. (eds.) Man-Machine Interactions. AISC, vol. 59, pp. 389–397. Springer, Heidelberg (2009)
4. Dustor, A., Kłosowski, P.: Biometric voice identification based on fuzzy kernel classifier. In: Kwiecień, A., Gaj, P., Stera, P. (eds.) CN 2013. CCIS, vol. 370, pp. 456–465. Springer, Heidelberg (2013)
5. Dustor, A., Kłosowski, P., Izydorczyk, J.: Influence of feature dimensionality and model complexity on speaker verification performance. In: Kwiecień, A., Gaj, P., Stera, P. (eds.) CN 2014. CCIS, vol. 431, pp. 177–186. Springer, Heidelberg (2014)
6. Dustor, A., Kłosowski, P., Izydorczyk, J.: Speaker recognition system with good generalization properties. In: International Conference on Multimedia Computing and Systems, ICMCS 2014, Marrakech, Morocco, April 2014, pp. 206–210. IEEE (2014)
7. Rabiner, L.R., Juang, B.H.: Fundamentals of Speech Recognition. Prentice Hall, NJ (1993)
8. Fazel, A., Chakrabartty, S.: An overview of statistical pattern recognition techniques for speaker verification. IEEE Circuits Syst. Mag. **11**(2), 62–81 (2011)
9. TIMIT corpus. https://catalog.ldc.upenn.edu/LDC93S1
10. Adamczyk, B., Adamczyk, K., Trawiński, K.: Zasób mowy ROBOT. Biuletyn Instytutu Automatyki i Robotyki WAT **12**, 179–192 (2000)
11. Linguistic data consortium. https://www.ldc.upenn.edu/

Stopping Criteria Analysis of the OMP Algorithm for Sparse Channels Estimation

Grzegorz Dziwoki$^{(\boxtimes)}$ and Jacek Izydorczyk

Institute of Electronics, Silesian University of Technology, Gliwice, Poland
{grzegorz.dziwoki,jacek.izydorczyk}@polsl.pl

Abstract. Wireless propagation environment utilised by broadband transmission systems usually has a sparse nature, i.e. only several isolated propagation paths are essential for information transfer. Receiver can recover the parameters of the particular paths using greedy, iterative algorithms that belong to the family of compressed sensing techniques. How to stop the iterative procedure, if no precise knowledge about the order of the channel sparsity is available in the receiver a priori, is a key question regarding a practical implementation of the method. The paper provides stopping criteria analysis of the Orthogonal Matching Pursuit (OMP) algorithm that is used as the core of the channel impulse response estimation method for Time-Domain Synchronous OFDM transmission system. There are investigated the residual error and the difference of successive residual errors of the OMP algorithm, as the possible metrics applied to stop the iteration procedure. Finally, a new heuristic stopping rule based on these two errors is proposed and numerically examined.

Keywords: Time domain estimation · Compressive sensing · Greedy methods · Orthogonal matching pursuit

1 Introduction

Channel impulse response of the wireless broadband transmission system can be often described by a sparse signal. This feature has been proved by many measurement experiments and is reflected in the developed channel models [1,2]. The several nonzero coefficients of the sparse impulse response can be effectively estimated using methods belonging to the relatively new Compressed Sensing (CS) theory [3]. There is required only a small set of the signal measurements $\mathbf{z}$ in the receiver's input to do so. The relation between the measurements and an unknown sparse signal (channel) $\mathbf{h}$ is described by the linear transformation with the measurement matrix Φ as follows:

$$\mathbf{z} = \Phi\mathbf{h}. \tag{1}$$

Although the system of linear equations in (1) is generally underdetermined, the CS methods can utilise knowledge about the signal sparsity to find only non-zero values with their positions in the sparse signal vector $\mathbf{h}$.

© Springer International Publishing Switzerland 2015
P. Gaj et al. (Eds.): CN 2015, CCIS 522, pp. 250–259, 2015.
DOI: 10.1007/978-3-319-19419-6_24

Among the CS reconstruction methods, the greedy ones have attracted widespread attention in the channel identification recently [4]. The Orthogonal Matching Pursuit (OMP) [5] and Compressive Sampling Matching Pursuit (CoSaMP) [6] belong to the ones of the highest popularity. Both of the mentioned methods work in an iterative manner under similar concept. The position of a new element in the sparse signal (or a group of elements in case of the CoSaMP) is estimated during each iteration and next its amplitude together with the amplitudes of the elements found before are calculated using the popular LS (Least Squares) algorithm.

The moment when the estimation algorithm should be stopped is an important issue for overall system performance including implementation complexity point of view. The exact sparsity order of the propagation channel is not available in the system a priori, because the real propagation environment is unpredictable and varies in time. On the other hand, a properly matched channel model can provide only supplementary information for the control of the estimation method. Therefore, metrics used in the design of the stopping criterion, should use internal data available during the iteration procedure. These information should allow to stop the estimation process when the distortion level in the system is low, that corresponds to finding the true value of the channel sparsity.

The goal of the paper is to present the numerical analysis which shows how the choice of the stopping criterion can influence the quality of the estimation procedure measured both in terms of the Symbol Error Rate (SER) and value of the algorithm residual error. The Time Domain Synchronous OFDM (TDS-OFDM) system and transmission through the wireless environment, described by the COST-207 TU-6 channel model [1] was picked as the test environment. The channel estimation procedure was driven by the OMP algorithm only. In the authors' opinion the results for the CoSaMP method will be similar because the core of both CS methods is based on the same calculation principle, which is the LS technique. The compressed sensing channel estimation techniques has been already evaluated in the TDS-OFDM systems in [7–9]. But, those research assumed that the CS-based estimation procedure knows the value of the channel sparsity order in advance.

The reminder of this paper is organised as follows. The Sect. 2 provides a sketch of the OMP algorithm with the main attention on the stopping criteria description. The simulation system model is presented in Sect. 3. The simulation scenarios are described in Sect. 4 and the final conclusion are in Sect. 5.

2   Stopping Criteria of the OMP Algorithm

The Orthogonal Matching Pursuit algorithm belongs to the family of greedy methods. It has attracted the widespread attention due to the simple and efficient implementation. Each iteration of the OMP procedure consists of two characteristic steps. Here, they are described in relation to the channel estimation problem. The first step is to determine a single delay of the propagation path and appends the result to the set of delays estimated in the previous iterations.

The second step is to estimate the new gains of the uncovered paths using the LS (Least Squares) method. The detailed description of the single OMP iteration loop may be found in [5]. The algorithm starts with the empty set of discovered channel paths, and the power of the residual error ε equals to power of the measurements $\mathbf{z}$. In each iteration loop the residual error is recalculated according to the formula:

$$\varepsilon_i = \mathbf{z} - \Phi\mathbf{h}_i \tag{2}$$

where $\mathbf{h}_i$ is the vector of the complex gains of the estimated channel impulse response in the i-th iteration.

The determination whether the algorithm should go to the next iteration or to stop the seeking procedure, is the main issue of the paper considerations. This research problem is worth of interest, because more number of the iterations means more time consumed for calculation, and in turn, the higher amount of time is related to an increase both in the general processing complexity and the higher power consumption. A natural way is to stop the OMP procedure when the sparsity order is reached, but the information about number of the propagation paths is rather unavailable in practical situation. Therefore, the known proposition is to use the power of the residual error ε to control the moment when the algorithm supposed to be stopped. The stopping threshold is usually $\mathbb{E}[\varepsilon^2] = \sigma^2$, where σ^2 is the power of noise in the channel. It seems to be a reasonable choice of the threshold value, because the noise is additive to the linear channel distortion. The interesting question is, how the power of the residual error varies as a function of iteration k. Does the chosen threshold value ensure the correct estimation of the channel impulse response, also for a time dynamic environment? An attempt to answers these questions is provided in Sect. 4.

Another possible way to control the iteration procedure is the authors' proposition to use the power of the difference of the successive residual errors. The difference of the residual errors can be described as follows:

$$\delta_i = \varepsilon_i - \epsilon_{i-1} = \Phi(\mathbf{h}_{i-1} - \mathbf{h}_i). \tag{3}$$

This metric informs how the residual error ε changes through the iterations. The small power values of δ indicates that a steady state of the channel estimation procedure has been reached and further iterations probably do not improve the system performance. A value of the stopping threshold for this metric was investigated during the simulations, which results are presented in Sect. 4.

3   TDS-OFDM Simulation Environment

The simulation environment designed for evaluation of the channel estimation procedure imitates the Time-Domain Synchronous OFDM system with transmission through the six paths wireless channel characterised by the COST-206 TU6 channel model. The TDS-OFDM is a type of multicarrier modulation which uses a pseudorandom noise sequence $\mathbf{c}$ transmitted in the guard time period as a training sequence for symbol synchronisation and channel estimation [10].

The OMP algorithm used in the channel estimation procedure requires, as every greedy method, higher amount of measurement samples than sparsity order, and it can be estimated using [5]

$$M - L > 2S\ln L \tag{4}$$

where M is length of the guard period, L is the maximum channel delay spread and S is the sparsity order. Figure 1 presents the TDS-OFDM time-domain symbol structure with the "IBI-free" region of the guard period indicated by the dark grey area in the received symbol. The important parameters of the simulated TDS-OFDM system are as follows:

- the transmitted symbol of the TDS-OFDM modulation consists of $N = 1024$ data samples $\mathbf{x}$ and $M = 128$ samples of the pseudorandom noise training sequence $\mathbf{c}$ in the guard period. No subcarriers are used as the pilot tones. The 16-QAM constellation is used for data coding in the subchannels. The total number of the OFDM symbols transmitted during one simulation cycle is 300 ($300 \times 1024 = 307\,200$ QAM data symbols for SER estimation);
- the COST-207 TU6 model of the time and frequency selective wireless channel for terrestrial propagation in an urban area with the delay spread about $5\,\mu s$ consists of $S = 6$ active paths. Adequately to the sampling frequency about $9\,MHz$ (DVB-T), the delays of the active paths can be approximated by the discrete time indices – $[0, 1, 4, 14, 21, 45]$;
- two cases of the channel time selectivity are considered – $f_d T = [0.1$ (fast fading), 0.02 (slow fading)$]$, where f_d is the Doppler spread and T is the OFDM symbol duration;
- no additional error correction coding is used;
- signal to noise ratios are $5\,[dB]$, $15\,[dB]$, $20\,[dB]$ and $25\,[dB]$.

The channel impulse response $\mathbf{h}$ estimated by the OMP algorithm, after the FFT processing, is further used to a frequency domain channel equalisation. The channel parameters are estimated just in time of the training sequence. Because the channel is considered doubly selective for the assumed time selectivity levels, the linear interpolation of the propagation paths gains within the duration of the data part of the OFDM symbol is performed [9].

4   Simulation Experiments

The investigation of the OMP-based channel impulse response estimation is performed under several simulation scenarios. The results of the previous simulations are used for some parameters adjustment of the next ones. The obtained results are discussed at once in the successive subsections.

4.1 $SER = f(SNR)$, and $SER = f(k)$

The aim of this scenario is to show how the channel estimation procedure affects the transmission performance expressed as the Symbol Error Rate (SER). The

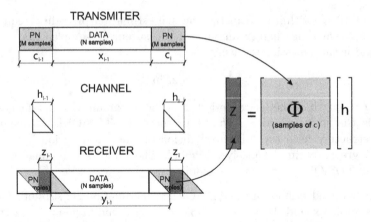

Fig. 1. The TDS-OFDM time-domain symbol structures at the transmitter and receiver with the featured samples of the training sequence used for channel reconstruction with the compressed sensing method

word "Symbol" in the metric name refers to the single element transmitted in every subchannel of the OFDM frequency domain block. The lines indicated by "CSI" on the graphs in Fig. 2 represent the absolute minimum of the SER for the given instance of the channel and was drawn for the perfect knowledge of the impulse response (perfect CSI) in every time sample of the data transmission.

The estimation procedure was stopped after the k iterations taken respectively from the subset of the integer numbers $I \in \{4, 6, 8, 10, 12, 18, 24\}$. These values have direct relation to the amount of nonzero coefficients in the estimated channel impulse response. The condition $k = S$, when the exact channel reconstruction is probable, is met only for second element of I. The Fig. 2 presents $SER = f(SNR)$ for the fast and slow fading channel. In both cases, if the number of iterations exceeds S, the obtained SER values for the estimated channel approaches to the perfect one. But an influence of k can be distinguished, and this phenomenon is clearly seen in Fig. 3. The number of iterations, too much higher than the channel sparsity order, generally slightly impairs the SER. It may not be a serious quality problem, because in real system an error correction coding may reduce this gap. But the excess number of iterations has adverse impact on the implementation complexity. Therefore, the upper bond for the number of iterations, that equals to the double value of the average sparsity S of the propagation environment, seems to be a good choice as the supplementary condition to stop the estimation procedure.

The considered cases for both types of time selective channels show no significant difference in the estimation efficiency with regard to the perfect CSI as a reference level. Hence, the next results discussion is limited to slow fading channel, though the experiments were performed for both.

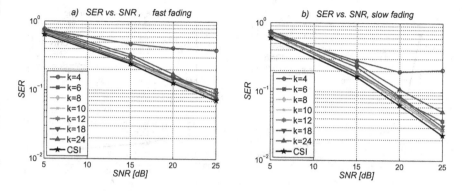

Fig. 2. SER = f(SNR) for different values of executed iterations of the OMP algorithm, (a) fast fading channel, (b) slow fading channel

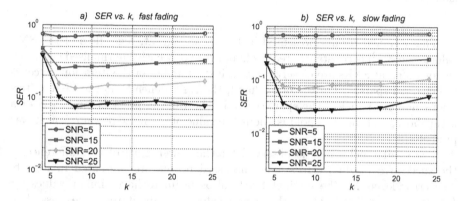

Fig. 3. SER = f(k) for different values of the SNR, (a) fast fading channel, (b) slow fading channel

4.2 $\mathbb{E}[\varepsilon^2] = f(k)$, and $\mathbb{E}[\delta^2] = f(k)$

Stopping the estimation process after the predefined number of iteration is not a convenient way of control. The sparsity driven threshold value, carefully chosen according to the expected channel features, might not guarantee neither high quality of the reconstructed channel (the algorithm stops too early) nor low complexity of the method (the algorithm stops too late).Therefore, another stopping condition, based on the metrics directly accessible inside the estimation procedure and well correlated with estimation quality, are suggested for a flexible control of the estimation procedure.

This subsection analyses the characteristics of the residual error power $\mathbb{E}[\varepsilon^2]$ and the power of the residuals difference $\mathbb{E}[\delta^2]$ in relation to their application for the stopping criteria design. For better comparison purposes, the values of the error powers are normalised by the noise power σ^2.

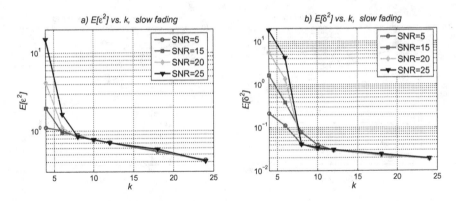

Fig. 4. $\mathbb{E}[\varepsilon^2] = f(k)$ (a), $\mathbb{E}[\delta^2] = f(k)$ (b) for different values of the SNR and the slow fading channel

The obtained results presented in Fig. 4a show that, regardless of the SNR values, the $\mathbb{E}[\varepsilon^2]$ is constantly reduced, and the consecutive normalised error levels for different SNRs are closely related, if the number of the iterations exceeds the channel sparsity. On the other hand, the $\mathbb{E}[\varepsilon^2]$ depends on the SNR if iterations number is below the sparsity level. The turning point is for $k = S$ or a little above, and the threshold is about the value of the noise power σ^2, as it was predicted.

The similar behaviour is observed for the power $\mathbb{E}[\delta^2]$ of the difference of the successive residual errors. The discussed results are presented in Fig. 4b. The error levels for the distinct SNR ratios begin to have similar values for k between S and $2S$. Then, the power values are about ten times less than in case of the $\mathbb{E}[\varepsilon^2]$. Using this metric to design the stopping condition, the $\mathbb{E}[\delta^2] = 0.1$ is suggested as the normalised threshold level. Although, the final number of coefficients of the estimated channel can be a little higher than the real amount of propagation paths, it can better protect the estimation procedure against the wrong delay paths selection, that can occur in some iteration steps.

Another useful observation concerns the much larger range of the $\mathbb{E}[\delta^2]$ values than it occurs in case of the power of the residual error. The metric $\mathbb{E}[\delta^2]$ crosses the threshold with the steepest slope while the iterations number increases, and additionally the bottom of this slope is much below of the predefined threshold. Both features are desirable with regard to how to stop the iteration procedure, because the sharp transition between high and low error values ensures the less sensitivity of the iteration stopping moment on the threshold value choice. It is strongly related to the range of the number of the estimated path delays consequently. Taking this conclusion into account, the better metric choice to control the estimation algorithm seems to be $\mathbb{E}[\delta^2]$. The next subsection presents the simulation results, which try to confirm that hypothesis.

4.3 Histograms of the Executed Iterations for the $\mathbb{E}[\varepsilon^2]$-Based, and the $\mathbb{E}[\delta^2]$-Based Stopping Criteria

The simulation results presented in this subsection concern a distribution analysis of the OMP algorithm iterations. The total number of the OFDM symbols transmitted during a single simulation cycle was 500. The investigations were performed for two distinct metric taken as the stopping criterion. The first one is the power of the residual error $\mathbb{E}[\varepsilon^2]$, and the second one is the power of the residuals difference $\mathbb{E}[\delta^2]$. The decision thresholds were adjusted according to result of the previous simulation scenarios.

Generally, the proposed threshold is proportional to the noise power σ^2:

$$\theta = m\sigma^2. \tag{5}$$

The simulations were performed for following values of m:

- $m = \{0.75; 1; 1.25\}$ for $\mathbb{E}[\varepsilon^2]$,
- $m = \{0.075; 0.1; 0.125\}$ for $\mathbb{E}[\delta^2]$,

and their results are presented in Fig. 5a, b in the form of histograms. The upper bound for the number of iterations was 24 (the quadruple of the channel sparsity).

In relation to the desired number of iterations, the OMP algorithm seems to be more sensitive on the threshold variations in case of the residual error based stopping criterion (or, it can be said conversely that, the OMP is more sensitive on the noise power fluctuations if the threshold value is fixed). For example, when the threshold value was below the noise power ($m = 0.75$), the executed iterations number of the OMP algorithm varies over broad range, regardless of the SNR. The similar feature can be noticed for all considered m values and the fixed value of SNR $= 5\,[\mathrm{dB}]$. The observed sensitivity of the OMP is reduced with the increasing values of the SNR, and finally it uses $k = S$ iterations in the most cases which is a desirable feature.

In order to protect the channel estimation method against noise power fluctuation the second metric – $\mathbb{E}[\delta^2]$ – was proposed. Then, the OMP procedure becomes less sensitive on the threshold and SNR variations (see Fig. 5b). The executed iterations number is concentrated about the single value, but it is slightly larger than the channel sparsity level, whereby it can be considered as shortcoming.

A solution which reconciles both independent approaches is their combination. The proposed stopping rule is defined as follows:

$$\text{OFF if } \begin{cases} k > 2S, & \text{any } \mathbb{E}[\varepsilon^2], \mathbb{E}[\delta^2] \\ \mathbb{E}[\varepsilon^2] < m\sigma^2, & \text{any } \mathbb{E}[\delta^2] \\ \mathbb{E}[\delta^2] < 0.1m\sigma^2, & \sigma^2 < \mathbb{E}[\varepsilon^2] < 2\sigma^2 \end{cases} \tag{6}$$

The second condition of (6) means that the estimation procedure stops in any case when the residual error is below the scaled value of noise power σ^2. But if the procedure get stuck, and the residual error is above the threshold (up to

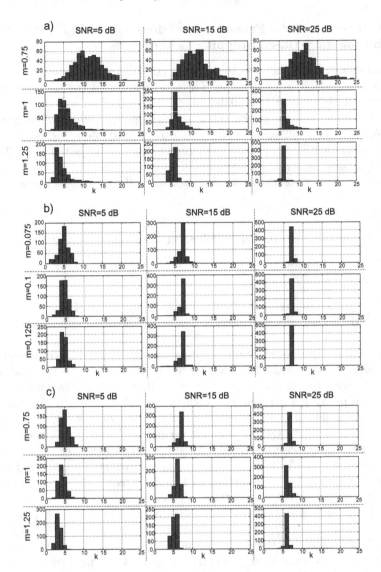

Fig. 5. Histogram of the executed iterations for threshold $\theta = m * \sigma^2$ (a), $\theta = 0.1m * \sigma^2$ (b), and combined threshold (c), where $m = \{0.75, 1, 1.25\}$ and SNR $= \{5, 15, 25\}$ [dB]

double value of the power noise), the $\mathbb{E}[\delta^2]$ controls the operation of the OMP algorithm (the third condition of (6)). The simulations were performed for $m = \{0.75; 1; 1.25\}$. The iteration distribution for this proposition is presented in Fig. 5c. The goals, which were less diversity of the iterations number and its precision with regard to channel sparsity reconstruction, seem to be fulfilled, although the stopping criterion (6) was developed only according to the simulation experiments and can not be considered optimal. Moreover, the number of iterations never reached the maximum value $k = 2S$ (the first condition of (6)). Nonetheless, the further investigations, supplemented with an analytical analysis, are recommended.

5   Conclusion

The paper has considered the stopping criteria analysis based on the residual error of the OMP algorithm. The power of the residual error and the power of the residuals difference were investigated. The results obtained through simulations show that the second one metric gives the more stable algorithm behaviour measured as the variability of the executed iterations. Next, the proposed combined solution additionally allows stopping the estimation procedure just after the iteration number that very often equals to the sparsity order of the transmission channel.

Acknowledgement. This work was supported by the Silesian University of Technology, Institute of Electronics under statutory research program in 2015.

References

1. Failli, M.: Digital land mobile radio communications COST 207. Technical report, European Commission (1989)
2. Guidelines for evaluation of radio transmission technologies for imt-2000. Technical report Rec. ITU-R M.1225, ITU (1997)
3. Candes, E., Wakin, M.: An introduction to compressive sampling. IEEE Signal Process. Mag. **25**(2), 21–30 (2008)
4. Berger, C., Wang, Z., Huang, J., Zhou, S.: Application of compressive sensing to sparse channel estimation. IEEE Commun. Mag. **48**(11), 164–174 (2010)
5. Tropp, J., Gilbert, A.: Signal recovery from random measurements via orthogonal matching pursuit. IEEE Trans. Inf. Theory **53**(12), 4655–4666 (2007)
6. Needell, D., Tropp, J.: Cosamp: iterative signal recovery from incomplete and inaccurate samples. Appl. Comput. Harmonic Anal. **26**(3), 301–321 (2009)
7. Dai, L., Wang, Z., Yang, Z.: Compressive sensing based time domain synchronous ofdm transmission for vehicular communications. IEEE J. Sel. Areas Commun. **31**(9), 460–469 (2013)
8. Zhu, X., Wang, J., Wang, Z.: Adaptive compressive sensing based channel estimation for tds-ofdm systems. In: 9th International Wireless Communications and Mobile Computing Conference (IWCMC), pp. 873–877 (2013)
9. Dziwoki, Grzegorz, Izydorczyk, Jacek, Szebeszczyk, Marcin: Time domain estimation of mobile radio channels for OFDM transmission. In: Kwiecień, Andrzej, Gaj, Piotr, Stera, Piotr (eds.) CN 2014. CCIS, vol. 431, pp. 167–176. Springer, Heidelberg (2014)
10. Wang, J., Yang, Z.X., Pan, C.Y., Song, J., Yang, L.: Iterative padding subtraction of the pn sequence for the tds-ofdm over broadcast channels. IEEE Trans. Consum. Electron. **51**(4), 1148–1152 (2005)

A Dynamic Energy Efficient Optical Line Terminal Design for Optical Access Network

Özgür Can Turna[1(✉)], Muhammed Ali Aydin[1], and Tülin Atmaca[2]

[1] Computer Engineering Department, Istanbul University, Istanbul, Turkey
{ozcantur,aydinali}@istanbul.edu.tr
[2] Laboratoire SAMOVAR, UMR 5157, IMT/Telecom SudParis, Evry, France
tulin.atmaca@telecom-sudparis.eu

Abstract. Computer networks are one of the major slices of the global energy consumption. Since 2009, a couple of standards have been developed for energy conservation in passive optical networks. These standards and most of the researches are based on improvements on Optical Network Unit side. In this study, a novel energy efficiency algorithm, which is based on coupling two Optical Line Terminal to reduce energy consumption in central office, is proposed. Our design employs optical switches and amplifiers to create a switch-box, which is under control of both Optical Line Terminal pairs.

Keywords: Passive optical network · Energy efficiency · Optical line terminal

1 Introduction

Energy consumption is one of the crucial problems for future of our world. In [1], the energy consumption projection is given for 2040. It is foreseen that energy consumption in world grow by 56 % from 2010 to 2040. Thus, in last decade industry and academia trend to develop more energy effective products. According to the study of M.Picavet from Ghent University in 2009 [2], the Information and Computing Technology (ICT) percentage of energy consumption in overall electronics is 8 %. In case of ICT, one of the considerable energy consumer is network devices which are distributed all over the world for Internet coverage. In [2], it is mentioned that access networks consume 70 % of total energy consumption of network devices in 2009. According to the projection of the energy consumption distribution in telecommunications network, the energy consumption of access network is considerably high for today and seem to be keeps its volume for future.

Passive Optical Network (PON) is the most promising solution for future access networks. Since, using optical technology in fix networks brings advantages as; reaching long-haul, less corruption on carried data, and far less being affected from environmental factors. Besides these advantages PON also gives the most favorable solution for energy efficiency. While PON is the best solution for energy

© Springer International Publishing Switzerland 2015
P. Gaj et al. (Eds.): CN 2015, CCIS 522, pp. 260–269, 2015.
DOI: 10.1007/978-3-319-19419-6_25

efficiency in access networks, the consumed energy can be reduced more by effective techniques, and/or devices.

FSAN and ITU-T made three surveys from 2006 to 2007 to gather requirement analysis for power-saving. According to these surveys, power saving has lesser priority compared to service quality, availability, and interface variability. In 2009, power conservation recommendations for GPON [3], and EPON [4] are released to cover sort of energy conservation methods in PON systems. For power consumption, G.Sup45 recommendation describes four different power saving methods; power shedding, dozing, deep-sleep, and fast/cyclic sleep. After releasing GPON recommendations, FSAN group and ITU-T go on working to develop the successor of GPON standard as NG-PONs. There are lots of ideas discussed for novel PON standardization. As a mid-term solution to meet service providers' demands, XG-PON [5] released. XG-PON, as an enhancement, inherits framing and management from GPON. For co-existence, XG-PON uses WDM in downstream and WDMA in upstream to share the same infrastructure with existing GPON implementations. XG-PON supports doze mode and cyclic sleep mode specifications in [3]. Applying any other power saving techniques left for the vendors' choice. In 2012, a novel approach Bit-Interleaving PON (Bi-PON) released by GreenTouch consortium [6]. GreenTouch is founded by a group of network companies and academic institutes to accomplish the purpose that ICT energy consumption in 2010 will be decreased factor of 1000 in 2015. Bi-PON aims to reduce the power consumption on end user devices (ONU) and reveal a success that energy consumption of ONU reduced from 3.5 W (XG-PON) to 0.5 W (Bi-PON) in average.

Central office equipment (OLT) consumes approximately 10–100 W energy according to the manufacturers [7–9], and [10]. There are a lot of ONU based studies for energy efficiency, where just a few propositions presented for energy conservation in OLT. Solutions for OLT regarding wavelength division are presented in [11], grouping OLTs are given in [12] and summary of some related studies are presented in [13]. When these solutions are examined, it is observed that these methods are mostly based on a major device renovation with novel technical devices or closing the device during some time period of day.

In follows we describe our OLT based energy efficiency design and algorithm. After, the performance of the design is presented under different conditions.

2   Dynamic Energy Efficiency Algorithm for OLT

In a PON system, OLT consumes considerable amount of energy which causes operational expenditure for service providers. Network equipment are not fully utilized whole runtime especially while serving home subscribers. Because home devices can be switched off, be idle or in usage with a quite light traffic (i.e. web surfing traffic). In such cases, by the use of different techniques some network equipment can be switched off or sent to sleep/deep-sleep mode. In PON's structure, there is no intermediate equipment that can be removed, and OLTs and ONUs must always be online for providing service. However, owing to

TDM/TDMA aspects, ONUs are actively in use less than 10 % of the time-line. Though, ONUs can be put into sleep mode while the ONU is not actively in use. On the contrary, OLT has to be active all the time to schedule and serve each ONU simultaneously. If we want to put an OLT into sleep, the subscribers must be served by another network element.

Our contribution provides an energy efficiency approach with a diminutive modification on central office systems and no alteration reflected to the ONU part. In our approach, two OLT cards are combined as a couple to handle each other traffic flows under low load. Thus, one OLT can be kept in deep-sleep mode till a heavy traffic volume occurs. Operating mode is switched dynamically according to the traffic pattern.

Optical amplifiers are used to elevate optical power to reach long-hauls and/or to recover the decay of passive splitter on divided optical power by two or more fraction. A summary of amplifiers used in PON systems are; xDFA (...-doped fiber amplifier), Raman amplifier, and SOA (semiconductor optical amplifier) is given in [14]. According to this study, a Raman amplifier consumes 0.5 W energy. Besides, decreasing the energy consumption of amplifiers is aimed by novel studies. In our proposition, one OLT's subscribers (ONUs) are for-warded to other one in power saving mode by optical switches. The proposed approach uses a switch-box. This component is electronically connected with OLTs for control plane and consists of 1 x 2 optical switches and an optical amplifier. The switch-box implementation is assumed to use far less power compared to an OLT. The amplifier is only in use when one OLT is put into sleep mode, in another words when it is in energy saving mode to satisfy the power budget for doubled fiber line split.

2.1 Developed Architecture

The developed architecture comprises two OLTs ($OLT - A$ for sleeping one, $OLT - B$ for master one) and a switch-box consists of five 1 x 2 optical switches and an amplifier. OLTs and switch-box are directly connected with dedicated communication line. The transition between normal and energy saving modes is decided according to the traffic load. If one of the OLTs (A) has a low traffic volume on both uplink and downlink, it requests B for passing to the energy saving mode. If B's traffic volume is under a reasonable amount, it responds positive and starts processes to put A into deep-sleep mode (A is completely shutdown except wake on by signal feature). Therefore, B must be able to take all responsibilities of A, and serves without any unsatisfactory delays. If B encounter a heavy traffic volume in uplink or downlink, than B wakes up and prepares A to serve as in usual way.

In system implementation, the following issues have to be considered. First, OLTs have to be capable of exchanging their synchronization and route table information with each other over a dedicated connection. It can be over an Eth-ernet, an USB or a BUS interface. Second, a switch-box that consists of optical switches and an amplifier must be implemented while keeping energy consumption at a very low level. Third, OLTs can be capable of controlling the switch-box

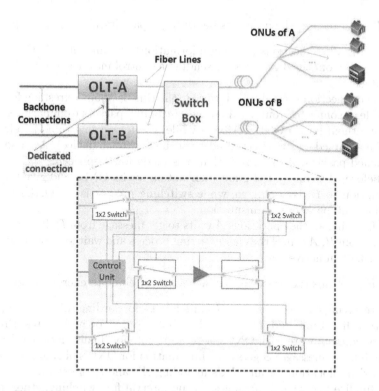

Fig. 1. General architectural design of OLT couple with switch-box

and buffering the traffic of backbone connection while other OLT is not ready
for switching process. The general structure and switch-box design are shown in
Fig. 1. Each OLT connects to the switch-box with a single fiber. The fiber coming from OLT is connected to a 1 x 2 optical switch. The fiber coming from the
ONUs side also connected to another 1 x 2 switch at the entrance of switch-box.
On normal mode, through input switches OLTs directly connected to its ONUs,
on energy saving mode both ONU set directed to middle line in switch-box and
it is directed to the active OLT by another 1 x 2 optical switch.

2.2   Proposed Algorithm for Control Plane

The proposed system works in two modes (Normal mode and Energy Saving
mode). In normal mode, switch-box always waits on stand-by. The amplifier is
not powered and optical switches are locked in their states. The system performs
as usual such as two separate OLTs are in use. To switch between these two
modes, a control plane is used to schedule and share information between OLTs.
In following example the two OLT pairs named as A and B. When A has less
traffic than a particular lower bound in both uplink and downlink, it informs
B from the dedicated link. If B responses positive then A prepares itself for
deep-sleep mode. Otherwise, A waits next load check interval for next query.

$OLT - A$ executes the processes below to prepare itself for deep-sleep mode;

- A prepares all connections (routing information) to transmit over B. Thus, A activates B from the direct link by sending a control message and its forwarding table.
- Till the connection is established over B (the time is about refreshing forwarding table of connected router), A receives incoming packets from backbone and schedule them for its ONUs. After a while, the forwarding table is refreshed (can be assumed or strictly checked from the backbone device). After sending all queued packets, A informs B that it is ready for deep-sleep.
- Just before, A clears all the grant information for its ONUs and give the sync information to B. By doing so, while switching process is on, ONUs belongs to A have no packets in transmission.
- After finishing all the operations A waits to get message from B. If no response returns from B, A cancel the energy saving process and wait for new commands and packets in active state.

$OLT - B$ performs the processes below before switching to energy saving mode;

- After receiving control message, B waits for A's forwarding table, then update its forwarding table and inform the backbone router for new routes. Prepare itself to communicate with ONUs of A, and waits for finalize message form A.
- Before finalize message, B gets sync information for ONU's of A and schedules bandwidth allocation for all ONUs including A's and its network.
- With finalize message, B arranges a time interval for switching, stops giving grants and stops downstream traffic for this interval, and turns the switch-box for forwarding packets to itself.
- While switching to energy saving mode, B stores any arrived packets that belongs to ONUs of A. After the switching procedure, these packets are scheduled with appropriate grant information.

When $OLT - B$ reaches to upper load limit, then it demands A to wake up to share the ONUs again and work in normal mode. After A receives wake-up signal, it powers up and sets its parameters (buffers, schedulers, counters etc.) and responses to B. By receiving this response, B knows that A can handle the incoming messages. B sends sync information to A and sets forwarding table of backbone router, itself, and A. After all updates, B sends finalization message to A. If preparations are finished as expected, A responds positive and B turns the switch-box in normal working mode. After that, B triggers A to start processing in normal mode with an ending control message.

3   Performance Evaluation

To evaluate the performance of our proposed design, we compared it with standard OLT implementation with OMNET++ 4.3 tool. We simulated a simple scenario where two OLT cards connected with aggregation node and a switch-box. The performance of the dynamic energy efficiency algorithm is quite related to the incoming traffic pattern. Transition between normal and energy saving modes occurs when the incoming traffic load varies. Classic traffic generation is not effective to observe the performance of the algorithm. Distribution algorithms in OMNET as Poisson or exponential are based on a fix load parameter and the generated packet distribution heaps up the given load value. Thus, traffic sources are configured to produce different loads in successive time intervals to easily analyze the performance of energy efficiency algorithm. The time intervals and load are exponentially random distributed. Traffic generator creates different loads in exponentially distributed time intervals. For the experimental results different traffic patterns were performed by changing the load and variation intervals. Parameters of the simulation environment are summarized in Table 1.

Table 1. Simulation parameters

Parameter	Symbol	Values
Bandwidth	β	1 Gbps
OLT buffer sizes		20 MB
Simulation times	t	10 000 s
Load check interval	LCI	10 ms
Packet sizes		1500 Byte
Load change interval	θ	3 s, 10 s
Optic switch-switching time	St	5 ms, 5 ns
Route table update time	Rt	25 ms, 5 ms

The performance of energy-efficiency algorithm has been investigated under different traffic pattern and bounds. Load change interval (θ) shows how long the traffic is generated based on the same load. In our simulations we examine two different θ value as 3 s and 10 s. θ equals to 3 s means that the load is very variable. Thus, incoming traffic is changed between low and high bounds of energy-efficiency algorithm. Low bound is the limit where an OLT can send sleep demand. High bound stands for when the master OLT sends wake up demand. The low and high bounds can affect the buffer fill and access delay. For performance evaluation we present three cases. Case 1: 90 % for high bound and 10 % for low bound, Case 2: 80 % for high bound and 10 % for low bound, Case 3: 90 % for high bound and 20 % for low bound. In each scenario, upper bound to accept the sleep demand is selected as 40 %. Firstly, the number of On-Off transitions, where St is 5 ms and Rt is 25 ms, is given in Fig. 2. θ has greater impact on the On-Off transitions in each scenario.

For the average load under 0.25, only Case 1 is investigated, since under low load values OLT couple stays in energy saving mode where one of the OLTs always stays in sleep mode. Under 0.17 average load, one OLT is always in sleep state. Same situation also occurs at high average load values. Thus, over

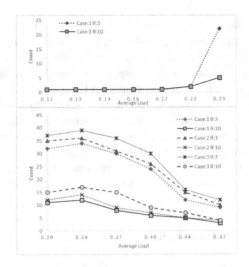

Fig. 2. On-Off transition under different load experiments (*St*: 5 ms, *Rt*: 25 ms)

0.50 average load is not investigated. Also, the other conditions are not given since they are comparably same. The number of On-Off transitions means more switching and route table update delays and results with an increase of access delay. When θ equals to 3 s, the maximum average of On-Off transition is resulted as 55 times in 10 000 s simulation duration.

In Figs. 3 and 4, total sleep duration of OLT couple is shown. These values are sum of both OLT's sleep durations in 10 000 s simulation time. The average load values under 0.23, one OLT stays 99 % in sleep mode. Thanks to dynamic scheme one OLT can be kept in sleep mode till the average load increase to 50 %. Load change interval of the traffic affects the algorithm performance in load values between 0.30 and 0.45. Arbitrary traffic means more transitions between sleep and active states. That seems to increase the sleep performance of OLT couple, however it also increases the access delay. For different case studies the results are very similar where θ taken as 10 s. Case 2 shows the lowest sleep performance and Case 3 shows the highest sleep performance, as expected, since the sleep decision bounds have great impact when the traffic load varies. The bounds of sleep algorithm has direct impact on sleep performance and access delay. During day time, Internet usage has peak (work hours) and idle (night) periods. Thus, it is better to select Case 2 (80-40-10) in peak periods, Case 3 (90-40-20) in idle periods, and Case 1 (90-40-10) in other periods. Though, the energy efficiency can be maximized while access delay values kept low. Switching operations in the designed architecture is far less than the switching in backbone networks.

Switching operation is done just for one time when the mode alters. Thus, the least power consuming optical switches can be selected (In the simulations switching time is taken 5 ms for the worst optical-switch performance and 5 ns for improved components). Also in our experiments the traffic generator creates

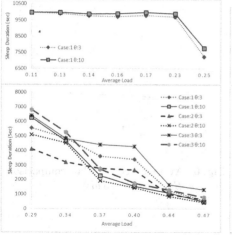

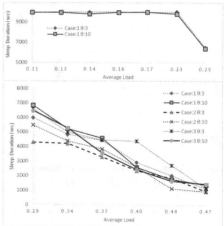

Fig. 3. Sleep duration of OLT couple (*St*: 5 ms, *Rt*: 25 ms)

Fig. 4. Sleep duration of OLT couple (*St*: 5 ns, *Rt*: 5 ms)

undulating load to see the performance under worst case scenarios. In this study the architecture is not implemented and tested on live conditions. In the literature and market analysis we see the existence and possibility to use optical amplifiers that can double the optical power on one wavelength with low energy consumption (i.e. 0.5 W). If we assume the extra components in switch-box uses 0.1 W and the least power consuming OLT card (10 W) is used in our design the OLT pairs use 20.1 W (20 for OLTs, 0.1 for switch-box) in active mode and 10.6 W (10 for OLT, 0.1 for switch-box and 0.5 for amplifier) in power saving mode. In this scenario if a system can stay 50 % in power saving mode the 48 % energy conservation can be succeeded. Considering this scenario the sleep time durations that are given in Figs. 3 and 4 for 10 000 s. simulation duration can directly show the energy conservation performance of our approach.

Figures 5 and 6 present average access delay performance of our algorithm compared to the system without energy-efficiency algorithm. The access delays are related to the time spent in energy saving mode, the number of On-Off transitions and the switching time. Thus, while $\theta = 3$ s, $St = 5$ ms, and $Rt = 25$ ms in Fig. 5, system has the maximum access delay results. The number of On-Off transitions is the critical parameter for the access delay increment. When the load is not highly variable, the energy-efficiency algorithm performs better. Case 2 has the lowest access delay performance. On the contrary, it is not as much energy-efficient as other cases. These results shows that changeable bounds according to the predicted traffic pattern can maximize the access delay versus energy-efficiency performance.

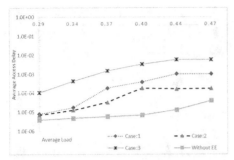 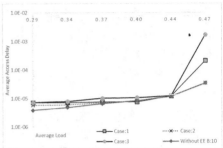

Fig. 5. Average access delay comparison (θ: 3 s, St: 5 ms Rt: 25 ms)

Fig. 6. Average access delay comparison (θ: 10 s, St: 5 ms Rt: 25 ms)

4   Conclusion

In this study, a novel OLT based energy efficiency technique in optical access network is presented. Our strategy is simply putting one of the OLTs into deep-sleep mode as much as possible while using standard TDM messaging procedure in PON. Our energy-efficiency algorithm's performance is compared with normal processing in case of access delay. Energy-efficiency algorithm can save 90 % of one OLTs energy consumption if the system average load is not high. When mostly 25 % of users are simultaneously using Internet, energy can be conserved whole day with our dynamic scheme. Our algorithm has the ability to conserve energy at peak hours of the daytime if any gap occurs between high loads, and excludes the performance problem caused by arbitrary increased traffic in idle hours (night). It does not bring too much structural changes and can be performed by a service provider without changing standard implementation of PON. Besides, our structure has no limitations on using ONU-based energy conservation approaches. Besides, this approach seems to be promising for implementation expenditures compared to WDM based solutions. As the negative aspects of our proposition, it requires a switch-box implementation and use of dedicated communication line between OLT pairs and switch-box. As a future study, the performance of the energy saving mode can be presented with different traffic patterns, and also dynamic vs. time-based schedule working modes can be compared. Afterwards, we aim to figure out 1 x N OLT switch-box design and evaluate its performance.

Acknowledgments. This work was partially supported by Scientific Research Projects Coordination Unit of Istanbul University (Project number 35709).

References

1. U.S. Energy Information Administration: International Energy Outlook 2013 (2013). http://www.eia.gov/forecasts/ieo/index.cfm
2. Mukherjee, B.: Energy Savings in Telecom Networks Energy Savings in Telecom Network (2011). http://www.ict.kth.se/MAP/FMI/Negonet/tgn-ons/documents/Energy%20efficiency%20in%20Access%20networks/bm-sbrc11-tutorial-1june11.pdf
3. ITU-T, G. Sup45: GPON power conservation (2009). http://www.itu.int/rec/T-REC-G.Sup45-200905-I/en
4. Mandin, J.: EPON Power saving via Sleep Mode (2008). http://www.ieee802.org/3/av/public/2008_09/3av_0809_mandin_4.pdf
5. ITU-T: G.987 Series: 10-Gigabit-capable passive optical network (XG-PON) systems: Definitions, abbreviations and acronyms (2010). http://www.itu.int/rec/T-REC-G.987/en
6. GreenTouch Consortium: BIPON: Bit-Interleaved Passive Optical Network Technology (2012). http://www.greentouch.org/index.php?page=Bi-PON
7. Grobe, K.: Performance, cost, and energy consumption in next-generation WDM-based access. In: Broadnets 2010, Athens (2010)
8. Ghazisaidi, N., Maier, M.: Techno-economic analysis of EPON and WiMAX for future fiber-wireless (FiWi) networks. Comput. Netw. **54**, 2640–2650 (2010)
9. Chowdhury, P.: Energy-Efficient Next-Generation Networks (E2NGN), Ph.D. thesis, University of California, Davis (2011)
10. Aleksic, S., Deruyck, M., Vereecken, W., Joseph, W., Pickavet, M., Martens, L.: Energy efficiency of femtocell deployment in combined wireless/optical access networks. Comput. Netw. **57**, 1217–1233 (2013)
11. Tokuhashi, K., Ishii, D., Okamoto, S., Yamanaka, N.: Energy saving optical access network based on hybrid Passive/Active architecture. In: The Institute of Electronics, Information and Communication Engineers, vol. 111(274), pp. 7–12 (2011). http://ci.nii.ac.jp/naid/110009465527/en/
12. Saliou, F., Chanclou, P., Genay, N., Laurent, F.: Energy efficiency scenarios for long reach PON central offices. In: Optical Fiber Communication Conference and Exposition (OFC/NFOEC), 2011 and the National Fiber Optic Engineers Conference, pp. 1–3 (2011)
13. Kani, J.: Power saving techniques and mechanisms for optical access networks systems. J. Lightwave Technol. **31**(4), 563–570 (2013)
14. Trojer, E., Dahlfort, S., Hood, D., Mickelsson, H.: Current and next-generation PONs: a technical overview of present and future PON technology. Ericsson Rev. **2**, 64–69 (2008)

Tracing of an Entanglement Level in Short Qubit and Qutrit Spin Chains

Marek Sawerwain[1]([⊠]) and Joanna Wiśniewska[2]

[1] Institute of Control and Computation Engineering, University of Zielona Góra, Licealna 9, 65-417 Zielona Góra, Poland
M.Sawerwain@issi.uz.zgora.pl
[2] Institute of Information Systems, Faculty of Cybernetics, Military University of Technology, Kaliskiego 2, 00-908 Warsaw, Poland
jwisniewska@wat.edu.pl

Abstract. The phenomenon of quantum entanglement is an important component of many protocols in the field of quantum computing. In this paper a level of entanglement in short qudit chains is evaluated. The evaluation is carried out with use of the CCNR criterion and the concurrence measure. There are also some explicit formulae describing the values of CCNR criterion and concurrence for exemplary short spin chains. Mentioned formulae may be used as an invariant for a protocol which describes a transfer of information in XY-like spin chains.

Keywords: Quantum information transfer · Qubit/qutrit chains · Entanglement · CCNR criterion · Concurrence · Numerical simulations

1 Introduction

A quantum entanglement [1] is a fundamental feature of quantum systems. A presence of quantum entanglement in an examined system indicates system's quantum nature. The entanglement is also present in spin chains which became one of the essential elements of quantum physics and quantum computing [2–4].

Naturally, the main function of spin chains is a transfer of information. A level of entanglement in a spin chain changes during the transmission of quantum state (i.e. quantum information). In spite of difficulties associated with formulating the entanglement levels, it is possible to evaluate the entanglement e.g. in spin chains. In this paper the level of entanglement is estimated for short qudit spin chains. The presented results may be used to verify if the process of transfer is correct, because it seems that the level of entanglement can be treated as an invariant for the transfer protocol in a spin chain.

The paper contains the following information: in Sect. 2 a form of Hamiltonian for performing a XY-like qudit transfer protocol is presented (including a short note concerning properties of the perfect transfer). There is also an algorithm describing the realization of transfer protocol with a γ condition playing a role of the invariant expressing a level of entanglement during the transfer.

© Springer International Publishing Switzerland 2015
P. Gaj et al. (Eds.): CN 2015, CCIS 522, pp. 270–279, 2015.
DOI: 10.1007/978-3-319-19419-6_26

In Sect. 3 the Computable Cross-Norm or Realignment (CCNR) criterion and the concurrence measure are presented.

Section 4 contains the results of the experiments for detecting entanglement and calculating the values of concurrence measure. This section presents the numerical test for the correctness of transfer in qubit and qutrit chains, as well. A summary and conclusions are presented in Sect. 5.

2 Hamiltonian for a Transfer Protocol in a Qudit Spin Chain

In this section a Hamiltonian H^{XY_d} will be defined. This Hamiltonian will be used to realize the perfect transfer of quantum information in qudit chains [5–7] for entanglement creation between chosen points of a chain. The form of Hamiltonian, given below, is naturally suitable for transfers discussed in this paper – i.e. for transmission in qubit and qutrit spin chains.

In a definition of the XY-like Hamiltonian for qudits' chain, given below, the Lie algebra's generator for a group $SU(d)$ was used, where $d \geq 2$ is to define a set of operators responsible for transfer dynamics. For clarity, the construction procedure of $SU(d)$ generators will be recalled – in the first step a set of projectors is defined:

$$(P^{k,j})_{v,\mu} = |k\rangle\langle j| = \delta_{v,j}\delta_{\mu,k}, \quad 1 \leq v, \mu \leq d. \tag{1}$$

The first suite of $d(d-1)$ operators from the group $SU(d)$ is specified as

$$\Theta^{k,j} = P^{k,j} + P^{j,k}, \quad \beta^{k,j} = -i(P^{k,j} - P^{j,k}), \tag{2}$$

and $1 \leq k < j \leq d$.

The remaining $(d-1)$ generators are defined in the following way

$$\eta^{r,r} = \sqrt{\frac{2}{r(r+1)}}\left[\left(\sum_{j=1}^{r}P^{j,j}\right) - rP^{r+1,r+1}\right], \tag{3}$$

and $1 \leq r \leq (d-1)$. Finally, the $d^2 - 1$ operators belonging to the $SU(d)$ group can be obtained.

The above construction of $SU(d)$ generator is used to create the XY-like Hamiltonian for qudits. Assuming that each qudit has the same freedom level $d \geq 2$ (the qudit is defined similar to qubit, however a computational base for qudits is expressed with d orthonormal vectors – in case of qubits, the base contains two orthonormal vectors):

$$H^{XY_d} = \sum_{(i,i+1)\in\mathcal{L}(G)} \frac{J_i}{2}\left(\Theta_{(i)}^{k,j}\Theta_{(i+1)}^{k,j} + \beta_{(i)}^{k,j}\beta_{(i+1)}^{k,j}\right), \tag{4}$$

where J_i is defined as follows: $J_i = \frac{\sqrt{i(N-i)}}{2}$ for $1 \leq k < j < d$ and $\Theta_{(i)}^{k,j}, \beta_{(i)}^{k,j}$ are $SU(d)$ group operators defined by (2) applied to the (i)-th and $(i+1)$-th qudit. The Hamiltonian (4) will be also called the transfer Hamiltonian.

The state transfers, studied in [8–10] use Hamiltonian H which have the following property

$$\left[H, \sum_{i=1}^{N} Z_{(i)}\right] = 0,$$ (5)

where Z represents the sign gate for qubits. This means that spins are preserved and dynamics generated by H is divided into series of subspaces denoted by the number of qubit in state $|1\rangle$ – see in [11]. In the case discussed here, it is not hard to show that

$$\left[H^{XY_d}, \sum_{i=1}^{N} \eta_{(i)}^{r,r}\right] = 0,$$ (6)

for $1 \leq r \leq (d-1)$, so the Eq. (6) generalizes the situation mentioned in the Eq. (5) – preserving spins and separating dynamics into subspaces.

It is necessary to add that an appropriate unitary operator for transfer operation is determined by the equation

$$U(a) = e^{-aH^{XY_d}},$$ (7)

where a represents evolution time and i represents imaginary unity. The symbol U represents the unitary operation which performs the transfer protocol.

Introducing a definition of the transfer protocol based on a Hamiltonian and an unitary operator allows to describe a transfer as an algorithm (or structural quantum program [12,13]). A very important issue is the use of γ condition as an invariant for the protocol. The invariant γ is based on a function calculating the level of entanglement.

Let $E(|\Psi(t)\rangle)$ be the function to determine a level of entanglement for a chain under perfect state transfer. The CCNR criterion, negativity and concurrence seem to be good candidates to play this role.

```
t ; time variable
N > 0 ; number of step
H_XY^d ; Hamiltonian for path L with l vertices)
U(t) = e^{-i(t/N)H_XY^d} ; unitary operator
|ψ(t_0)⟩ ; initial state of chain
i := 0;
{γ : E(|ψ(t_i)⟩) ≅ E(|Ψ(t)⟩) ∧ i < N}
while i < N do
begin
      |ψ(t_{i+1})⟩ := U(t)|ψ(t_i)⟩
      i := i + 1
      {γ : E(|ψ(t_i)⟩) ≅ E(|Ψ(t)⟩)} ∧ i < N
end ;
{γ : E(|ψ(t_i)⟩) ≅ E(|Ψ(t)⟩) ∧ i = N}
```

Remark 1. Calculating the level of entanglement in multiple qubit, qutrit and, particularly, qudit systems is a problem which is still unsolved for mixed quantum states (however it is possible to give lower bound for concurrence for mixed states [14]). Due to this fact, in the notation of invariant γ for perfect state protocol (algorithm) – $\gamma : E(|\psi(t_i)\rangle) \cong E(|\Psi(t)\rangle)$ – sign $\cong$ expresses that the level of entanglement is comparable with used measure (e.g. CCNR criterion or concurrence).

In this paper the transfer protocol is used to realize the transmission of information through the path which length equals l. The process of transfer involves an unknown qudit state (where d is a freedom level):

$$|\psi\rangle = \alpha_0|0\rangle + \alpha_1|1\rangle + \ldots + \alpha_{d-1}|d-1\rangle \quad \text{where} \quad \alpha_i \in \mathbb{C} \text{ and } i < d. \quad (8)$$

Figure 1 shows the scheme of transfer protocol's realization. The whole process is divided into discrete steps which are realized by operator U.

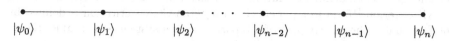

$$|\psi_0\rangle \qquad |\psi_1\rangle \qquad |\psi_2\rangle \qquad |\psi_{n-2}\rangle \qquad |\psi_{n-1}\rangle \qquad |\psi_n\rangle$$

Fig. 1. The picture of quantum information transfer's realization in a spin chain for a single unknown qudit state $|\psi\rangle$ in the path of length l. The state $|\psi_0\rangle$ represents the initial state and state $|\psi_n\rangle$ denotes the final state obtained after execution of transfer protocol.

3 The CCNR Criterion and Concurrence

In this part a basic information about CCNR criterion and concurrence will be recalled. The mentioned methods may be used as measures to estimate a level of entanglement in a bipartite system. The CCNR criterion uses two auxiliary operations called vectorization and realignment.

The vectorization of a given matrix $A = [a_{ij}]_{m \times n}$ is represented by the following operation: vec : $A_{m \times n} \to \mathbb{C}^{mn}$:

$$\text{vec}(A) = [a_{11}, \ldots, a_{m1}, a_{12}, \ldots, a_{m2}, \ldots, a_{1n}, \ldots, a_{mn}]. \quad (9)$$

The realignment is based on the vectorization of matrix and applies to the scalar product of two matrices:

$$R(A \otimes B) = \text{vec}(A)^T \cdot \text{vec}(B). \quad (10)$$

The realignment of matrix's elements – the initial matrix dimensions are 2×2 – in a space $H = H_A \otimes H_B$ may be illustrated by the example:

$$\rho = \begin{pmatrix} a_{11} & a_{12} & a_{13} & a_{14} \\ a_{21} & a_{22} & a_{23} & a_{24} \\ a_{31} & a_{32} & a_{33} & a_{34} \\ a_{41} & a_{42} & a_{43} & a_{44} \end{pmatrix}, \quad R(\rho) = \begin{pmatrix} a_{11} & a_{21} & a_{12} & a_{22} \\ a_{31} & a_{41} & a_{32} & a_{42} \\ a_{13} & a_{23} & a_{14} & a_{24} \\ a_{33} & a_{43} & a_{34} & a_{44} \end{pmatrix}. \quad (11)$$

The operation of realignment is realized similarly for larger matrices, e.g. for matrices 3×3:

$$
R(\rho) = \begin{pmatrix}
a_{11} \, a_{21} \, a_{31} & a_{12} \, a_{22} \, a_{32} & a_{13} \, a_{23} \, a_{33} \\
a_{41} \, a_{51} \, a_{61} & a_{42} \, a_{52} \, a_{62} & a_{43} \, a_{53} \, a_{63} \\
a_{71} \, a_{81} \, a_{91} & a_{72} \, a_{82} \, a_{92} & a_{73} \, a_{83} \, a_{93} \\
a_{14} \, a_{24} \, a_{34} & a_{15} \, a_{25} \, a_{35} & a_{16} \, a_{26} \, a_{36} \\
a_{44} \, a_{54} \, a_{64} & a_{45} \, a_{55} \, a_{65} & a_{46} \, a_{56} \, a_{66} \\
a_{74} \, a_{84} \, a_{94} & a_{75} \, a_{85} \, a_{95} & a_{76} \, a_{86} \, a_{96} \\
a_{17} \, a_{27} \, a_{37} & a_{18} \, a_{28} \, a_{38} & a_{19} \, a_{29} \, a_{39} \\
a_{47} \, a_{57} \, a_{67} & a_{48} \, a_{58} \, a_{68} & a_{49} \, a_{59} \, a_{69} \\
a_{77} \, a_{87} \, a_{97} & a_{78} \, a_{88} \, a_{98} & a_{79} \, a_{89} \, a_{99}
\end{pmatrix} . \tag{12}
$$

The realignment allows to specify the criterion which is able to detect and track the entanglement's behaviour during the transfer process. This method is called CCNR criterion [15] in this paper. Generally, the CCNR criterion is defined by the fact that if the matrix ρ_{AB} of a bipartite $m \times n$ system is separable, then:

$$
||\rho_{AB}^R|| \leq 1, \tag{13}
$$

in the case $||\rho_{AB}^R|| > 1$ state ρ_{AB} is entangled. Of course ρ^R stands for the completion of realignment operation on the state ρ.

Remark 2. Generally, the CCNR criterion is not sufficient to detect all the cases of entanglement in any quantum system. However, it is sufficient to correctly detect the entanglement in qudit spin chains.

It should be also pointed out that the value of CCNR criterion may be calculated using Singular Value Decomposition (SVD):

$$
||\rho_{AB}^R|| = \sum_{i=1}^{q} \sigma_i \left(\rho_{AB}^R \right) \tag{14}
$$

where σ_i represents a singular value of ρ_{AB}^R and $q = \min(m^2, n^2)$.

The correctness of CCNR criterion may be amplified by using the states of subsystems, what was shown in [16]:

$$
||(\rho_{AB} - \rho_A \otimes \rho_B)^R|| \leq \sqrt{(1 - \mathrm{Tr}\,(\rho_A^2))(1 - \mathrm{Tr}\,(\rho_B^2))}. \tag{15}
$$

If the above condition is fulfilled it means that the state ρ_{AB} is separable (otherwise the state is entangled).

The concurrence measure is used to calculate a level of entanglement, as well. Concurrence [14] for pure state $|\psi_{AB}\rangle$ in bipartite system $d \otimes d$ is defined as

$$
\mathcal{C}(|\psi_{AB}\rangle) = \sqrt{2\left(1 - \mathrm{Tr}\,(\rho_A^2)\right)}. \tag{16}
$$

4   Tracing of Entanglement

In this section the results of calculating a level of entanglement with use of CCNR criterion and concurrence are presented. Figure 2 shows the values of CCNR criteria calculated according to the Eqs. (14) and (15). The process of transfer was realized for the state $|+\rangle$, which for qubits is $|+\rangle_2$ state and for qutrits is expressed as $|+\rangle_3$ state:

$$|+\rangle_2 = \frac{1}{\sqrt{2}}(|0\rangle + |1\rangle) \quad \text{or} \quad |+\rangle_3 = \frac{1}{\sqrt{3}}(|0\rangle + |1\rangle + |2\rangle). \tag{17}$$

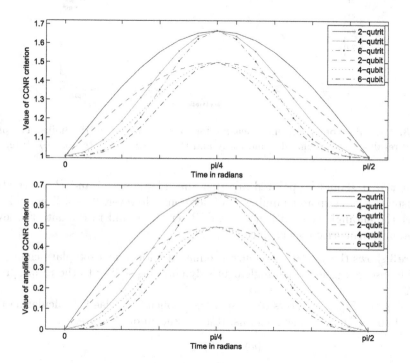

Fig. 2. The entanglement detection with use of CCNR criterion and amplified CCNR criterion for short qubit and qutrit chains. The presented values were calculated numerically and the transferred state was $|+\rangle$ state.

The value of CCNR criterion was calculated according to the Eq. (14). In case of amplified CCNR criterion, the Eq. (15) was transformed to:

$$\left\| (\rho_{AB} - \rho_A \otimes \rho_B)^R \right\| - \sqrt{(1 - \mathrm{Tr}\,(\rho_A^2))(1 - \mathrm{Tr}\,(\rho_B^2))} > 0. \tag{18}$$

If the above inequality is true, then the analysed state is entangled.

Naturally, during the transfer process, the level of entanglement initially increases and then decreases. All analysed short spin chains were divided into two subsystems: A and B.

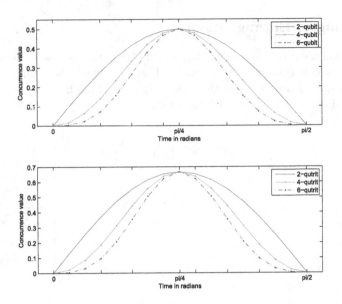

Fig. 3. The values of concurrence measure for short qubit and qutrit chains. The presented results were calculated numerically and the transferred state was $|+\rangle$ state.

Remark 3. It should be pointed out that the subsystems A and B are treated as systems with a greater number of dimensions. However, it is still possible to detect the entanglement with use of CCNR criterion and to evaluate the level of entanglement with concurrence.

Figure 3 shows the values of concurrence measure for some exemplary chains. It can be observed that the entanglement's dynamics is similar to the results from Fig. 2.

Using the Eq. (16) allows to present the explicit formulae for calculating the level of entanglement in spin chains. If an unknown qubit state:

$$|\psi\rangle = \alpha|0\rangle + \beta|1\rangle, \tag{19}$$

is transferred through the short path ($l = 2$), then the level of entanglement in this 2-qubit chain can be expressed as:

$$\mathcal{C}_{d=2}^{l=2}(a) = \frac{1}{4}\left(4\alpha^4 + 3\beta^4 + 8\alpha^2\beta^2\cos(2a) + \beta^4\cos(4a)\right) \tag{20}$$

where a represents the duration of the transfer process.

For a 2-qutrit chain the transfer of an unknown state:

$$|\psi\rangle = \alpha|0\rangle + \beta|1\rangle + \gamma|2\rangle, \tag{21}$$

can be precisely described by the concurrence:

$$\mathcal{C}_{d=3}^{l=2}(a) = \frac{1}{4}\left(4\alpha^4 + 3(\beta^2+\gamma^2)^2 + 8\alpha^2\left(\beta^2+\gamma^2\right)\cos(2a) + \left(\beta^2+\gamma^2\right)^2\cos(4a)\right) \tag{22}$$

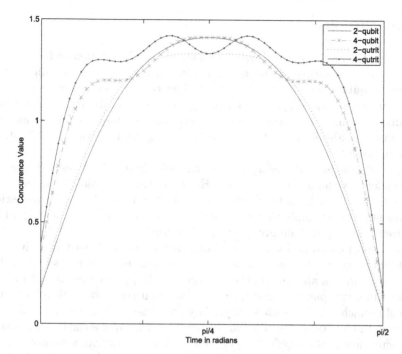

Fig. 4. The values of concurrence measure, for short spin chains, obtained with use of the formulae (20) and (22)

The dynamics of changes for concurrence values depends on the values $\cos(2a)$ and $\cos(4a)$ – both in qubit and qutrit chains. The constants preceding $\cos(2a)$ and $\cos(4a)$ correspond to the probability amplitudes. A sum of probability amplitudes' values equals one what is easy to proof when for some a values: $\cos(2a) = \cos(4a) = 1$. Furthermore the values of amplitudes:

$$(\alpha^2 + \beta^2)^2 = 1, \quad (\alpha^2 + \beta^2 + \gamma^2)^2 = 1 \tag{23}$$

both for qubit and qutrit chains of any length, because of a trace calculation in (16). An exemplary value of concurrence measure for 4-qutrit chain is:

$$C_{d=3}^{l=4}(a) = \frac{1}{64}(c_0 + (c_1 + c_2)\cos 2a + (c_3 + c_4 + c_5)\cos 4a$$
$$+ (c_6 + c_7)\cos 6a + (c_8 + c_9 + c_{10})\cos 8a + (c_{11} + c_{12} + c_{13})\cos 12a) \tag{24}$$

where $\sum_i c_i = 1$ and the c_i coefficients are composed of the probability amplitudes' values of transferred state.

Figure 4 presents the values of concurrence measure for explicit formulae. The changes in the level of entanglement, illustrated as local minima, were not detectable with numerical calculations (see Fig. 2). These minima correspond to the different levels of entanglement between separate chain's parts, e.g. between adjacent qubits/qutrits.

5 Conclusions

Calculating the level of entanglement with use of CCNR criterion and, especially, with concurrence proved to be a good solution for tracking the information transfer in qubit and qutrit spin chains. The concurrence measure indicates unambiguously the details of transfer process and due to this fact, it can be used as an invariant in the algorithmic description of transfer protocol for an unknown qubit state and also for an unknown qutrit state (generalization for qudit states is also available).

Naturally, the presented analysis concerns only short spin chains and it should be extended for chains of any length. However, this extension is in fact only estimating the upper bound for concurrence value in a spin chain. The numerical experiments seem to imply that the length $l = 2$ points out the upper bound for concurrence in a spin chain given by the Hamiltonian (4).

It is important to add that the concurrence measure – apart from its analytical functions, e.g. to track the level of entanglement, as it was shown in the paper – is at present also used in the experiments [17], where the entanglement is produced in a two-photon system. In [18] the concurrence was utilized for spin-1/2 chain which can be realized physically with use of Benzene molecules [8]. Therefore, the functions described in the paper may be applied as a tool to verify the concurrence in the physical experiments where e.g. qutrits were used.

The other problem is calculating the analytical form of the spin chain's state. This state should be expressed as a density matrix, taking into account the additional environmental influence [19]. The measure described in [14] is an interesting candidate to be a very useful tool for tracking entanglement in spin chains with XY-like dynamics in general.

Acknowledgments. We would like to thank for useful discussions with the *Q-INFO* group at the Institute of Control and Computation Engineering (ISSI) of the University of Zielona Góra, Poland. We would like also to thank to anonymous referees for useful comments on the preliminary version of this paper. The numerical results were done using the hardware and software available at the "GPU μ-Lab" located at the Institute of Control and Computation Engineering of the University of Zielona Góra, Poland.

References

1. Horodecki, R., Horodecki, P., Horodecki, M., Horodecki, K.: Quantum entanglement. Rev. Mod. Phys. **81**, 865–942 (2009)
2. Hirvensalo, M.: Quantum Computing. Springer, Berlin (2001)
3. Nielsen, M.A., Chuang, I.L.: Quantum Computation and Quantum Information. Cambridge University Press, New York (2000)
4. Klamka, J., Węgrzyn, S., Znamirowski, L., Winiarczyk, R., Nowak, S.: Nano and quantum systems of informatics. Bull. Pol. Acad. Sci.: Tech. Sci. **52**(1), 1–10 (2004)
5. Jafarizadeh, M.A., Sufiani, R., Taghavi, S.F., Barati, E.: Optimal transfer of a d-level quantum state over pseudo-distance-regular networks. J. Phys. A Math. Theor. **41**, 475302 (2008)

6. Paz-Silva, G.A., Rebić, S., Twamley, J., Duty, T.: Perfect mirror transport protocol with higher dimensional quantum chains. Phys. Rev. Lett. **102**, 020503 (2009)
7. Bayat, A.: Arbitrary perfect state transfer in d-level spin chains. Phys. Rev. A **89**(6), 062302 (2014)
8. Bose, S.: Quantum communication through an unmodulated spin chain. Phys. Rev. Lett. **91**, 207901 (2003)
9. Bose, S.: Quantum communication through spin chain dynamics: an introductory overview. Contemp. Phys. **48**, 13–30 (2007)
10. Vinet, L., Zhedanov, A.: How to construct spin chains with perfect state transfer. Phys. Rev. A **85**, 012323 (2012)
11. Kay, A.: A review of perfect state transfer and its application as a constructive tool. Int. J. Quantum Inf. **8**(4), 641–676 (2010)
12. Sawerwain, M., Gielerak, R.: Natural quantum operational semantics with predicates. Int. J. Appl. Math. Comput. Sci. **18**(3), 341–359 (2008)
13. Klamka, J., Gawron, P., Miszczak, J., Winiarczyk, R.: Structural programming in quantum octave. Bull. Pol. Acad. Sci. Tech. Sci. **58**(1), 77–88 (2010)
14. Chen, K., Albeverio, S., Fei, S.M.: Concurrence of arbitrary dimensional bipartite quantum states. Phys. Rev. Lett. **95**, 040504 (2005)
15. Rudolph, O.: Some properties of the computable cross-norm criterion for separability. Phys. Rev. A **67**, 032312 (2003)
16. Zhang, C.J., Zhang, Y.S., Zhang, S., Guo, G.C.: Entanglement detection beyond the computable cross-norm or realignment criterion. Phys. Rev. A **77**, 060301(R) (2008)
17. Almutairi, K., Tanaś, R., Ficek, Z.: Generating two-photon entangled states in a driven two-atom system. Phys. Rev. A **84**, 013831 (2011)
18. Wang, Z.H., Wang, B.S., Su, Z.B.: Entanglement evolution of a spin-chain bath coupled to a quantum spin. Phys. Rev. B **79**, 104428 (2009)
19. Gawron, P., Klamka, J., Winiarczyk, R.: Noise effects in the quantum search algorithm from the viewpoint of computational complexity. Int. J. Appl. Math. Comput. Sci. **22**(2), 493–499 (2012)

A Novel Multicast Architecture
of Programmable Networks

Jakub Kiciński and Michał Hoeft[✉]

Faculty of Electronics, Telecommunications and Informatics, Gdańsk University
of Technology, 11/12 Gabriela Narutowicza Street, 80-233 Gdańsk, Poland
kubakici@wp.pl, michal.hoeft@eti.pg.gda.pl

Abstract. In the paper a multicast architecture for programmable networks based on separation of group management and network control
tasks is proposed. Thanks to this separation, services which want to
make use of multicast communications no longer have to implement low-
level network functionalities and their operation is greatly simplified.
Abstracting service's view of the network into a fully connected cloud
enables us to transparently implement within the network complex features such as host mobility without any disturbances to the operation
of the service or it's clients. Aspects of the multicast architecture such
as signaling, addressing, form of data delivery, security and cooperation
with legacy applications are discussed; precise requirements for interfaces
specified.

Keywords: SDN · Multicast · Network architecture

1 Introduction

Multicast is a group communication model (1-to-N or N-to-N) which allows for
simpler and more efficient use of network resources. It is appropriate for many
multimedia applications (e.g. teleconferencing, IPTV, games) and applications
which have to distribute identical copies of data to multiple hosts (common in
data-center scenarios). Increasingly multicast is also employed to build overlay
networks [1]. Unfortunately, delivering messages to a group of network end-nodes
in an optimal manner, either with minimal bandwidth usage or optimizing any
other metric, is an inherently difficult task. Computationally it's equivalent to a
NP-hard problem of finding minimal Steiner tree [2] and requires coordination
of state in a large part of the network to operate properly. For these reasons,
multicast delivery remains a mostly unsolved problem in existing networks. Lack
of established, single, well-defined protocol implemented in all modern networks
discourage software developers from adopting this communication model. Multicast remains today a territory of network engineers.

Devices in traditional IP networks operate and make decisions autonomously
in accordance with their own configuration and control capabilities. To enable
non-trivial features various protocols are defined. They allow for cooperation

© Springer International Publishing Switzerland 2015
P. Gaj et al. (Eds.): CN 2015, CCIS 522, pp. 280–292, 2015.
DOI: 10.1007/978-3-319-19419-6_27

between separate devices and facilitate exchange of information. However, the set and scope of protocols implemented is arbitrarily decided by manufacturers hindering interoperability. The full-independence of devices in traditional networks leads to a formation of a network control-plane which is physically and logically distributed. Because control information is exchanged using the same channels as data, a trade-off between information propagation speed and possibility of network overload has to be made. Network overload is particularly likely when state is disseminated using broadcast communication. On the other hand, slow convergence is unacceptable in mobile scenarios in which user expects no service disturbance during handoff. Furthermore, in distributed environment, multicast traffic distribution tree has to be constructed by autonomous units which prohibits the use of many state-of-the-art algorithms [3]. It also significantly increases the complexity of network equipment.

Software Defined Networking (SDN) is a disruptive approach to networking which assumes decoupling of control- and data-planes. In SDN all control is handled by a single centralized unit appropriately named SDN Controller (SDN-C). Network devices, named SDN Forwarding Engines (SDN-FE), are configured to provide all information about their state and occurring events to the controller, thus giving it a full and up-to-date view of the network. Data-plane processing depends on the controller to specify explicit handling rules.

Figure 1 presents the overall architecture of Software Defined Networks. Name Software Defined Networking comes from the fact that controllers are software constructs usually written in high-level programming languages. Devices with external API designed for use with a SDN controller form a Programmable Network. Interface naming model is controller-centric the interface between SDN-C and SDN-FE is called Southbound Interface (SBI). OpenFlow is the most popular SBI protocol implemented in many software and hardware switches. Communication between SDN-C and SDN-FE happens on a Management Network (MN) which is logically or physically separated from the data

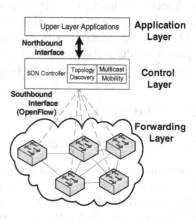

Fig. 1. SDN network architecture

network (dashed lines in Fig. 1). External Upper Layer Applications (ULA) can control the network through interaction with the controller over the Northbound Interface (NBI).

From the short description above it can be deduced that SDN greatly simplifies implementation of multicast communication. Implementation of multicast in traditional networks is hampered by the distributed control plane. In SDN, thanks to the centralization, SDN-C possesses full knowledge about network state hence it's able to make better decisions consequently achieving both better resilience [4] and higher efficiency [5–8].

Regardless of the advantages of SDN, it must be said that implementing a multicast application with a SDN network is currently harder than in the legacy networks. This situation is caused by the fact that SDN controllers often do not provide multicast routing by default and application developers must themselves take care of tasks such as distribution tree computation and management using low-level controller API. The main contribution of this paper is an architecture which makes it easy for application developers to use multicast communication in SDN networks without requiring any network-specific knowledge. This is achieved by taking care of the low-level aspects of group communication in the controller and exposing a high-latitude Northbound Interface for applications.

Abstracting network operation and clearly separating application logic from network tasks make it possible, for the network, to evolve freely without any need to adapt existing applications. Flexibility of SDN-approach is particularly useful in mixed wired/wireless scenarios. It has been reported that common-sense rules of constructing multicast distribution trees do not always hold true in this environment [9]. Wireless scenario is of particular interest to us as multimedia streams increasingly have to be delivered to users with mobile terminals. Enabling user mobility, in this paper, is considered to be an important requirement for described architecture.

Section 2 provides reader with the related work. Description of the current multicast architecture of IP networks is presented in Sect. 2.1. Propositions of multicast management in SDN networks are discussed in Sect. 2.2. Section 3 introduces and describes proposed architecture. Section 4 presents a proof-of-concept implementation and the results of performed evaluations. The paper is finished with conclusions in Sect. 5.

2   Related Works

2.1   IP-Based Multicast

There are five elements which comprise any multicast communication scheme. Fundamentally, there must be *(i)* a source of traffic, *(ii)* a group of clients to which this traffic has to be delivered in an efficient way and *(iii)* a set of network devices responsible for the delivery. Remaining two elements are communication protocols *(iv)* for routing and *(v)* for client signaling. Modern IP networks use predominantly Protocol Independent Multicast (PIM) for communication between routers. Unlike earlier solutions, it does not depend on any particular protocol being used to establish unicast routes. PIM can operate in Dense Mode (PIM-DM) [10] or Sparse Mode (PIM-SM) [11].

Choice of multicast routing protocol is transparent to the listeners. Specification of PIM also does not obligate any particular client signaling method yet it is generally assumed that IP clients use Internet Group Management Protocol (IGMP) [12] or MLD [13][1] to communicate their subscriptions to the first-hop

---

[1] Because IGMP [12] and MLD [13] are mostly equivalent, in the context of this work whenever referring to IGMP the reference to MLD is implied.

routers of their subnet. IGMP allows clients to join or leave a group and routers to query group's membership on an attached subnet.

In IP networks, multicast traffic always has a Class D destination address and source address of the original source. In case of PIM-SM the original datagram may be encapsulated in a unicast tunnel on the path between source's Designated Router and Rendezvous Point. Dedicated Multicast Forwarding Information Base (MFIB) at the routers contains information about forwarding rules for different (source, group) pairs. Because the group address is visible to the clients, they must configure their IP stack appropriately to be able to receive data. Layer 2 forwarding devices such as Ethernet switches use IP-signaling snooping techniques (e.g. IGMP snooping, PIM snooping) to derive their forwarding rules and achieve efficient operation in multicast scenarios [14].

IP multicast architecture leaves client-source communication fully in the range of responsibilities of the end-node applications. Sources have no capability to query or control the group membership. Before IGMPv3 introduced the per-source subscriptions, there had been also no way for the clients to restrict which sources they trust and groups could have been flooded with useless data.

Security has always been one of the weakest points of classical IP multicast architecture. Neither client signaling nor routing protocols provide means of authentication and it is recommended by the standards that security-conscious users employ generic IPSec methods [15]. Sources cannot restrict which hosts will be able to receive their data. If client authorization is necessary it is usually enforced through encryption of the data stream and providing decryption keys only to the legitimate users. Apart from high computational cost encryption may cause additional delays unwelcome in real-time applications. It also cannot prevent DoS attacks e.g. congesting the distribution network or spoofing join requests to flood end-nodes.

Administratively Scoped IP Multicast [16] defines ranges of addresses for which forwarding is confined to a specific topological region (link/site/organization). This mechanism helps to prevent multicast traffic from leaking outside of intended segment of the network and partially alleviate the need to filter this traffic at the edges. It also provides more logical control of data delivery than scoping based on limited TTL value. However, region-based policies are very coarse-grained and require trustworthy relationship between the connected clients and the infrastructure elements.

2.2 Multicast in SDN Networks

Many publications show advantages and new capabilities of programmable networks when compared to the ossified IP architecture. Zou et al. [6] used SDN to accomplish simple user authentication. Source IP and MAC addresses of the IGMP Join message were used to validate whether user is allowed to become a member of the requested group.

Kotani et al. [4] showed a multicast OpenFlow controller with fast switching between distribution trees to enable quick failover for link and switch failures.

Iyer et al. [7] presented a centralized algorithm which can compute optimal distribution trees for common data-center network topologies in polynomial time. Using this algorithm would be impossible with distributed control plane. Implementation presented in the paper also ensured that traffic is distributed evenly throughout the infrastructure and enforced a simple access policy.

In above-mentioned cases, researchers used designs where entire multicast functionality was implemented inside the SDN controller itself.

Marcondes et al. proposed a CastFlow system [5] to deliver multicast in OpenFlow-enabled networks which require fast group join/leave operations. The implementation featured a NOX controller enhanced with multicast routing capabilities and a separate network management entity which configured virtual testbed and the multicast group. In a sense, this ad hoc architecture is the closest to our proposal because of separation between group management and routing tasks.

Zou et al. [6] created a system capable of delivering multicast traffic on a SDN network with multiple controllers. In their design, controllers share full information about network topology, all multicast groups, clients, and forwarding rules. Controller receiving a client signal conducts tree recomputation for the whole network, configures locally attached switches and broadcasts rule updates to other controllers.

Tajik and Rostami [8] evaluated alternative delivery modes to achieve better performance in a wireless scenario. They used OpenFlow to dynamically decide when to use multicast delivery and when to duplicate frames and send them over the radio as unicast.

3   Proposed Multicast Architecture for SDN

This paper presents a new SDN architecture which redesigns multicast communication taking into account all capabilities and complexities of modern networks including end-node mobility. An overview picture of the proposed architecture is presented in Fig. 2. It shows a simple 3-layer model of a SDN network together with end-nodes. The lowest layer corresponds to the end-nodes participating in the communication and the highest contains the control logic. End-nodes represent clients connected to the network – they can act as both listeners and sources of the multicast traffic. End nodes use non-fixed network connections which are managed by the Mobility module of the SDN Controller. In the presented approach, mobility is a service of the SDN-C and manages connections and abstracts the way clients connect to the network from applications which are not interested in such details. Mobility may be required both for physical hosts e.g. in a wireless environment to provide seamless handoffs and in a virtualized environment for a Virtual Machine (VM) migration. Network applications are implemented at application layer. In the proposed approach, a high-altitude multicast control application called Upper Layer Multicast Application (ULMA) is included. This application is responsible for group management and control tasks. It is separated from the applications used by multicast sources and clients

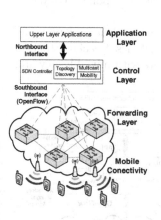

Fig. 2. Topology of the proposed design

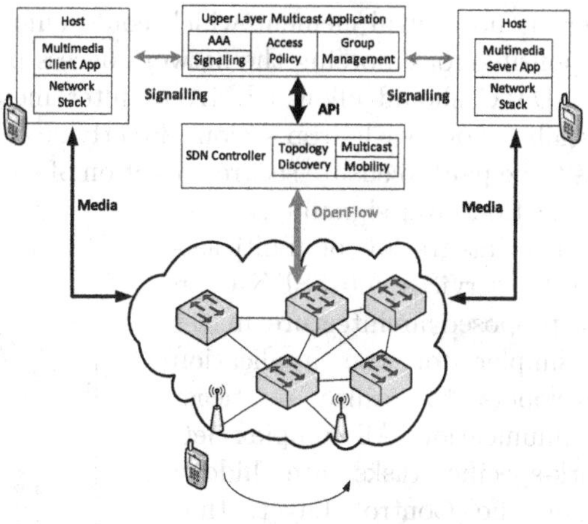

Fig. 3. Multicast architecture

(see Fig. 3). Moreover ULMA does not need to be aware of mobility or any other aspect of network topology.

SDN Controller in our architecture is extended for multicast operation. It handles all of the basic tasks such as topology discovery and flow management. Moreover, it contains multicast specific services like: legacy signaling snooping, source discovery and group communication awareness. It is responsible for providing efficient and reliable routing for unicast and multicast data. Advanced functionalities such as fast reroute and redundant trees to protect from link failures can be offered [4]. SDN-C may also accept QoS parameter specification for both unicast and multicast traffic. In case of multicast flows, parameters may include factors which would influence selection of a distribution tree construction algorithm like required bandwidth (trees should be highly optimized) or tree calculation time (faster algorithms). If network supports client mobility Controller is responsible for handling it, without interferences with applications traffic and without assist from ULMA.

Upper Layer Multicast Application (ULMA) contains high-level application logic and is responsible for the control of clients and sources of multicast traffic. From the point of view of the SDN-C it is just a network application which makes use of the multicast API. ULMA handles *join* and *leave* requests from the group members and ensure that only legitimate sources can transmit data. ULMA passes this information to SDN-C which computes distribution trees and install the necessary flows at the switches. ULMA does not have any knowledge about the network topology or location of the listeners. All this information remains at SDN-C, and ULMA only has to identify the involved hosts in an unambiguous way e.g. by providing their host name or IP address.

The enhanced security in our model comes mostly from the fact that ULMA has full control over the group membership, while in traditional IP multicast there is no central element which could enforce membership policy. ULMA is responsible for enforcing the network access policies.

SDN-C should allow ULMA to filter messages from malicious sources by installing flows with drop actions directly on the edge switches. Special attention has to be paid to ensure secure operation of environments with mixed legacy and application layer signaling.

The integration of multicast-specific services into SDN-C in the proposed architecture makes it simpler for the application developers to utilize multicast communication. All complex network-specific tasks are hidden inside the Control Layer, thus there is no need to re-implement them for each application. The clearly defined API could facilitate ULMAs to work with SDN-C from different vendors. Clear responsibility separation also eases the management and makes it simpler to upgrade and innovate inside one part of the stack without disturbing the existing elements. Figure 4 demonstrates how network events are handled. Case (a) shows a new client requesting access to the group as a listener while case (b) presents handling of mobile client handoff. Note that in (b) ULMA is not involved in the process.

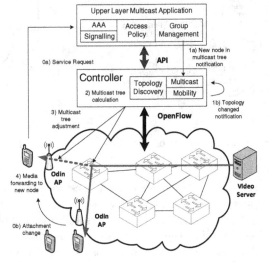

Fig. 4. SDN network architecture

3.1 Signaling

To implement client-ULMA signaling existing application layer protocols such as HTTP, SIP, XMPP or even an entirely custom solution designed from scratch can be used.

Choice of the protocols is unrestricted because all signaling is replaced by end-to-end communication between the client and ULMA.

It should be highlighted, that unlike other researchers [4,6] the proposed solution does not force client-ULMA communication to use the Management Network (MN). In the introduction, it was stated that mixing control and data messages on a single logical link is undesirable and cited it as a disadvantage of IP multicast. However, in our solution from the perspective of forwarding devices the client-ULMA messages are equal to application traffic because they require no additional processing.

Burdening MN with client signaling makes an unjustified use of this precious resource. The fast and reliable management communication is crucial to high

Quality of Services of the entire network. Depending on the implementation Management Network can have limited bandwidth (network or CPU-to-forwarding engine bandwidth in switches) and require extra processing (encapsulation, encryption in case of TLS). Quality of Service for signaling, necessary to achieve appropriate Grade of Service (connection setup time etc.), can be configured separately in the same way it is provided to sensitive and critical applications. Explicit configuration is more robust and precise than relying on characteristics of MN. Client signaling should also account for the possibility of an occasional frame loss, hence there is no need for privileged treatment.

3.2   Delivery

Traffic delivery configuration lies entirely in the authority of the SDN-C. Programmable networks offer choice of the form in which datagrams are delivered to the end stations, specifically the destination address. OpenFlow has the ability to modify IP addresses of the datagram since early version of the protocol. Using this feature edge switches can be instructed to rewrite the destination address with clients' unicast addresses making the multicast functionality completely *transparent* to clients. The authors called this mode *Transparent Multicast Delivery* (TMD) as opposed to commonly used mode, in this parer named *Legacy Multicast Delivery* (LMD). In LMD addresses are not modified by network elements thus clients have to configure their IP-stacks to receive the group traffic. TMD removes the dependency on IP addressing completely. As a result clients do not need to be aware of the addresses used by the communication group and can identify content based on a completely independent scheme. This also removes the theoretical limit on the number of groups in the network. Thanks to TMD, client applications need less privileges and can be executed in more restricted environment e.g. in the web browser.

In an example of multicast streaming application for mobile nodes (Fig. 4), RTP-encapsulated video frames are distributed throughout the network with a multicast IP address. However, in proposed approach, the multicast address can be replaced with the client's unicast address at the last-hop switch. This was beneficial in two ways. Firstly, our main use case presumes that most clients are mobile and use Wi-Fi connections. Multicast traffic in wireless networks is usually transmitted at a low data rate and can be delayed until Delivery Traffic Indication Message (DTIM) if some of the clients are in power saving mode. There is also no acknowledge mechanism for multicast frames in IEEE 802.11. Rewriting destination address to the unicast address allows us to avoid all those problems [8]. Secondly, the client application does not have to perform any multicast-specific configuration of the IP stack of its host and required no privileges hence it could also be implemented as a web page or in similarly restricted environment.

3.3   Controller Interfaces

Although several projects have tackled multicast in a SDN environment, they usually implement all multicast management functionality on the side of a topology-aware upper-layer application itself and not inside the SDN controller. In this paper, authors would like to argue that it would be beneficial if group communication was implemented inside the SDN-C and accessed by applications via a high-latitude Northbound Interface.

Adding a specialized multicast module to the proposed design as an element residing between ULMA and SDN-C is also a viable solution. However, it might be necessary for the multicast module to interact with low-level details and elements of the network controller such as configuration of the client's wireless connection. Because of the tight coupling between SDN-C and multicast module, multicast module is better suited to become a high-level part of SDN controller rather than a completely independent unit.

Proposed Northbound Interface must make it possible to perform following tasks: (i) group creation and configuration, (ii) membership management, (iii) source management, (iv) legacy signaling, (v) information query. The API can be implemented in any suitable form e.g. in a REST architecture style. Creation and basic configuration of the group has to be performed to inform SDN-C about desired delivery mode (LDM vs. TDM) and to configure traffic and signaling QoS. ULMA may also define access policies appropriate for the applications. Membership management consists a set of actions to add and remove listeners from the group. Note that SDN-C may refuse the add action on the basis of admission control. The API should allow ULMA to configure the source signaling policy and manage the list of authorized sources. If source signaling is implicit, SDN-C has to inform ULMA whenever new source appears. Even with explicit signaling (sources announce their presence to ULMA before transmission) ULMA may still be interested in notifications when an unauthorized source tries to transmit to the group. List of authorized sources is similar to group membership and also requires add and remove operations. It is also anticipated that ULMA may need to query group information and statistics from SDN-C for AAA services.

SDN cannot fully replace legacy IP-based multicast unless it offers some form of inter-network interfaces. Naturally if networks are connected over Layer 3 routers, legacy routing protocols can be exploited with little or no effort. However, if two SDN networks are interfacing with each other controllers could communicate directly to exchange multicast information and build more optimal distribution trees.

4   Evaluation of Proposed Solution

4.1   Implementation

Figure 5 presents the testbed implementation of proposed system. Our use case refers to real-life streaming scenarios e.g. IPTV broadcasting. The main software part, the SDN Controller was based on the Floodlight framework. As a controller module providing mobility management a Odin [17] framework was used. It served as a Wi-Fi mobility layer but had to be extended to trigger automatic multicast tree reconstruction after listener handoff. Evaluated Upper-Layer Multicast Application was a GStreamer-based [18] video streamer. Desktop client application, streamer and Upper-Layer Multicast Application were written in C++ completely independently from the Floodlight SDN-C framework which was written in Java. As it is stated in Sect. 3.2, the proposed architecture mechanisms do not depend on specific network layer protocols, thus for our evaluation IPv4 was used.

The client application requested access to the video stream from the controllers which then reconfigured the SDN network so that new listener was attached to the multicast distribution tree. All active listeners had to periodically renew their subscription.

Sources of the traffic were limited to two predefined servers on our network, no other hosts were allowed to transmit to the group. Only one of the sources could be streaming at any given time. The second stand-by source was kept for reliability. Choice of the active source was left in the hands of the ULMA – it could detect source failure or request a manual switch e.g. for the purpose of scheduled maintenance of servers.

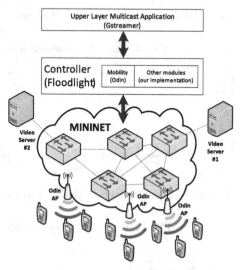

Fig. 5. Testbed implementation

Video sources had their clocks synchronized using Precision Time Protocol (PTP) [19] which allowed source switches to be performed seamlessly. The switch action was scheduled by the ULMA according to both wall-clock and media-clock which enabled new source to begin transmission from the point in the video at which the previous source has finished thus making the transition invisible to the end-user.

A simple Minimal Spanning Tree-based heuristic is used to find the distribution tree. Finding the optimal tree is a NP-hard problem known as the Steiner Tree problem [2] and solving it efficiently was outside of the scope of our work. However, any algorithm can be easily implemented and utilized in our solution.

The client application communicates directly with the ULMA which then passed the membership information to the SDN-C. Apart from group creation and tear-down ULMA – Controller API included only member add/remove and source switch actions. JSON as the data format for serialization and deserialization of objects passed over the interface was used. ULMA was not aware of the network topology and client mobility.

4.2 Results

Manual tests of client mobility with physical Wi-Fi Access Points (AP) and listener running on a laptop have been conducted. Any changes in frame loss

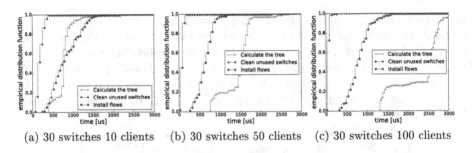

(a) 30 switches 10 clients   (b) 30 switches 50 clients   (c) 30 switches 100 clients

Fig. 6. Time distributions with varying number of clients

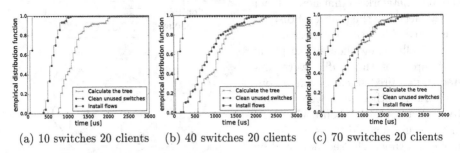

(a) 10 switches 20 clients   (b) 40 switches 20 clients   (c) 70 switches 20 clients

Fig. 7. Time distributions with varying number of switches

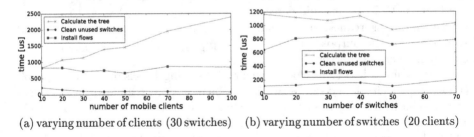

(a) varying number of clients (30 switches)   (b) varying number of switches (20 clients)

Fig. 8. The average time of multicast tree calculation, cleaning of unused switches and installation of new flows

during handoff and source switch were not observed. Traffic inspection at the APs revealed that tree reconstruction did not cause any frame loss inside the SDN network. Because our video source generated frames at the rate of 30 fps it had big inter-frame gaps (ca. 30 ms) hence our experiments were repeated with traffic generated using the *trafgen* [20]. Again no frame loss was observed.

To further evaluate our design with different number of clients and switches, our physical network was connected to a emulated large SDN network based on Mininet [21]. It allowed us to scale the test topology to 100 clients or 70 switches. Test topologies were generated by creating a path of required number of switches – n and then adding $2n$ extra links between randomly selected switches. Beginning with a path was necessary to ensure our network maintained connectivity. Afterwards, clients were attached to random switches. Because changing the streaming source requires SDN-C to rebuild the whole tree not only a part of it (which is the case during client actions) source switching was used to measure the worst-case tree setup time. The execution time of software paths in the controller which were responsible for tree reconstruction were captured using code instrumentation.

Figures 6 and 7 show empirical distributions of servicing various stages of tree update process. The lines with x markers refer to execution time of the tree-building heuristic – the purely algorithmic part of our implementation. The square and the dot markers represent time required for the controller to issue appropriate OpenFlow messages accordingly to fix the tree and to remove flows from the switches which no longer take part in the communication. The entire process have never taken more than 10 ms to complete, counting from the moment then triggering action was observed.

Figure 8a and b show the mean values of execution times. It can be seen that number of switches does not have a noticeable impact on execution times when the number of clients remains constant. When number of clients increases our algorithm takes longer to calculate the tree but the OpenFlow communication times again do not change. It should be noted that there are some outliers in the computation times which are probably related to the fact that our SDN-C (Floodlight) was written in Java.

5 Conclusions

In the paper, a novel approach of multicast and mobility management for Software Defined Networks is proposed. The advantages of such solutions are presented and compared with traditional multicast utilization in legacy IP networks. The new architecture, its elements components and communication interfaces are defined. Proposed solution was evaluated in a testbed reflecting to real-world streaming scenarios like IPTV broadcasting. Even for large networks including 30 switches and 100 mobile nodes, a multicast tree is calculated and install on all switches in less than 10 ms. Performed measurements proofed that proposed approach can be successfully implemented for multimedia applications and the described separation of responsibilities has no influence on the multicast performance.

References

1. Nakagawa, Y., et al.: A management method of IP multicast in overlay networks using openflow. In: Proceedings of the First Workshop on Hot Topics in Software Defined Networks, HotSDN 2012, pp. 91–96. ACM, New York (2012)
2. Imase, M., Waxman, B.: Dynamic steiner tree problem. SIAM J. Discrete Math. 4(3), 369–384 (1991)
3. Huang, L.-H., et al.: Scalable steiner tree for multicast communications in software-defined networking. arXiv preprint arXiv:1404.3454 (2014)

4. Kotani, D., et al.: A design and implementation of openflow controller handling IP multicast with fast tree switching. In: 2012 IEEE/IPSJ 12th International Symposium on Applications and the Internet (SAINT), pp. 60–67. July 2012
5. Marcondes, C., et al.: Castflow: clean-slate multicast approach using in-advance path processing in programmable networks. In: 2012 IEEE Symposium on Computer and Communication (ISCC), pp. 94–101. July 2012
6. Zou, J., et al.: Design and implementation of secure multicast based on SDN. In: 2013 5th IEEE International Conference on Broadband Network & Multimedia Technology (IC-BNMT), pp. 124–128. IEEE (2013)
7. Iyer, A., et al.: Avalanche: data center multicast using software defined networking. In: 2014 6th International Conference on Communication Systems and Networks (COMSNETS), pp. 1–8. January 2014
8. Tajik, S., Rostami, A.: Multiflow: enhancing IP multicast over IEEE 802.11 WLAN. In: Wireless Days (WD), 2013 IFIP, pp. 1–8. November 2013
9. Thaler, D.: Evolution of the IP model. IETF, RFC6250, May 2011
10. Nicholas, J., et al.: Protocol Independent Multicast - Dense Mode (PIM-DM): Protocol specification (revised). IETF, RFC3973, January 2005
11. Fenner, B., et al.: Protocol Independent Multicast - Sparse Mode (PIM-SM): Protocol specification (revised). IETF, RFC4601, August 2006
12. Kouvelas, I., et al.: Internet Group Management Protocol, Version 3. IETF, RFC3376, October 2002
13. Vida, R., Costa, L.: Multicast Listener Discovery Version 2 (MLDv2) for IPv6. IETF, RFC3810, June 2004
14. Christensen, M.J., et al.: Considerations for Internet Group Management Protocol (IGMP) and Multicast Listener Discovery (MLD) snooping switches. IETF, RFC4541, May 2006
15. Atwood, J.W., et al.: Authentication and confidentiality in Protocol Independent Multicast Sparse Mode (PIM-SM) link-local messages. IETF, RFC5796 (2010)
16. Meyer, D.: Administratively scoped ip multicast. IETF, RFC2365, Jul 1998
17. Suresh, L., et al.: Towards programmable enterprise WLANs with Odin. In: Proceedings of the 1st Workshop on Hot Topics in SDN, pp. 115–120. ACM (2012)
18. Gstreamer. http://gstreamer.freedesktop.org
19. IEEE Standard for a precision clock synchronization protocol for networked measurement and control systems. In: IEEE Std 1588–2008 (Revision of IEEE Std 1588–2002), pp. c1–269. July 2008
20. netsniff-ng toolkit. http://netsniff-ng.org
21. Lantz, B., et al.: A network in a laptop: rapid prototyping for software-defined networks. In: Proceedings of the 9th ACM SIGCOMM Workshop on Hot Topics in Networks, Hotnets-IX, pp. 19:1–19:6. ACM (2010)

Mobile Offloading Framework: Solution for Optimizing Mobile Applications Using Cloud Computing

Henryk Krawczyk, Michał Nykiel$^{(\boxtimes)}$, and Jerzy Proficz

Faculty of Electronics, Telecommunications and Informatics,
Gdansk University of Technology, Narutowicza 11/12, 80-233 Gdansk, Poland
hkrawk@eti.pg.gda.pl, {mnykiel,jerp}@task.gda.pl

Abstract. Number of mobile devices and applications is growing rapidly in recent years. Capabilities and performance of these devices can be tremendously extended with the integration of cloud computing. However, multiple challenges regarding implementation of these type of mobile applications are known, like differences in architecture, optimization and operating system support. This paper summarizes issues with mobile cloud computing and analyzes existing solutions in this field. A new innovative approach consisting of an application model and a mobile offloading framework are considered and adopted for practical applications.

Keywords: Cloud computing · Mobile · Network · Optimization · Model · Framework · Offloading

1 Introduction

The capabilities of mobile devices such as smartphones, tablets and wearables are increasing at very fast pace. Average processing power of a smartphone increased almost 4 times from 2011 to 2014 [1]. Mobile network connection speeds more than doubled in 2013 – the average mobile network downstream speed was 526 kbps in 2012 and almost 1.4 Mbps in 2013 [2]. It's estimated that by 2018 there will be over 7.4 billion mobile devices with 3G or 4G connection speed and global mobile data traffic will exceed 15 exabytes per month. This is all result of an increased demand from users, expecting continuous access to Internet services, multimedia, social networks, etc.

Arguably the most important issue in mobile devices currently is battery life. While CPU (Central Processing Unit) power, memory size, screen size and number of sensors increased significantly, we're not seeing any noticeable increase in battery capacity. For example the first iPhone, released in 2007, had a 5180 mWh battery and the iPhone 5 s model, released in 2013, have a 5966 mWh battery. The huge increase in processing capability over the 6 years has come with only a 15 % increase in battery capacity.

© Springer International Publishing Switzerland 2015
P. Gaj et al. (Eds.): CN 2015, CCIS 522, pp. 293–305, 2015.
DOI: 10.1007/978-3-319-19419-6_28

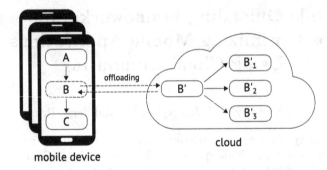

Fig. 1. The concept of offloading a part (component B) of mobile application consisting of three components A, B and C from a mobile device to the cloud

Although processing power of smartphones and tablets are increasing very rapidly, they are still far behind desktop computers or even laptops. Low performance on tasks such as image and video processing, 3D modeling or processing large amounts of numerical data prevents users from replacing their PCs with smaller and cheaper devices.

In recent years these problems have been addressed by integrating cloud computing with mobile devices [3]. Thanks to a new generation of mobile networks the applications are able to transmit large amounts of data to the cloud-hosted services. The concept of using remote servers and resources to extend capabilities of smartphones and tablets enables computationally intensive tasks to run efficiently while minimizing battery usage.

Mobile devices have many features and applications nowadays, from taking pictures, recording videos and playing games to advanced augmented reality software. With the computing power and storage space of the cloud these features may be further extended and new applications are possible. This concept is known in literature as computation offloading and refers to transferring certain processing tasks from devices with hardware limitations to external machines with more computational power [4]. Figure 1 presents the concept of using cloud for offloading of data processing from mobile device. Application component B can be executed in the cloud as parallel tasks B'_1, B'_2 and B'_3. Two advantages of offloading this component will be reduced execution time and less battery usage on mobile device.

However, there are still many unresolved issues with mobile cloud computing and existing solutions don't provide solutions for all of them. A new elastic mobile application model together with an innovative mobile offloading framework that could be adopted for practical applications is presented in this paper. Some examples of applications that could greatly benefit from this approach are multimedia processing [5], mobile gaming [6] and augmented reality [7,8] applications. The cloud offloading could be implemented in many others to improve battery life and performance of mobile devices.

Table 1. Comparison of existing solutions

	Run-time optimization	Application modifications	OS modifications	Other
CloneCloud	Performance	None	Significant	—
Weblets	Composite	Significant	None	Requires using one of supported application patterns
μCloud	None	Significant	None	—
Cloudlet	None	None	Significant	Requires low-latency server, private cloud
eXCloud	None	None	None	—
MAUI	Composite	Small	None	Based on outdated OS
ThinkAir	Energy/ performance	Small	None	Dedicated compiler, offloading only if both objectives are met
Cuckoo	None	Small	None	—

2   Existing Solutions

Multiple frameworks and models designed for integration of cloud and mobile devices exist. They differ in objective, as some of them are designed to maximize performance of an application, others try to optimize energy usage of mobile device by offloading computational-heavy tasks to cloud. There are also several solutions that try to achieve multiple objectives at once. The most popular solutions and differences between them are presented in Table 1.

CloneCloud [9] model is based on cloned version of mobile device's operating system running as a virtual machine on remote server in the cloud. When resource intensive task is started on the mobile device the execution is paused and process state is transferred to the clone. Partitioning component decides in run-time when to offload process to the cloud. Migration is fully automatic and doesn't require any changes in source code by application programmer. However, it requires significant changes in the operating system of mobile device, specifically a new implementation of Dalvik VM [10] in Android. Furthermore, the CloneCloud model optimizes only application performance and doesn't consider energy consumption or transfer cost.

Another cloud integration model assumes that mobile application is built from multiple loosely coupled components – weblets [11]. An application composed from weblets is called elastic application and supports three topology patterns: replication, splitting and aggregation. Elastic application optimization is performed in run-time using cost model, which is based on various parameters. However, authors don't provide any details regarding implementation of the optimization algorithm and example applications use only predefined configurations. Another disadvantage of weblet model is that the application must be built from scratch using provided SDK (Software Development Kit).

In μCloud [12] a mobile application is composed from multiple heterogeneous components. Building an application is reduced to orchestrating existing components to an execution graph. The biggest disadvantage of this model is that optimization must be performed a priori by the developer and no run-time profiling is proposed by the authors. Additionally, new development process requires existing application to be completely rewritten.

Cloudlet [13] concept is based on migrating whole mobile operating system to resource-rich server. When user starts a computationally intensive task virtual machine on mobile device is paused and transferred to remote computer. The main advantage of this model is that it allows any existing mobile application to be migrated without changes in implementation. However, significant modification must be made to mobile operating system. Furthermore, migration of whole virtual machine or even small VM overlay, as authors propose, requires high speed connection and low-latency servers.

Extensible cloud or eXCloud [14] relies on migration of JVM (Java Virtual Machine) stack frames to the cloud. Specialized pre-processor modifies byte code of the application before it is loaded by the JVM. Among the advantages of this solution is that it doesn't require any changes to existing application implementation or mobile operating system. The biggest disadvantage is that eXCloud only performs migration when resources on mobile device are insufficient and doesn't take into account performance or cost of the execution.

MAUI [15] is a framework that is able to optimize execution time and energy use by offloading individual methods to the cloud. Programmer is required to annotate methods allowed for remote execution and the profiler determines in run-time if migration is beneficial. Modifications that must be implemented in mobile application are relatively small, which is one of the biggest advantages of MAUI model. However, MAUI is based on an old .NET Framework version used in outdated Windows Mobile 6.5.

ThinkAir [16] use an Android emulator that allows to execute mobile applications on machines with x86 architecture. Decision whether code should be offloaded to the cloud is based on energy consumption, execution time or both. One of the shortcomings of ThinkAir model is that it doesn't take into account data transfer cost or network speed. Additionally programmer is required to use a dedicated compiler which could be not up-to-date with latest Android version and may not support new features or even fail to compile some applications.

Cuckoo [17] is a very simple offloading framework that was designed to be simple to use for programmers. Application developer provides two implementations of resource-intensive methods: local and remote. Among the advantages of Cuckoo framework is that it doesn't require any changes in operating system. However, offloading decisions in this model are based solely on remote server availability and doesn't consider performance gain, energy usage or migration cost.

3 Proposed Solution

Research and analysis of existing solutions lead to conclusion, that the most common flaws of aforementioned models and frameworks are as follows.

1. Implementation requires use of completely new architecture, patterns or compiler, hence existing applications must be rewritten from scratch.
2. Run-time optimization is not implemented or takes into account only single parameter like performance or battery consumption.
3. The framework requires changes in mobile operating system, making it difficult to use in practical applications without support from manufacturer of the OS.

Considering these issues authors decided that there is a need to design new model of integration between mobile devices and the cloud, together with a programming framework supporting computation offloading. Solution proposed in this paper tries to address aforementioned problems, basing on experiences and results from existing models.

3.1 Design Goals

First of all, proposed model must be elastic and flexible to allow development of different kinds of mobile applications – both simple and complex. Programmer should be able to use various architecture patterns and not be enforced to design application in one way. That implies that any programming interface should be minimal and integrated seamlessly into existing mobile system SDK (Software Development Kit).

Application optimization should be context-aware and should be performed in run-time, as parameters of network, device status or cloud availability could change in any time. Framework should monitor the execution of an application and collect all necessary data like execution time, CPU, memory, network and energy usage, both in mobile device and in the cloud. Optimization method should work on abstract application model and be able to minimize multiple objectives at once.

To maximize potential practical applications of the framework it shouldn't require any changes to the mobile operating system. Any provided tools, compilers or APIs (Application Programming Interface) must be designed as plugins, extensions or libraries, in such a way that future updates of the system wouldn't break existing applications.

3.2 Application Model

To build a feasible mobile application framework the most popular types of applications should be analyzed and used as a foundation to create an universal application model. As a result of an extensive research the following models were identified.

1. Sequential model – control flow in application is sequential, the output from one component is the input of exactly one other component. Simple image processing is an example of this model.
2. Parallel model – same as sequential but the output from one component may be passed to multiple components executing in parallel. Complex image filters, processing of video streams or rendering of 3D scene are good representations of this model.
3. Complex model – components of the application exchange data in arbitrary way, as in any data processing network that could be represented as a DAG (directed acyclic graph). A good example of this model would be an artificial intelligence algorithm.

Because the sequential and parallel models are special cases of the complex model, the later could be used as a general mobile application model. The model consists of multiple components that contains blocks of data processing instructions. It can be represented in form of a directed graph, where nodes are the application components and edges illustrate data flow. Some of these components could be executed either on mobile device or on remote server, other are constrained to the device because they require user input, specific device data or mobile operating system environment. It is also possible that component must be executed in the cloud, for example to exchange data with multiple users.

Mobile application cost model is based on proposed application model. Figure 2 demonstrates a sample application and cost models. If N is a set of components and E is a set of connections between two components the model is defined as follows:

$$M = (N, E, c_\mathrm{m}, c_\mathrm{c}, c_\mathrm{t}, t_\mathrm{m}, t_\mathrm{c}, t_\mathrm{t}) \tag{1}$$

$$E \subseteq \{(n_i, n_j) : n_i \in N \land n_j \in N \land n_i \neq n_j\} \tag{2}$$

$$c_\mathrm{m} : N \to \mathbb{R}, \quad c_\mathrm{c} : N \to \mathbb{R}, \quad c_\mathrm{t} : E \to \mathbb{R}, \tag{3}$$

$$t_\mathrm{m} : N \to \mathbb{R}, \quad t_\mathrm{c} : N \to \mathbb{R}, \quad t_\mathrm{t} : E \to \mathbb{R}. \tag{4}$$

The model contains four functions over set of components:

- execution time on mobile device $t_\mathrm{m}(n)$,
- execution time in the cloud $t_\mathrm{c}(n)$,
- cost of executing component on mobile device $c_\mathrm{m}(n)$,
- cost of executing component in the cloud $c_\mathrm{c}(n)$.

For purpose of this paper it is assumed that the cost is equal to energy used during component execution by the device or by remote server. Components that are constrained to the device have infinite cloud cost and cloud execution time, analogically for the cloud constrained components. Two functions over the set of edges (communication) are used in the cost model:

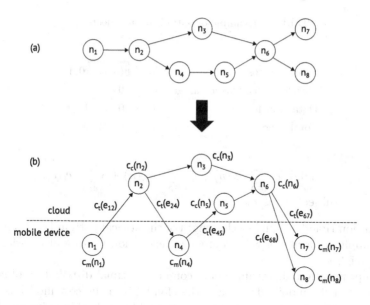

Fig. 2. Example of the application (a) and cost model (b)

- transfer time $t_t(e)$,
- transfer cost $c_t(e)$.

Authors assume that time and cost of data transfer between two components executed in the same environment (both on the device or both in the cloud) is negligible and equals zero. If N_m is a subset of components executed on the mobile device and N_c is a subset of components offloaded to the cloud then total cost c of running the application could be calculated:

$$N_c \subset N, \quad N_m = N \setminus N_c \tag{5}$$

$$E_{cm} = \{(n_c, n_m) : (n_c, n_m) \in E \land n_c \in N_c \land n_m \in N_m\} \tag{6}$$

$$E_{mc} = \{(n_c, n_m) : (n_c, n_m) \in E \land n_c \in N_c \land n_m \in N_m\} \tag{7}$$

$$E_t = E_{cm} \cup E_{mc} \tag{8}$$

$$c = \alpha \sum_{n \in N_c} c_c(n) + \beta \sum_{n \in N_m} c_m(n) + \gamma \sum_{e \in E_t} c_t(e). \tag{9}$$

Introducing the α, β and γ coefficients to the equation allows to adjust the cost model to different objectives. Table 2 presents different optimization objectives that can be achieved by assigning different values to the coefficients.

3.3   Optimization Problem

The goal is to find a subset of components to be executed in the cloud that minimizes the sum of total execution cost and transfer cost, while simultaneously preserving total execution time t below a certain threshold t_{max}:

Table 2. Example of optimization objectives

	α	β	γ
Monetary cost	$\langle 0, 1 \rangle$	$\langle 0, 1 \rangle$	$\langle 0, 1 \rangle$
Mobile device battery usage	1	0	0
Data transfer	0	0	1
Total energy use	$\langle 0, 1 \rangle$	$\langle 0, 1 \rangle$	0

$$\min_{N_c \in \mathbb{P}(N)} \alpha \sum_{n \in N_c} c_c(n) + \beta \sum_{n \in N_m} c_m(n) + \gamma \sum_{e \in E_t} c_t(e).$$
$$\text{subject to} \, t \leqslant t_{max} \tag{10}$$

Execution time t is equal to the length of the critical path in DAG representing the application model with component execution times and transfer times used as weights.

The optimization algorithm should compute optimal distribution of components between the mobile device and the cloud, given the cost model as input. Minimization of the cost without constraints could be solved in polynomial time, because the optimization problem could be reduced to minimum cut problem [18]. However, when the execution time constraint is considered the problem becomes NP-hard in general. In special cases of sequential and parallel application models the problem could still be solved in polynomial time. For general approach a genetic algorithm has proved to provide good results for sparse graphs, i.e. application models where the number of edges is significantly lower than square number of components.

4   Mobile Offloading Framework

This paper introduces Mobile Offloading Framework (or MOFF) as an innovative contribution to the field of mobile cloud computing. The idea of MOFF is to provide lightweight middleware both for the cloud and mobile devices that allows dynamic optimization of mobile application cost by offloading parts of data processing to the cloud. Initial implementation of the framework is based on Android operating system and allows of development of mobile applications using both in Java and C++ language. MOFF supports Android version 4.0 and newer which currently covers over 75 % of the mobile market [19].

The idea of MOFF Framework is to enable each component to be executed on mobile device or in the cloud. Application developer is given a programming library that supports development of components in Java or C++ programming language. Developer can prepare one general implementation or different implementations of the same component for mobile and cloud execution to take advantage of resource-rich remote machines. Compiled components are embedded in mobile application package (APK) together with Mobile Service needed for run-time monitoring and communication with the cloud. All components are also deployed to the Cloud Service which supports execution in the cloud.

After implementation of all required components the application developer must orchestrate them and provide the application model which is defined as a JSON document. For example, a definition of the application model from Fig. 2 would look like this:

```
{
    components: [{
        "id": "n1",
        "output": [
            "n2"
        ]
    },{
        "id": "n2",
        "output": [
            "n3", "n4"
        ]
    },
    /* ...definitions of components n3, n4, n5... */
    {
        "id": "n6",
        "output": [
            "n7", "n8"
        ]
    },{
        "id": "n7"
    },{
        "id": "n8"
    }]
}
```

Typical sequence of events in mobile application execution using Mobile Offloading Framework is shown in Fig. 3. When users starts the mobile application (1) Mobile Service gathers information about the device and current execution context (like network connection type, battery level, etc.) via various Android APIs and Linux kernel modules. This data is sent together with application model definition to the Cloud Service (2).

In order to perform accurate optimization the cost model must be constructed on top of the application model. The Cloud Service queries the database for historic data regarding execution cost and time on mobile device and in the cloud for each component (3). The Cloud Service tries to use data for the specific execution context or average from similar contexts in case that the specific one does not exist in the database yet. When the cost model is constructed (i.e. functions of cost and time have values for each component and communication between them) optimization of application model can be performed by genetic algorithm. The result of optimization is a partition of the components set into two subsets: components that should be executed on mobile device and components that should be executed in the cloud.

Results of the optimization are sent back to the mobile device (4) and the application executes components that are assigned to it (5). The Mobile Service

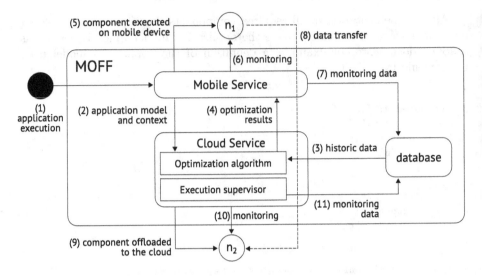

Fig. 3. Sequence of events in application execution with Mobile Offloading Framework. Components n_1 and n_2 are a part of mobile application presented in Fig. 2

must constantly monitor execution time and cost of every component (6). It is important to notice that cost and time functions could be different for every device and depend on current execution context, i.e. network type, battery level, device load, etc. Monitoring data is stored in centralized database (7), therefore historical data from previous executions and other devices could be used as a good basis for future optimizations.

When application tries to execute component that should be offloaded to the cloud Mobile Service communicate with Cloud Service and transfers input data over the network (8). The Cloud Service launches required component on remote server (9), supervises the execution (10) and stores monitoring data in database (11). When component processing is finished the output data is transferred back to the mobile device. The offloading process is transparent for the application.

MOFF Framework introduces a few tweaks to further improve performance of the offloading process. First obvious improvement is to execute two or more consecutive cloud components in one batch rather than sending data to the mobile device after each one. Secondly, the components that are executed in the cloud are started immediately after the optimization is performed and are waiting for input data from the mobile device. This approach helps to decrease delay between sending the data and receiving results from the component because the system process is already initialized.

The Mobile Service includes also some basic fail-over mechanisms. When network connection is unavailable and optimization could not be performed all components are executed on the device if possible. If connection is lost during application execution the Mobile Service takes care of restarting the component locally.

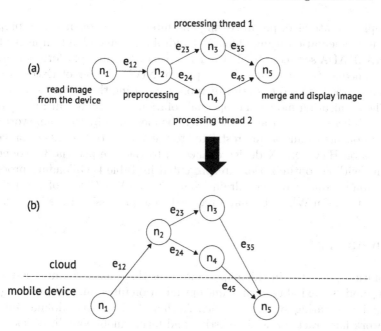

Fig. 4. The application model (a) and optimal component distribution (b) of the image processing application

Table 3. The values of cost [mWh] and time [s] functions measured by MOFF

	c_m	c_c	t_m	t_c		c_t	t_t
n_1	0.04	∞	0.09	∞	e_{12}	0.12	4.06
n_2	0.73	0	1.47	0.31	e_{23}	0.06	4.21
n_3	4.31	0	8.72	1.86	e_{24}	0.06	4.21
n_4	4.31	0	8.72	1.86	e_{35}	0.05	3.22
n_5	0.07	∞	0.08	∞	e_{45}	0.05	3.21

Table 4. Test results of image processing application with and without optimization

	single image		20 images	
	cost [mWh]	time [s]	cost [mWh]	time [s]
not optimized	9.46	10.36	189.21	207.22
optimized	0.33	9.62	6.67	192.41

5   Case Study

MOFF was used to implement sample mobile application for processing images. It consisted of five components: reading an image from the device, preprocessing, applying filters using two parallel threads and displaying processed image.

Two implementations of preprocessing and filtering components were prepared: mobile implementation using OpenCV4Android [20] and cloud implementation using KASKADA services [21]. KASKADA is a distributed platform for processing multimedia streams realized as a part of NIWA Center of Excellence [22]. The framework was used to optimize application using the battery saving objective. The application model and optimal solution is presented in Fig. 4. Table 3 contains values of the cost and time functions measured by the framework. Note that components n_1 and n_5 are restricted to the device. Test results of processing 20 photos on HTC One X device connected to the internet via 3G connection with and without optimization are presented in Table 4. Offloading processing to the cloud reduced battery drain from 189.21 mWh (2.8 % of total battery capacity) to 6.67 mWh (0.1 %) and reduced total processing time by 15 s.

6 Summary

The solution presented in this paper addresses three most common issues with existing models: flexibility of run-time optimization function, programming effort and mobile operating system support. Authors believe that Mobile Offloading Framework has practical use in existing and future mobile applications.

Regarding future research the framework should be more extensively tested by developing more advanced mobile applications. Additionally, more complex execution monitoring could be implemented to improve optimization results in more complicated applications, perhaps by including static code analysis. Support for different mobile operating systems is also worth considering.

In the near future Mobile Offloading Framework will be deployed to the cloud based on Tryton supercomputer that is a part of NIWA Center of Excellence project [22]. Tryton is located in Academic Computer Centre in Gdansk (CI TASK) an is one of the fastest supercomputers in Europe with over 1.2 PFLOPS computing power, 2600 CPUs and 31000 cores. Powered by this infrastructure MOFF could be easily adopted by application developers and scientist to create efficient mobile applications.

Acknowledgments. This work was carried out as a part of the Center of Excellence in Scientific Application Development Infrastructure "NIWA" project, Operational Program Innovative Economy 2007–2013, Priority 2 "Infrastructure area R&D".

References

1. Chitkara, R.: Application processors: driving the next wave of innovation. In: Mobile Technologies Index, PwC (2012)
2. Cisco visual networking index: global mobile data traffic forecast update 2013–2018 (2014)
3. Khan, A.R., Othman, M., Madani, S.A., Khan, S.U.: A survey of mobile cloud computing application models. In: Communications Surveys and Tutorials, vol.16, no. 1, pp. 393–413. IEEE (2014)

4. Kumar, K., Liu, J., Lu, Y.-H., Bhargava, B.: A survey of computation offloading for mobile systems. In: Mobile Networks and Applications, vol. 18, no. 1, pp. 129–140. Springer (2013)

5. Satyanarayanan, M.: Mobile computing: the next decade. In: Proceedings of the 1st ACM Workshop on Mobile Cloud Computing & Services: Social Networks and Beyond (MCS 2010), pp. 1–6. ACM, New York (2010)

6. nVidia GRID game service. http://shield.nvidia.com/grid-game-streaming/

7. Google glass. http://www.google.com/glass/

8. Azuma, R.T.: A survey of augmented reality. In: Presence: Teleoperators and Virtual Environments, vol. 6, no. 4, pp. 355–385 (1997)

9. Chun, B.G., Ihm, S., Maniatis, P., Naik, M., Patti, A.: CloneCloud: elastic execution between mobile device and cloud. In: Proceedings of the Sixth Conference on Computer systems, pp. 301–314. ACM (2011)

10. ART and Dalvik. http://source.android.com/devices/tech/dalvik/

11. Zhang, X., Jeong, S., Kunjithapatham, A., Gibbs, S.: Towards an elastic application model for augmenting computing capabilities of mobile platforms. In: Mobile Networks and Applications, vol. 16, no.3, pp. 270–284. Springer, New York (2011)

12. March, V., Gu, Y., Leonardi, E., Goh, G., Kirchberg, M., Lee, B.S.: μCloud: towards a new paradigm of rich mobile applications. In: Procedia Computer Science, vol. 5, pp. 618–624. Elsevier (2011)

13. Satyanarayanan, M., Bahl, P., Caceres, R., Davies, N.: The case for VM-based cloudlets in mobile computing. In: Pervasive Computing, vol. 8, no. 4, pp. 14–23. IEEE (2009)

14. Ma, R.K.K., Lam, K.T., Wang, C.L.: eXCloud: Transparent runtime support for scaling mobile applications in cloud. In: 2011 International Conference on Cloud and Service Computing, pp. 103–110. IEEE (2011)

15. Cuervo, E., Balasubramanian, A., Cho, D.K., Wolman, A., Saroiu, S., Chandra, R., Bahl, P.: MAUI: making smartphones last longer with code offload. In: Proceeding of the 8th International Conference on Mobile Systems, Applications, and Services, pp. 49–62. ACM (2010)

16. Kosta, S., Aucinas, A., Hui, P., Mortier, R., Zhang, X.: Unleashing the power of mobile cloud computing using ThinkAir. In: arXiv preprint arXiv:1105.3232 (2011)

17. Kemp, R., Palmer, N., Kielmann, T., Bal, H.: Cuckoo: a computation offloading framework for smartphones. In: Gris, M., Yang, G. (eds.) MobiCASE 2010. LNICST, vol. 76, pp. 59–79. Springer, Heidelberg (2012)

18. Hao, J.X., Orlin, J.B.: A faster algorithm for finding the minimum cut in a directed graph. J. Algorithms 17(3), 424–446 (1994). Elsevier

19. Smarthphone OS market share, Q3 (2014). http://www.idc.com/prodserv/smartphone-os-market-share.jsp

20. OpenCV4Android. http://opencv.org/platforms/android.html

21. Krawczyk, H., Proficz, J.: KASKADA - multimedia processing platform architecture. In: Proceedings of the International Conference on Signal Processing and Multimedia Applications, SIGMAP 2010. IEEE (2010)

22. NIWA center of excellence. http://www.niwa.gda.pl

Multi-server Queueing System $MAP/M/N^{(r)}/\infty$ Operating in Random Environment

Chesoong Kim[1]($\boxtimes$), Alexander Dudin[2], Sergey Dudin[2], and Olga Dudina[2]

[1] Sangji University, Wonju, Kangwon 220-702, Korea
dowoo@sangji.ac.kr
[2] Belarusian State University, 4 Nezavisimosti Ave., 220030 Minsk, Belarus
dudin@bsu.by,dudin85@mail.ru, dudina_olga@email.com

Abstract. A multi-server queueing system with an infinite buffer operating in the Markovian random environment is analyzed. Under the fixed state of the random environment, the arrival flow is described by the Markovian arrival process. Customers in the buffer may be impatient and leave the system. The number of available servers, the arrival process, the rate of customers' service, and the impatience intensity depend on the state of the random environment. Behavior of the system is described by the multi-dimensional asymptotically quasi-Toeplitz Markov chain. The ergodicity condition is derived. Algorithm for computation of the stationary distribution is provided. The main performance measures of the system are calculated.

Keywords: Multi-server queue · Markovian arrival process · Random environment · Stationary state distribution

1 Introduction

Random factors can significantly impact on the operation of telecommunication systems. Such factors can affect the system bandwidth, the characteristics of the arrival flow of customers, the quality of data transmission, etc. Examples of random factors may be artificial and natural disturbances, different levels of noise in the transmission channel, change of the distance between wireless transmitters, the parallel transmission of the high priority information, various malfunctions and equipment failures, and the effects of weather, etc. Particularly, random factors have strong influence on wireless networks. Accounting the effects of random factors on the system operation is a very important task in the construction of adequate mathematical models for calculation of performance measures of telecommunication systems.

Mathematical methods of the theory of queuing systems operating in the random environment (RE) allow us to create adequate stochastic models of modern telecommunication systems, taking into account the influence of random factors. The RE is assumed to be a random process with a finite state space,

P. Gaj et al. (Eds.): CN 2015, CCIS 522, pp. 306–315, 2015.
DOI: 10.1007/978-3-319-19419-6_29

independent of the queuing system. For the fixed state of the RE the system operates as a classical queueing system. However, the parameters of the system (the arrival process, the distribution of the service time, etc.) instantly change its values with the change of the state of the RE. For background information and an overview of the present state of the art in the study of queueing systems operating in the RE, the reader is referred to the papers [1–5], and as well as references therein.

In this paper, we analyze a multi-server queueing system with an infinite buffer operating in the RE. The main contribution of our paper is the assumption that the number of available servers depends on the state of the RE. To the best of our knowledge, in all existing papers devoted to analysis of queues operating in the RE, the influence of the RE on the number of available servers is not considered. Unreliable multi-server queues are a very special example of queues with the random number of available servers, however such queues are also not considered in literature in assumption that the arrival and service processes parameters depend on the RE.

The queueing model under study is described by a multi-dimensional Markov chain with an infinite state space and inhomogeneous state behavior. We use results for the asymptotically quasi-Toeplitz Markov chains, see [6], to derive the ergodicity condition and calculate the system performance measures.

2   Mathematical Model

We consider a multi-server queueing system with an infinite buffer operating in the Markovian RE. The structure of the system under study is presented in Fig. 1.

Behavior of the system depends on the state of the RE. The RE is given by the stochastic process r_t, $t \geq 0$, which is an irreducible continuous time Markov chain with the state space $\{1, 2, \ldots, R\}$ and the infinitesimal generator H.

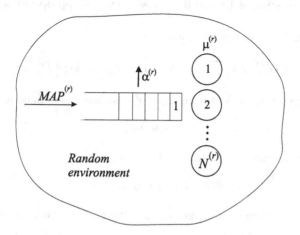

Fig. 1. Queueing system under study

Under the fixed state r of the RE the number of working servers is $N^{(r)}$, $r = \overline{1, R}$. Without loss of generality we assume that the number of available servers increases with the increase of the number of the RE states, i.e., $0 \leq N^{(1)} \leq N^{(2)} \leq \cdots \leq N^{(R)}$.

The customers arrive to the system according to the switching Markovian arrival process (MAP). This means the following. The arrival of customers is directed by the stochastic process ν_t, $t \geq 0$, with the finite state space $\{0, 1, \ldots, W\}$. Under the fixed state r of the RE this process behaves as an irreducible continuous time Markov chain. Under the fixed state r of the RE, the sojourn time of the chain ν_t, $t \geq 0$, in the state ν is exponentially distributed with the positive finite parameter $\lambda_\nu^{(r)}$, $r = \overline{1, R}$. When the sojourn time in the state ν expires, with probability $p_0^{(r)}(\nu, \nu')$ the process ν_t jumps to the state ν' without generation of customers and with probability $p_1^{(r)}(\nu, \nu')$ it jumps with generation of customer, $\nu, \nu' = \overline{0, W}$, $r = \overline{1, R}$.

Behavior of the MAP under the fixed state r of the RE is completely characterized by the matrices $D_0^{(r)}$, $D_1^{(r)}$, defined by the entries

$$
\begin{aligned}
\left(D_1^{(r)} \right)_{\nu, \nu'} &= \lambda_\nu^{(r)} p_1^{(r)}(\nu, \nu'), \ \nu, \ \nu' = \overline{0, W}, \\
\left(D_0^{(r)} \right)_{\nu, \nu} &= -\lambda_\nu^{(r)}, \ \nu = \overline{0, W}, \\
\left(D_0^{(r)} \right)_{\nu, \nu'} &= \lambda_\nu^{(r)} p_0^{(r)}(\nu, \nu'), \ \nu, \nu' = \overline{0, W}, \ \nu \neq \nu', \ r = \overline{1, R}.
\end{aligned}
$$

The matrix $D^{(r)}(1) = D_0^{(r)} + D_1^{(r)}$ represents the generator of the process ν_t, $t \geq 0$, under the fixed value r, $r = \overline{1, R}$.

Under the fixed state r, $r = \overline{1, R}$, of the RE the average arrival rate $\lambda^{(r)}$ is given as $\lambda^{(r)} = \boldsymbol{\theta}^{(r)} D_1^{(r)} \mathbf{e}$, where $\boldsymbol{\theta}^{(r)}$ is the invariant vector of the stationary distribution of the Markov chain ν_t, $t \geq 0$. The vector $\boldsymbol{\theta}^{(r)}$ is the unique solution to the system $\boldsymbol{\theta}^{(r)} D^{(r)}(1) = \mathbf{0}$, $\boldsymbol{\theta}^{(r)} \mathbf{e}_{\bar{W}} = 1$. Here $\mathbf{e}_{\bar{W}}$ is a column vector of size $\bar{W} = W + 1$ consisting of 1's and $\mathbf{0}$ is a row vector of appropriate size consisting of zeroes.

Let us introduce the following matrices:

$$
\tilde{D}_1 = \mathrm{diag} \left\{ D_1^{(r)}, r = \overline{1, R} \right\}, \ \tilde{D}_0 = H \otimes I_{\bar{W}} + \mathrm{diag} \left\{ D_0^{(r)}, r = \overline{1, R} \right\}.
$$

The average intensity λ of input flow of customers is defined as

$$
\lambda = \boldsymbol{\theta} \tilde{D}_1 \mathbf{e},
$$

vector $\boldsymbol{\theta}$ is the unique solution to the following system:

$$
\boldsymbol{\theta}(\tilde{D}_0 + \tilde{D}_1) = \mathbf{0}, \ \boldsymbol{\theta}\mathbf{e} = 1.
$$

The squared coefficient of variation c_{var} of intervals between successive arrivals is given as

$$
c_{\mathrm{var}} = 2\lambda\boldsymbol{\theta}(-\tilde{D}_0)^{-1}\mathbf{e} - 1.
$$

The coefficient of correlation c_{cor} of two successive intervals between arrivals is given as

$$c_{\text{cor}} = \left(\lambda \boldsymbol{\theta}(-\tilde{D}_0)^{-1} \tilde{D}_1 (-\tilde{D}_0)^{-1} \mathbf{e} - 1 \right) / c_{\text{var}}.$$

We assume that during the epochs of the transitions of process r_t, $t \geq 0$, the states of the process ν_t, $t \geq 0$, do not change, only the intensities of the further transitions of this process change. For more information about the MAP see [7].

If there is a free server during an arbitrary customer arrival epoch, this customer is accepted to the system and starts service. If all servers are busy during an arbitrary customer arrival epoch, this customer goes to the buffer.

In case when the transition of the RE leads to the reduction of the number of servers we suppose that, at first, the number of free servers decreases, and if that is not enough, then service of the corresponding number of customers is terminated. We assume that all customers whose service was terminated due to reducing the number of servers return to the buffer.

If the state of the RE is changed in such a way as the number of available servers increases, we suppose that the corresponding number of customers that wait in the buffer occupy the free servers.

The customers in the buffer are assumed to be impatient. Under the fixed state r of the RE, each customer leaves the buffer after an exponentially distributed with the parameter $\alpha^{(r)}$, $\alpha^{(r)} \geq 0$, $r = \overline{1,R}$, time due to the lack of service.

The probability of finishing service of the customer in an interval of infinitesimal length $(t, t + \Delta t)$ under the fixed state r of the RE is equal to $\mu^{(r)} \Delta t + o(\Delta t)$, $r = \overline{1,R}$.

3   Process of System States and Stationary Distribution

Let

- i_t, $i_t \geq 0$, be the number of customers in the system,
- r_t, $r_t = \overline{1,R}$, be the state of the RE,
- ν_t, $\nu_t = \overline{0,W}$, be the state of the underlying process of the MAP during the epoch t, $t \geq 0$.

The Markov chain $\xi_t = \{i_t, r_t, \nu_t\}$, $t \geq 0$, is the regular irreducible continuous time Markov chain.

Let us introduce the following notations:

- I is the identity matrix and O is a zero matrix of appropriate dimension;
- $\text{diag}\{A_1, \ldots, A_l\}$ is the block diagonal matrix with the diagonal blocks $A_1, \ldots, A_l$;
- C_l is the square matrix of size $l + 1$ defined as follows $C_l = \text{diag}\{0, 1, \ldots, l\}$;
- $\otimes$ indicates the symbol of Kronecker product of matrices;
- E_l^- is the square matrix of size $l + 1$ with all zero entries except the entries $(E_l^-)_{k,k-1}$, $k = \overline{1,l}$, which are equal to 1;

- E_l^+ is the square matrix of size $l+1$ with all zero entries except the entries $(E_l^+)_{k,k+1}$, $k = \overline{0, l-1}$, which are equal to 1;
- $\hat{I}_l$ is the square matrix of size $l+1$ with all zero entries except the entry $(\hat{I}_l)_{l,l} = 1$.

Let us enumerate the states of the Markov chain ξ_t in the lexicographic order of the components (i, r, ν) and refer to the set of states of the chain having values (i, r) of the first two components of the Markov chain as macro-state (i, r).

Let Q be the generator of the Markov chain ξ_t, $t \geq 0$.

Lemma 1. *The generator Q has the following block-three-diagonal structure:*

$$
Q = \begin{pmatrix}
Q_{0,0} & Q_{0,1} & Q_{0,2} & O & O \cdots \\
Q_{1,0} & Q_{1,1} & Q_{1,2} & O & O \cdots \\
O & Q_{2,1} & Q_{2,2} & Q_{2,3} & O \cdots \\
\vdots & \vdots & \vdots & \vdots & \vdots & \ddots
\end{pmatrix}.
$$

The non-zero blocks $Q_{i,j}$, $i, j \geq 0$, have the following form:

$$
\begin{aligned}
Q_{i,i} &= (Q_{i,i})_{r,r'}, r, r' = \overline{1, R}, \\
(Q_{i,i})_{r,r} &= \left(-\mu^{(r)}i + (H)_{r,r}\right) I_{\bar{W}} + D_0^{(r)}, \quad i = \overline{0, N^{(r)}}, \\
(Q_{i,i})_{r,r} &= \left[-\left(\mu^{(r)}N^{(r)} + (i - N^{(r)})\alpha^{(r)}\right) + (H)_{r,r}\right] I_{\bar{W}} + D_0^{(r)}, i > N^{(r)}, \\
(Q_{i,i})_{r,r'} &= (H)_{r,r'} I_{\bar{W}}, r, r' = \overline{1, R}, \\
Q_{i,i+1} &= \operatorname{diag}\left\{ D_1^{(r)}, r = \overline{1, R} \right\}, \quad i \geq 0, \\
Q_{i,i-1} &= \operatorname{diag}\left\{ \mu^{(r)}i I_{\bar{W}}, r = \overline{1, R} \right\}, \quad i = \overline{1, N^{(r)}}, \\
Q_{i,i-1} &= \operatorname{diag}\left\{ \left(\mu^{(r)}N^{(r)} + (i - N^{(r)})\alpha^{(r)}\right) I_{\bar{W}}, r = \overline{1, R} \right\}, \quad i > N^{(r)}.
\end{aligned}
$$

Remark 1. The Markov chain ξ_t, $t \geq 0$, belongs to the class of continuous-time asymptotically quasi-Toeplitz Markov chains $(AQTMC)$, see [6].

Let us consider the case when $\alpha^{(r)} > 0$ at least for one state r of the RE, $r = \overline{1, R}$.

As follows from [6], a necessary condition for the existence of a stationary distribution of $AQTMC$ ξ_t, $t \geq 0$, is expressed in terms of the matrices Y_0, Y_1 and Y_2 defined by

$$
Y_0 = \lim_{i \to \infty} R_i^{-1} Q_{i,i-1}, \quad Y_1 = \lim_{i \to \infty} R_i^{-1} Q_{i,i} + I, \quad Y_2 = \lim_{i \to \infty} R_i^{-1} Q_{i,i+1}
$$

where the matrix R_i is the diagonal matrix with diagonal entries defined as modules of the corresponding diagonal entries of the matrix $Q_{i,i}$, $i \geq 0$.

It is easy to verify that for the Markov chain describing the considered system the matrices Y_0, Y_1 and Y_2 have the following form:

$$
\begin{aligned}
Y_0 &= \operatorname{diag}\left\{ \tilde{\Omega}_1, \ldots, \tilde{\Omega}_R \right\}, \\
Y_1 &= \begin{pmatrix}
\tilde{Q}_{1,1} & \tilde{Q}_{1,2} & \cdots & \tilde{Q}_{1,R} \\
\tilde{Q}_{2,1} & \tilde{Q}_{2,2} & \cdots & \tilde{Q}_{1,R} \\
\vdots & \vdots & \ddots & \vdots \\
\tilde{Q}_{R,1} & \tilde{Q}_{R,2} & \cdots & \tilde{Q}_{R,R}
\end{pmatrix}, \\
Y_2 &= \operatorname{diag}\left\{ Z_1, \ldots, Z_R \right\},
\end{aligned}
$$

where

$$\tilde{Q}_{r,r'} = K_r(Q_{N,N})_{r,r'} + \delta_{r-r',0}I_{\bar{W}}, \quad \text{if } \alpha^{(r)} = 0,$$
$$\tilde{Q}_{r,r'} = O, \quad \text{if } \alpha^{(r)} > 0, \, r, r' = \overline{1,R},$$
$$\tilde{Z}_r = K_r D_1^{(r)}, \quad \text{if } \alpha^{(r)} = 0,$$
$$\tilde{Z}_r = O, \quad \text{if } \alpha^{(r)} > 0, \, r, r' = \overline{1,R},$$
$$\tilde{\Omega}_r = N^{(r)}\mu^{(r)}K_r, \quad \text{if } \alpha^{(r)} = 0,$$
$$\tilde{\Omega}_r = I_{\bar{W}} \quad \text{if } \alpha^{(r)} > 0, \, r = \overline{1,R},$$

$$K_r = (\mu^{(r)}N^{(r)}I_{\bar{W}} + \Sigma_0 - (H)_{r,r}I_{\bar{W}})^{-1}, \, r = \overline{1,R},$$

$\delta_{r,0}$ indicates Kronecker delta.

Here $\Sigma_0^{(r)}$ is the diagonal matrix the diagonal entries of which are defined as the corresponding diagonal entries of the matrix $-D_0^{(r)}$.

Theorem 1. *If customers are impatient $(\alpha^{(r)} > 0)$ at least for one state r of the RE, then the Markov chain ξ_t is ergodic for any set of parameters of the queueing system under study.*

Proof. Let $L = \{l_1, l_2, \ldots, l_S\}$ be the set of the states of the RE for which $\alpha^{(l)} > 0$, $l \in L$. It can be seen that, in this case, the matrix $Y_0 + Y_1 + Y_2$ is reducible and by coordinated permutation of rows and columns can be represented as

$$Y_0 + Y_1 + Y_2 = \begin{pmatrix} Y_{1,1} & Y_{1,2} \\ O & Y_{2,2} \end{pmatrix}$$

where
$Y_{1,1}$ is the matrix obtained from the matrix Y by discarding block rows and columns with numbers l, $l \in L$,
$Y_{1,2}$ is the non-zero matrix obtained from the matrix Y by discarding block rows with numbers l, $l \in L$, and block columns with numbers r, $r \notin L$,
$Y_{2,2} = I_{\bar{W}S}$.
To prove Theorem 1 we need the following lemma.

Lemma 2. *Suppose that for any stochastic vector $\bar{y}$ the following inequality holds true:*

$$\bar{y}Y_0^{(2)}\mathbf{e} > \bar{y}Y_2^{(2)}\mathbf{e}, \tag{1}$$

where the matrices $Y_0^{(2)}$ and $Y_2^{(2)}$ are obtained from the matrices Y_0 and Y_2 by discarding block rows and columns with numbers r, $r \in R \setminus L$. Then the Markov chain ξ_t, $t \geq 0$, is ergodic.

Proof. Proof of the lemma steps up from the results of [6]. If the matrix $Y_{2,2}$ is irreducible, then the existence of a stationary distribution requires that the inequality

$$\mathbf{z}Y_0^{(2)}\mathbf{e} > \mathbf{z}Y_2^{(2)}\mathbf{e}, \tag{2}$$

holds true where the vector $\mathbf{z}$ is the unique solution of the system:

$$\mathbf{z}Y_{2,2} = \mathbf{z}, \ \mathbf{z}\mathbf{e} = 1. \tag{3}$$

If condition (1) holds true for any stochastic vector $\bar{\mathbf{y}}$, then it holds for the vector $\mathbf{z}$ as well.

In the case of the reducible matrix $Y_{2,2}$, this matrix has to be transformed to a canonical form and for the existence of a stationary distribution of the chain ξ_t, $t \geq 0$, it is required that the inequalities like (2) hold true for each irreducible block of the matrix $Y_{2,2}$. By representations of the vector $\bar{\mathbf{y}}$ in form $\bar{\mathbf{y}} = \{\mathbf{0}, \mathbf{z}_l, \mathbf{0}\}$, where $\mathbf{z}_l$ is the unique solution of the system like (3) for l-th irreducible block of the matrix $Y_{2,2}$, it is easy to see that all the required inequalities hold true if condition (1) holds true.

We continue the proof of the theorem. It is easy to see that the matrix $Y_2^{(2)} = O$, and the matrix $Y_{2,2} = Y_0^{(2)}$. Consequently, inequality (1), which can be rewritten as

$$\bar{\mathbf{y}}Y_0^{(2)}\mathbf{e} = \bar{\mathbf{y}}Y_{2,2}\mathbf{e} = \bar{\mathbf{y}}\mathbf{e} = 1 > 0 = \bar{\mathbf{y}}Y_2^{(2)}\mathbf{e},$$

holds true for any vector $\bar{\mathbf{y}}$. Hence, the Markov chain ξ_t, $t \geq 0$, is ergodic. Thus the theorem is proved.

Now, let us consider the case when $\alpha^{(r)} = 0$ for all states r, $r = \overline{1, R}$, of the RE. One can see, that in this case, the blocks of the generator for $i > N^{(R)}$ have the following form:

$$Q_1 = Q_{i,i} = (Q_{i,i})_{r,r'}, r, r' = \overline{1, R},$$

$$(Q_{i,i})_{r,r} = -\mu^{(r)} N^{(r)} I_{\bar{W}} + D_0^{(r)} + (H)_{r,r} I_{\bar{W}}, \quad i > N^{(R)},$$

$$(Q_{i,i})_{r,r'} = (H)_{r,r'} I_{\bar{W}}, \ r, \ r' = \overline{1, R},$$

$$Q_2 = Q_{i,i+1} = \text{diag}\left\{D_1^{(r)}, r = \overline{1, R}\right\}, \quad i \geq 0,$$

$$Q_0 = Q_{i,i-1} = \text{diag}\left\{\mu^{(r)} N^{(r)} I_{\bar{W}}, r = \overline{1, R}\right\}, \quad i > N^{(R)},$$

and do not depend on i. So, the Markov chain ξ_t, $t \geq 0$, belongs to the class of continuous-time quasi-Toeplitz Markov chains ($QTMC$) or $M/G/1$-type Markov chains, see [8].

As follows from [8], the necessary and sufficient condition for the ergodicity of the $QTMC$ is the fulfillment of the following inequality

$$\mathbf{x}Q_0\mathbf{e} > \mathbf{x}Q_2\mathbf{e} \tag{4}$$

where the vector $\mathbf{x}$ is the unique solution to the system

$$\mathbf{x}(Q_0 + Q_1 + Q_2) = \mathbf{0}, \ \mathbf{x}\mathbf{e} = 1.$$

Theorem 2. *If* $\alpha^{(r)} = 0$ *for all states of the RE* r, $r = \overline{1, R}$, *then the Markov chain* ξ_t, $t \geq 0$, *is ergodic if the following inequality holds true:*

$$\sum_{r=1}^{R} \delta_r \mu^{(r)} N^{(r)} > \lambda$$

where δ_r, $r = \overline{1, R}$, *are the components of the vector* $\boldsymbol{\delta}$ *that is the vector of the stationary distribution of the RE and is defined as the unique solution of the system:* $\boldsymbol{\delta} H = \mathbf{0}$, $\boldsymbol{\delta} \mathbf{e} = 1$.

Proof. Let us consider the matrix $\bar{Q} = Q_0 + Q_1 + Q_2$ which has the form

$$\bar{Q} = H \otimes I_{\bar{W}} + \text{diag}\left\{ D_0^{(r)} + D_1^{(r)}, r = \overline{1, R} \right\}.$$

It is easy to see, that

$$\bar{Q} = \tilde{D}(1).$$

Hence the vector $\mathbf{x}$ coincides with $\boldsymbol{\theta}$. So, the right side of inequality (4) can be rewritten as

$$\mathbf{x} Q_2 \mathbf{e} = \boldsymbol{\theta} Q_2 \mathbf{e} = \boldsymbol{\theta} \text{diag}\left\{ D_1^{(r)}, r = \overline{1, R} \right\} \mathbf{e} = \lambda.$$

The left hand side of inequality (4) can be rewritten as

$$\boldsymbol{\theta} Q_0 \mathbf{e} = \boldsymbol{\theta} \text{diag}\left\{ \mu^{(r)} N^{(r)} \otimes I_{\bar{W}}, r = \overline{1, R} \right\} \mathbf{e}$$
$$= \left(\tilde{\boldsymbol{\theta}}_1, \ldots, \tilde{\boldsymbol{\theta}}_R \right) \text{diag}\left\{ \mu^{(r)} N^{(r)} I_{\bar{W}}, r = \overline{1, R} \right\} \mathbf{e},$$

where $\tilde{\boldsymbol{\theta}}_r = \left(P\{r_t = r, \nu_t = 0\}, \ldots, P\{r_t = r, \nu_t = W\} \right).$

It can be shown that

$$\left(\text{diag}\left\{ \mu^{(r)} N^{(r)} \otimes I_{\bar{W}}, r = \overline{1, R} \right\} \right) \mathbf{e} = \left(\mu^{(1)} N^{(1)}, \ldots, \mu^{(R)} N^{(R)} \right)^T \otimes \mathbf{e}_{\bar{W}}.$$

So,

$$\boldsymbol{\theta} Q_0 \mathbf{e} = \left(\tilde{\boldsymbol{\theta}}_1, \ldots, \tilde{\boldsymbol{\theta}}_R \right) \text{diag}\left\{ \mu^{(r)} N^{(r)} \otimes I_{\bar{W}}, r = \overline{1, R} \right\} \mathbf{e}$$
$$= \left(\tilde{\boldsymbol{\theta}}_1 \mu^{(1)} N^{(1)} \mathbf{e}_{\bar{W}} + \cdots + \tilde{\boldsymbol{\theta}}_R \mu^{(R)} N^{(R)} \mathbf{e}_{\bar{W}} \right). \tag{5}$$

Taking into account that $\tilde{\boldsymbol{\theta}}_r \mathbf{e} = \delta_r$, the right hand side of equality (5) can be rewritten in the form $\sum_{r=1}^{R} \delta_r \mu^{(r)} N^{(r)}$. Thus the theorem is proved.

If the ergodicity condition holds true, the following limits (stationary probabilities) exist:

$$\pi(i, r, \nu) = \lim_{t \to \infty} P\{i_t = i, r_t = r, \nu_t = \nu\}, \quad i \geq 0, \quad r = \overline{1, R}, \quad \nu = \overline{0, W}.$$

Let us form row vectors π_i as follows:

$$\pi(i,r) = (\pi(i,r,0), \pi(i,r,1), \ldots, \pi(i,r,W)), \quad r = \overline{1,R},$$

$$\pi_i = (\pi(i,1), \pi(i,2), \ldots, \pi(i,R)), \quad i \geq 0.$$

It is well known that the probability vectors π_i, $i \geq 0$, satisfy the following system of linear algebraic equations:

$$(\pi_0, \pi_1, \ldots)Q = \mathbf{0}, \quad (\pi_0, \pi_1, \ldots)\mathbf{e} = 1. \tag{6}$$

System (6) is infinite and cannot be directly solved on computer. It can be solved by means of the numerically stable algorithm which is presented in [9].

4   Performance Measures of the System

Having computed the vectors of the stationary probabilities π_i, $i \geq 0$, it is possible to compute a variety of the performance measures of the system.

The stationary distribution of the number of the customers in the system is

$$\lim_{t \to \infty} P\{i_t = i\} = \pi_i \mathbf{e}, \quad i \geq 0.$$

The average number of customers in the system is $L = \sum_{i=1}^{\infty} i\pi_i \mathbf{e}$.

The average number of busy servers is

$$N_{\text{server}} = \sum_{i=0}^{\infty} \sum_{r=1}^{R} \min\left\{i, N^{(r)}\right\} \pi(i,r)\mathbf{e}.$$

The average number of customers in the buffer is

$$N_{\text{buffer}} = \sum_{i=0}^{\infty} \sum_{r=1}^{R} \max\left\{0, i - N^{(r)}\right\} \pi(i,r)\mathbf{e}.$$

The intensity of output flow of customers is

$$\lambda_{\text{out}} = \sum_{i=0}^{\infty} \sum_{r=1}^{R} \min\left\{i, N^{(r)}\right\} \mu^{(r)} \pi(i,r)\mathbf{e}.$$

The loss probability of an arbitrary customer is $P^{(\text{loss})} = 1 - \frac{\lambda_{\text{out}}}{\lambda}$.

5   Conclusion

In this paper a multi-server queueing system with an infinite buffer operating in the Markovian RE is analyzed. The total number of servers, as well as the arrival process, the rate of service of customers, and the impatience intensity depend on the state of the RE. Using the results for the asymptotically quasi-Toeplitz

Markov chains we derive the ergodicity condition and provide the algorithm for computing the stationary distribution. The main performance measures of the system are presented. Results can be applied for capacity planning and performance evaluation of a variety of computer networks, e.g., for analysis of intellectual transportation systems. The number of available servers in adjusted areas along the route of a vehicle defines the state of the RE and the changes of the RE are caused by the motion of a vehicle. Application to analysis of the systems of cognitive radio is also transparent. Here the change of available channels for the service of cognitive users is defined by the change of the number of the active licensed users, see [10,11].

Acknowledgements. This research was supported by Basic Science Research Program through the National Research Foundation of Korea (NRF) funded by the Ministry of Education (Grant No. 2014K2A1B8048465) and by Belarusian Republican Foundation of Fundamental Research (Grant No. F15KOR-001).

References

1. Kim, C.S., Dudin, A.N., Klimenok, V.I., Khramova, V.V.: Erlang loss queueing system with batch arrivals operating in a random environment. Comput. Oper. Res. **36**, 674–967 (2009)
2. Kim, C.S., Klimenok, V., Mushko, V., Dudin, A.: The BMAP/PH/N retrial queueing system operating in Markovian random environment. Comput. Oper. Res. **37**, 1228–1237 (2010)
3. Wu, J., Liu, Z., Yang, G.: Analysis of the finite source $MAP/PH/N$ retrial G-queue operating in a random environment. Appl. Math. Model. **35**, 1184–1193 (2011)
4. Yang, G., Yao, L.G., Ouyang, Z.S.: The $MAP/PH/N$ retrial queue in a random environment. Acta Math. Appl. Sin. **29**, 725–738 (2013)
5. Kim, C.S., Dudin, A., Dudin, S., Dudina, O.: Analysis of $MMAP/PH_1, PH_2/N/\infty$ queueing system operating in a random environment. Int. J. Appl. Math. Comput. Sci. **24**(3), 485–501 (2014)
6. Klimenok, V.I., Dudin, A.N.: Multi-dimensional asymptotically quasi-Toeplitz Markov chains and their application in queueing theory. Queueing Syst. **54**, 245–259 (2006)
7. Lucantoni, D.: New results on the single server queue with a batch Markovian arrival process. Commun. Stat. Stoch. Models **7**, 1–46 (1991)
8. Neuts, M.: Matrix-Geometric Solutions in Stochastic Models - An Algorithmic Approach. Johns Hopkins University Press, Baltimore (1981)
9. Dudina, O., Kim, C., Dudin, S.: Retrial queueing system with Markovian arrival flow and phase-type service time distribution. Comput. Ind. Eng. **66**, 360–373 (2013)
10. Sun, B., Lee, M.H., Dudin, S.A., Dudin, A.N.: Analysis of multiserver queueing system with opportunistic occupation and reservation of servers. Math. Probl. Eng. **2014**, 1–13 (2014). ID 178108
11. Sun, B., Lee, M.H., Dudin, S.A., Dudin, A.N.: $MAP + MAP/M_2/N/\infty$ queueing system with absolute priority and reservation of servers. Math. Probl. Eng. **2014**, 1–15 (2014). ID 813150

A Model of Erlang's Ideal Grading
with Multi-service Traffic Sources

Sławomir Hanczewski[1]([✉]), Maciej Sobieraj[1], and Joanna Weissenberg[2]

[1] Chair of Communication and Computer Networks,
Poznan University of Technology, Poznan, Poland
{slawomir.hanczewski,maciej.sobieraj}@put.poznan.pl
http://nss.et.put.poznan.pl
[2] Physics and Technical Sciences Institute of Mechanics and Applied
Computer Science, Kazimierz Wielki University, Bydgoszcz, Poland
joanna@weissenberg.pl

Abstract. The paper presents a new model of Erlang's Ideal Grading
with multi-service traffic sources. Traffic sources of this kind correspond
better with actual devices utilized by network users. As the proposed
model is an approximate one, the results obtained with its help have been
compared to the results of a computer simulation. The comparison has
confirmed the correctness of the model's theoretical assumptions. The
new model may successfully be used for modeling modern communication
systems.

Keywords: Erlang's ideal grading · BPP traffic · Multi-service sources

1   Introduction

In present-day communication systems, service integration occurs not only at the
network level, but also at the level of end devices (terminals) utilized by network
users. While network-level integration has been known and analyzed for more
than 30 years [1,2], the services integration at the level of end devices is relatively new. The best example of a system involving multi-service end devices is
the 2G/3G/4G mobile network. Nowadays, a smartphone not only handles voice
calls, but also enables (often simultaneously) web browsing, sending e-mails,
or multimedia content streaming (Fig. 1). The classic approach to multi-service
modeling telecommunication systems assumes the existence of single-service traffic sources. Each source could generate a single type of calls. In works teaching
about the analysis and modeling of communication systems, e.g., in [3,4] the proposed models used exactly that approach. Currently, it needs to be updated to
include multi-service end devices used in the network. Some models taking into
account multi-service traffic sources have started to appear in the literature. For
instance, in [5–7] the authors proposed models for multi-service networks. In [8,9]
is presented a models of of full availability group (FAG) and limited availability
group (LAG) with reservation and with threshold mechanism, respectively.

© Springer International Publishing Switzerland 2015
P. Gaj et al. (Eds.): CN 2015, CCIS 522, pp. 316–325, 2015.
DOI: 10.1007/978-3-319-19419-6_30

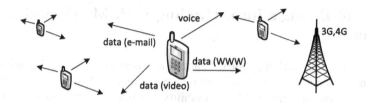

Fig. 1. Multi-service terminals (traffic sources) in mobile phone network

In parallel to the studies on models with multi-service traffic sources, research is conducted with reference to other issues in traffic engineering. For instance, one may mention the works by [10] which proposed models for queueing systems with multi-rate traffic. On the other hand, [11] presented a general model of telecommunication systems, based on a convolutional algorithm, and the authors of [12] attempted to solve the problem of modeling systems with elastic and adaptive traffic.

Erlang's Ideal Grading is one of the oldest models of communication systems. The structure of this group and its precise mathematical model for single-service traffic was elaborated by Erlang as early as in 1917 [13,14]. Initially, the model was used, though to a limited extent, to model the outgoing directions of electromechanical telephone exchanges. The limitation was caused by computation complexity (calculation of the factorial function). In the absence of easily available computation tools, a much simpler Palm-Jacobaeus' formula [15] was used more frequently, though it was less accurate. Next, the EIG model was also used in the efficient availability method [16] which enabled the calculation of the blocking probability in single-service switching networks. Various modifications of this method allowed the modeling of fields with both unicast [3,17], and multicast connections [18]. When networks with service integration appeared in the 1980s, EIG has again stopped being used for the analysis and modeling of telecommunication systems. The elaboration of an EIG model with multi-mate traffic [19,20] has restored EIG's popularity and made it useful in modeling modern telecommunication systems. For example, in [21], a model was proposed for systems with reservation; in [22], the authors attempted to model the WCDMA radio interface, and [23] proposed VoD system models and in [24,25] proposed model of systems with overflow traffic. These models, however, assume single-service traffic sources. This paper proposes a model of EIG with BPP-type (Binomial-Poisson-Pascal) traffic generated by multi-service traffic sources. As the proposed model is an approximation, the results obtained with its help have been compared to the results of a digital simulation. The comparison proved a high accuracy of the proposed method. The new model may successfully be used for modeling modern communication systems.

The rest of the paper is organized as follows. Section 2 presents the model of EIG with multi-service traffic sources and with differentiated availability. Section 3 presents the sample results. The paper is concluded in Sect. 4.

2   Model Erlang's Ideal Grading with Multi-service Sources

The structure of the Erlang's Ideal Grading is described with the following three parameters: capacity V AUs[1] (allocation units), availability d AUs and number of, so called, load groups g. The availability parameter determines the number of allocation units of the grading to which traffic sources have access. The load group is composed of traffic sources that have access to the same AUs.

Let us consider Erlang's Ideal Grading with multi-rate traffic and with differentiated availability. Assume that the group is offered m ($\mathbb{M} = \{1, 2, \ldots, m\}$) classes of calls generated by multi-service sources. Each class c ($c \in \mathbb{M}$) is described by the following parameters: the number t_c of demanded AUs to set up new connection, the intensity μ_c of exponential distribution of service time and the availability d_c. The number of load groups for class c, in the EIG, is equal to the number of possible choices d_c AUs from among all V AUs:

$$g_c = \binom{V}{d_c}. \tag{1}$$

This means that two load groups differ in at least one AU. An interesting property of EIG is the dependence of transitions between adjacent states of the service process in such a group exclusively on its structure. As a consequence, the conditional probability of transition ($\sigma_c(n)$) can be determined in a combinatorial way [13,14,20]. Figure 2 shows a simple model of the EIG with the capacity $V = 3$ AU's. The group services two classes of calls with the availability $d_1 = 2$ AUs, $d_2 = 3$ AUs. Hence, the number of load groups for relevant call classes is equal to $g_1 = 3$ and $g_2 = 1$.

In the EIG model considered in the paper, it is assumed that calls are generated by s-service traffic sources. Each such source can generate s calls at the same time (one for each supported class of calls). We also assume that traffic sources, depending on the type, can generate calls of BBP streams (Binomial-Poisson-Pascal) [27]. The traffic sources were grouped into sets. Let $\mathbb{Z}_{T,s_z}$ mark the z-th set of traffic sources, where T is one of the considered traffic streams ($T = Er, En, Pa$). This set includes all traffic sources generating the same calls: $\mathbb{C}_{T,s_z} = (1, 2, \ldots, C_{T,s_z})$.

The total traffic offered by class c is the sum of the traffic generated by traffic sources belonging to individual sets of sources. In order to determine what part of the total traffic is generated by a given set of sources, the ω_{T,s_z}^c parameter has been introduced, fulfilling the following condition:

$$\bigwedge_{c \in \mathbb{M}} \sum_{s=1}^{m} \left(\sum_{z=1}^{C_{Er,s_z}} \omega_{Er,s_z}^c + \sum_{z=1}^{C_{En,s_z}} \omega_{En,s_z}^c + \sum_{z=1}^{C_{Er,s_z}} \omega_{Pa,s_z}^c \right) = 1. \tag{2}$$

---

[1] The allocation unit (AU) defines a certain, and basic for a given system, bit rate expressed in bps [26]. Back in Erlang times [13] the equivalent to AU was a link with the capacity that would make a transmission of one call (voice service) possible.

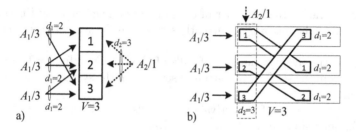

Fig. 2. Erlang's Ideal Grading with different availability [20] (a) offered traffic distribution, (b) concept of availability

The parameter ω^c_{T,s_z} is equal to 0 if a given set $\mathbb{Z}_{T,s_z}$ of sources is not able to generate calls of class c.

Knowing determined the contribution of the sources of particular types in generating calls of class c and the total intensity λ_c of generating calls of this class in the system, we can determine the value of particular types of traffic offered (Erlang, Engset, and Pascal) by s-service traffic sources.

Let us first consider Erlang's traffic streams. The distinguishing feature of this type of traffic is that the intensity of the inflow of new calls is independent of the number of serviced calls (i.e., sources from which calls are currently serviced). Therefore, the average value of the traffic offered by class c generated by this type of traffic sources from set $\mathbb{Z}_{Er,s_z}$ is equal to:

$$A_{Er,s_z,c} = \gamma_{Er,s_z,c}/\mu_c, \tag{3}$$

where $\gamma_{Er,s_z,c}$ is the arrival rate of calls of class c generated by sources that belong to the set $\mathbb{Z}_{Er,s_z}$:

$$\gamma_{Er,s_z,c} = \lambda_c \omega^c_{Er,s_z}. \tag{4}$$

In the case of Engset traffic sources, the intensity of the arrival of new calls of particular traffic classes decreases with the increase in the occupancy state of the system, while in the case of Pascal sources the intensity of the arrival of new calls of particular traffic classes increases with the increase of the occupancy state of the system. The values $A_{Er,s_z,c}(n)$ and $A_{Er,s_z,c}(n)$ for traffic offered respectively by Engset traffic sources from the set $\mathbb{Z}_{En,s_z}$ and Pascal sources from the set $\mathbb{Z}_{Pa,s_z}$, generating calls of class c in state n busy AUs, can be determined by the following formulas:

$$A_{En,s_z,c}(n) = [N_{En,s_z} - y_{En,s_z,c}(n)]\gamma_{En,s_z,c}/\mu_c, \tag{5}$$

$$A_{Pa,s_z,c}(n) = [S_{Pa,s_z} + y_{Pa,s_z,c}(n)]\gamma_{Pa,s_z,c}/\mu_c, \tag{6}$$

where:

N_{En,s_z} is the number of Engset traffic sources from the set $\mathbb{Z}_{En,s_z}$,
S_{Pa,s_z} is the number of Pascal traffic sources from the set $\mathbb{Z}_{Pa,s_z}$,

$y_{En,s_z,c}(n)$ is the average number of calls of class c generated by Engset traffic sources from the set $\mathbb{Z}_{En,s_z}$, currently serviced in the system in the occupancy state n,

$y_{Pa,s_z,c}(n)$ is the average number of calls of class c generated by Pascal traffic sources from the set $\mathbb{Z}_{Pa,s_z}$, currently serviced in the system in the occupancy state n,

$\gamma_{En,s_z,c}$ is the intensity of the arrival of new calls of class c generated by a single idle Engset traffic source from the set $\mathbb{Z}_{En,s_z}$:

$$\gamma_{En,s_z,c} = \frac{\lambda_c \omega_{En,s_z}^c}{N_{En,s_z} - y_{En,s_z,c}(n)}, \tag{7}$$

$\gamma_{Pa,s_z,c}$ is the intensity of the arrival of new calls of class c generated by a single idle Pascal traffic source from the set $\mathbb{Z}_{Pa,s_z}$:

$$\gamma_{Pa,s_z,c} = \frac{\lambda_c \omega_{Pa,s_z}^c}{S_{Pa,s_z} + y_{Pa,s_z,c}(n)}. \tag{8}$$

The occupancy distribution in Erlang's Ideal Grading can be determined on the basis of the generalized Kaufman-Roberts formula [1,2]. In [20] appropriate formula for traffic BPP was proposed. Taking into account multi-service traffic sources [28], this formula can be rewritten in the following form:

$$
\begin{aligned}
n[P_n]_V &= \sum_{s=1}^{m} \sum_{z=1}^{C_{Er,s}} \sum_{c=1}^{C_{Er,s_z}} A_{Er,s_z,c} t_c \sigma_c(n - t_c)[P_{n-t_c}]_V \\
&+ \sum_{s=1}^{m} \sum_{z=1}^{C_{En,s}} \sum_{c=1}^{C_{En,s_z}} A_{En,s_z,c}(n - t_c) t_c \sigma_c(n - t_c)[P_{n-t_c}]_V \\
&+ \sum_{s=1}^{m} \sum_{z=1}^{C_{Pa,s}} \sum_{c=1}^{C_{Pa,s_z}} A_{Pa,s_z,c}(n - t_c) t_c \sigma_c(n - t_c)[P_{n-t_c}]_V,
\end{aligned} \tag{9}
$$

where:

$[P_n]_V$ is the probability of n AUs being busy in the EIG with the capacity of V AUs,

$C_{T,s}$ is the number of sets of s-service sources $(T = Er, En, Pa)$,

$\sigma_c(n)$ is the conditional probability of transition for a traffic stream of class c in occupancy state n in the group [21]:

$$\sigma_c(n) = 1 - \sum_{x=d_c-t_c+1}^{k} \binom{d_c}{x}\binom{V-d_c}{n-x} \Big/ \binom{V}{n}, \tag{10}$$

where:

$k = n - t_c$, if $(d_c - t_c + 1) \le (n - t_c) < d_c$,
$k = d_c$, if $(n - t_c) \ge d_c$.

Note that the knowledge of the distribution $[P_n]_V$ is necessary to determine the parameters $y_{En,s_z,c}(n)$ and $y_{Pa,s_z,c}(n)$ [29]:

$$y_{En,s_z,c}(n) = \begin{cases} A_{En,s_z,c}(n-t_c)[P_{n-t_c}]_V/[P_n]_V & \text{for } n \le V \\ 0 & \text{for } n > V \end{cases} \tag{11}$$

$$y_{Pa,s_z,c}(n) = \begin{cases} A_{Pa,s_z,c}(n-t_c)[P_{n-t_c}]_V/[P_n]_V & \text{for } n \le V \\ 0 & \text{for } n > V \end{cases} \tag{12}$$

However, to determine the distribution $[P_n]_V$, the knowledge of the value of the parameters $y_{En,s_z,c}(n)$ and $y_{Pa,s_z,c}(n)$, necessary to determine the value of traffic generated by Engset and Pascal sets of traffic sources (Eqs. (5) and (6)), is needed. Equations (11) and (12) and (9) form then a system of confounding equations. In order to solve a given system of confounding equations, we will apply the iterative method proposed in [30]. To achieve that, let us assume that the distribution $\left[P_n^{(l)}\right]_V$ is the occupancy distribution determined in the l-th iteration:

$$n\left[P_n^{(l)}\right]_V = \sum_{s=1}^{m} \sum_{z=1}^{C_{Er,s}} \sum_{c=1}^{C_{Er,s_z}} A_{Er,s_z,c}^{(l)} t_c \sigma_c(n-t_c)\left[P_{n-t_c}^{(l)}\right]_V$$

$$+ \sum_{s=1}^{m} \sum_{z=1}^{C_{En,s}} \sum_{c=1}^{C_{En,s_z}} A_{En,s_z,c}^{(l)}(n-t_c)t_c\sigma_c(n-t_c)\left[P_{n-t_c}^{(l)}\right]_V$$

$$+ \sum_{s=1}^{m} \sum_{z=1}^{C_{Pa,s}} \sum_{c=1}^{C_{Pa,s_z}} A_{Pa,s_z,c}^{(l)}(n-t_c)t_c\sigma_c(n-t_c)\left[P_{n-t_c}^{(l)}\right]_V \tag{13}$$

where $A_{En,s_z,c}^{(l)}$ and $A_{Pa,s_z,c}^{(l)}$ determine the value of offered traffic in the l-th iteration by appropriate Engset and Pascal sources that belong to the set No. z of the set of s-service sources:

$$A_{En,s_z,c}^{(l)}(n) = \left[N_{En,s_z} - y_{En,s_z,c}^{(l-1)}(n)\right]\gamma_{En,s_z,c}/\mu_c, \tag{14}$$

$$A_{Pa,s_z,c}^{(l)}(n) = \left[N_{Pa,s_z} - y_{Pa,s_z,c}^{(l-1)}(n)\right]\gamma_{Pa,s_z,c}/\mu_c, \tag{15}$$

whereas $y_{En,s_z,c}^{(l-1)}$ and $y_{Pa,s_z,c}^{(l-1)}$ determine the average number of serviced calls of class c generated by traffic sources that belong to the sets $\mathbb{Z}_{En,s_z}$ and $\mathbb{Z}_{Pa,s_z}$, respectively:

$$y_{En,s_z,c}^{(l)}(n) = \begin{cases} A_{En,s_z,c}^{(l)}(n-t_c)\left[P_{n-t_c}^{(l)}\right]_V / \left[P_n^{(l)}\right]_V & \text{for } n \le V \\ 0 & \text{for } n > V \end{cases} \tag{16}$$

$$y_{Pa,s_z,c}^{(l)}(n) = \begin{cases} A_{Pa,s_z,c}^{(l)}(n-t_c)\left[P_{n-t_c}^{(l)}\right]_V / \left[P_n^{(l)}\right]_V & \text{for } n \le V \\ 0 & \text{for } n > V \end{cases} \tag{17}$$

Table 1. Details of experiments

Experiment 1	
V	64
service classes	$m = 3$, $t_1 = 1$ AUs, $d_1 = 30$ AUs, $\mu_1^{-1} = 1$, $t_2 = 2$ AUs, $d_2 = 40$ AUs, $\mu_2^{-1} = 1$, $t_3 = 6$ AUs, $d_3 = 60$ AUs, $\mu_1^{-1} = 1$
traffic sources	$\mathbb{C}_{Er,3_1} = \{1, 2, 3\}$, $\omega_{Er,3_1}^1 = 0.5$, $\omega_{Er,3_1}^2 = 0.2$, $\omega_{Er,3_1}^3 = 0.3$, $\mathbb{C}_{En,2_1} = \{1, 2\}$, $\omega_{En,2_1}^1 = 0.5$, $\omega_{En,2_1}^2 = 0.5$, $N_{En,2_1} = 50$, $\mathbb{C}_{Pa,2_2} = \{2, 3\}$, $\omega_{Pa,2_2}^2 = 0.3$, $\omega_{Pa,2_2}^3 = 0.7$, $S_{Pa,2_2} = 50$

Experiment 2	
V	30
service classes	$m = 3$, $t_1 = 1$ AUs, $d_1 = 10$ AUs, $\mu_1^{-1} = 1$, $t_2 = 2$ AUs, $d_2 = 30$ AUs, $\mu_2^{-1} = 1$, $t_3 = 3$ AUs, $d_3 = 20$ AUs, $\mu_1^{-1} = 1$
traffic sources	$\mathbb{C}_{Er,3_1} = \{1, 2, 3\}$, $\omega_{Er,3_1}^1 = 0.4$, $\omega_{Er,3_1}^2 = 0.4$, $\omega_{Er,3_1}^3 = 0.3$, $\mathbb{C}_{En,2_1} = \{1, 2\}$, $\omega_{En,2_1}^1 = 0.6$, $\omega_{Er,2_1}^2 = 0.3$, $N_{En,2_1} = 50$, $\mathbb{C}_{Pa,2_2} = \{2, 3\}$, $\omega_{Pa,2_2}^2 = 0.3$, $\omega_{Pa,2_2}^3 = 0.7$, $S_{Pa,2_2} = 50$

Experiment 3	
V	30
service classes	$m = 3$, $t_1 = 1$ AUs, $d_1 = 10$ AUs, $\mu_1^{-1} = 1$, $t_2 = 1$ AUs, $d_2 = 20$ AUs, $\mu_2^{-1} = 1$, $t_3 = 1$ AUs, $d_3 = 30$ AUs, $\mu_1^{-1} = 1$
traffic sources	$\mathbb{C}_{En,3_1} = \{1, 2, 3\}$, $\omega_{En,3_1}^1 = 1$, $\omega_{En,3_1}^2 = 1$, $\omega_{En,3_1}^3 = 1$, $N_{En,3_1} = 50$

Experiment 4	
V	64
service classes	$m = 5$, $t_1 = 1$ AUs, $d_1 = 50$ AUs, $\mu_1^{-1} = 1$, $t_2 = 4$ AUs, $d_2 = 20$ AUs, $\mu_2^{-1} = 1$, $t_3 = 4$ AUs, $d_3 = 40$ AUs, $\mu_1^{-1} = 1$, $t_4 = 6$ AUs, $d_4 = 64$ AUs, $\mu_4^{-1} = 1$, $t_5 = 2$ AUs, $d_3 = 45$ AUs, $\mu_5^{-1} = 1$
traffic sources	$\mathbb{C}_{Er,5_1} = \{1, 2, 3, 4, 5\}$, $\omega_{Er,5_1}^1 = 1$, $\omega_{Er,5_1}^2 = 0.2$, $\omega_{Er,5_1}^3 = 0.3$, $\omega_{Er,5_1}^4 = 0.3$, $\omega_{Er,5_1}^5 = 0.3$, $\mathbb{C}_{En,2_1} = \{2, 3\}$, $\omega_{En,2_1}^2 = 0.7$, $\omega_{Er,2_1}^3 = 0.8$, $N_{En,2_1} = 50$, $\mathbb{C}_{Pa,2_2} = \{4, 5\}$, $\omega_{Pa,2_2}^4 = 0.7$, $\omega_{Pa,2_2}^5 = 0.7$, $S_{Pa,2_2} = 50$

Finally, the blocking probability for calls of class c can be determined on the basis of the following formula:

$$E_c = \sum_{n=d_c-t_c+1}^{V} [1 - \sigma_c(n)][P(n)]_V. \tag{18}$$

3   Numerical Examples

The proposed model of Erlang's Ideal Grading with multi-service sources is an approximate one. In order to verify its accuracy, the results obtained with its use were compared with digital simulation results. For this purpose, an EIG simulator has been worked out and implemented in C++. In each step of the simulation, 5 series were conducted, with the length of each single series being set to 100 000 calls of the class demanding the highest number of AUs. This allowed the confidence interval to be determined at 95 %, at least one order of magnitude

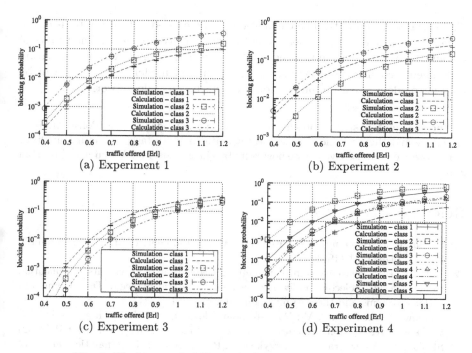

Fig. 3. Blocking probability in EIG with multi-service sources

less than the values of the results of the simulation. The blocking probability is presented in graphs in relation to the mean value of traffic a offered to a single AU in the EIG group:

$$a = \sum_{c=1}^{m} A_c/V. \qquad (19)$$

Table 1 presents the specification of the experiments the results of which are shown in Fig. 3a–d. The presented results display a high level of accuracy, confirming the correctness of the theoretical assumptions adopted. In the worst case, the difference between the simulation results and the results of the analytical model does not exceed 17%. All presented results confirm that blocking probability in EIG is strictly depends on availability parameter. It should be noted that Experiment 3 (classes 1, 2, 3) and Experiment 4 (classes 2 and 3) demonstrate the influence of the availability value on the blocking probability for classes demanding an identical number of AUs to be served (the lower the availability, the higher value of the blocking probability). It is also worth noting that the model accuracy is not dependent on the number of classes served.

4 Conclusions

The paper proposes a new method for a determination of the blocking probability in Erlang's Ideal Grading with multi-service traffic sources and with differentiated availability. The calculations to be carried out on the basis of the formulas presented in the paper are not complex or complicated. The results of a comparison of the analytical calculations with the results of the simulation experiments confirm fair accuracy of the proposed model. The model can be used for modeling telecommunications systems e.g. mobile networks.

Acknowledgements. The presented work has been funded by the Polish Ministry of Science and Higher Education within the status activity task "Struktura, analiza i projektowanie nowoczesnych systemów komutacyjnych i sieci telekomunikacyjnych" in 2015.

References

1. Kaufman, J.: Blocking in a shared resource environment. IEEE Trans. Commun. **29**(10), 1474–1481 (1981)
2. Roberts, J.: A service system with heterogeneous user requirements - application to multi-service telecommunications systems. In: Performance of Data Communications Systems and their Applications, pp. 423–431 (1981)
3. Głąbowski, M., Stasiak, M.: Point-to-point blocking probability in switching networks with reservation. Annales Des Télécommunications **57**(7–8), 798–831 (2002)
4. Stasiak, M., Wiewióra, J., Zwierzykowski, P., Parniewicz, D.: Analytical model of traffic compression in the UMTS network. In: Bradley, J.T. (ed.) EPEW 2009. LNCS, vol. 5652, pp. 79–93. Springer, Heidelberg (2009)
5. Głąbowski, M., Sobieraj, M.: Modelling of network nodes with threshold mechanisms and multi-service sources. In: 16th International Telecommunications Network Strategy and Planning Symposium, pp. 1–7 (2014)
6. Stasiak, M., Sobieraj, M., Weissenberg, J., Zwierzykowski, P.: Analytical model of the single threshold mechanism with hysteresis for multi-service networks. IEICE Trans. Commun. **E95–B(1)**, 120–132 (2012)
7. Sobieraj, M., Stasiak, M., Zwierzykowski, P.: Model of the threshold mechanism with double hysteresis for multi-service networks. In: Kwiecień, A., Gaj, P., Stera, P. (eds.) CN 2012. CCIS, vol. 291, pp. 299–313. Springer, Heidelberg (2012)
8. Głąbowski, M., Sobieraj, M., Stasiak, M.: Modelling limited-availability groups with BPP traffic and bandwidth reservation. In: 5th Advanced International Conference on Telecommunications, pp. 89–94 (2009)
9. Głąbowski, M., Sobieraj, M., Stasiak, M.: A full-availability group model with multi-service sources and threshold mechanisms. In: 8th International Symposium on Communication Systems. Networks Digital Signal Processing, pp. 1–5 (2012)
10. Hanczewski, S., Stasiak, M., Weissenberg, J.: A queueing model of a multi-service system with state-dependent distribution of resources for each class of calls. IEICE Trans. Commun. **E97–B(8)**, 1592–1605 (2014)
11. Kaliszan, A., Głąbowski, M., Stasiak, M.: Generalised convolution algorithm for modelling state-dependent systems. Circuits, Devices Syst. IET **8**(5), 378–386 (2014)
12. Moscholios, I., Vardakas, J., Logothetis, M., Boucouvalas, A.: Congestion probabilities in a batched poisson multirate loss model supporting elastic and adaptive traffic. Annales Des Télécommunications **68**(5–6), 327–344 (2013)

13. Brockmeyer, E., Halstrøm, H.L., Jensen, A.: The life and works of A.K. Erlang. In: Transactions of the Danish Academy of Technical Sciences, no. 2. Copenhagen Telephone Comany (1948)
14. Lotze, A.: History and development of grading theory. In: 5th International Teletraffic Congress, pp. 148–161 (1967)
15. Palm, C.: Nagra foljdsatser urde Erlang'ska formlerna. Tekniska meddelanden frün Kungliga Telegrafstyrdsen (1–3) (1943)
16. Ershova, E., Ershov, V.: Digital Systems for Information Distribution. Radio and Communications, Moscow (1983). [in Russian]
17. Stasiak, M.: Blocage interne point a point dans les reseaux de connexion. Annales Des Télécommunications **43**(9–10), 561–575 (1988)
18. Hanczewski, S., Stasiak, M.: Point-to-group blocking in 3-stage switching networks with multicast traffic streams. In: Dini, P., Lorenz, P., Souza, J.N. (eds.) SAPIR 2004. LNCS, vol. 3126, pp. 219–230. Springer, Heidelberg (2004)
19. Stasiak, M.: An approximate model of a switching network carrying mixture of different multichannel traffic streams. IEEE Trans. Commun. **41**(6), 836–840 (1993)
20. Głąbowski, M., Hanczewski, S., Stasiak, M., Weissenberg, J.: Modeling Erlang's ideal grading with multirate BPP traffic. Math. Probl. Eng. **2012**, 35 (2012). Article ID 456910
21. Stasiak, M., Hanczewski, S.: Approximation for multi-service systems with reservation by systems with limited-availability. In: Thomas, N., Juiz, C. (eds.) EPEW 2008. LNCS, vol. 5261, pp. 257–267. Springer, Heidelberg (2008)
22. Stasiak, M., Głąbowski, M., Hanczewski, S.: The application of the Erlang's ideal grading for modelling of UMTS cells. In: 8th International Symposium on Communication Systems. Networks Digital Signal Processing, pp. 1–6 (2012)
23. Hanczewski, S., Stasiak, M.: Performance modelling of video-on-demand systems. In: 17th Asia-Pacific Conference on Communications, pp. 784–788 (2011)
24. Głąbowski, M., Hanczewski, S., Stasiak, M.: Erlang's ideal grading in DiffServ modelling. IEEE Africon **2011**, 1–6 (2011)
25. Głąbowski, M., Hanczewski, S., Stasiak, M.: Modelling of cellular networks with traffic overflow. Math. Probl. Eng. **2015**, 15 (2015). Artical ID 286490
26. Roberts, J., Mocci, V., Virtamo, I. (eds.): Broadband Network Teletraffic, Final Report of Action COST 242. Commission of the European Communities. Springer, Berlin (1996)
27. Głąbowski, M., Sobieraj, M., Stasiak, M.: Blocking probability calculation in UMTS networks with bandwidth reservation, handoff mechanism and finite source population. In: International Symposium on Communications and Information Technologies, pp. 433–438 (2007)
28. Głąbowski, M., Sobieraj, M., Stasiak, M.: Analytical modelling of multi-service systems with multi-service BBP traffic sources. In: 2nd European Teletraffic Seminar, pp. 1–6 (2013)
29. Głąbowski, M., Kaliszan, A., Stasiak, M.: Modeling product-form state-dependent systems with BPP traffic. Perform. Eval. **67**(3), 174–197 (2010)
30. Głąbowski, M.: Modelling of state-dependent multirate systems carrying BPP traffic. Annales des Télécommunications **63**(7–8), 393–407 (2008)

Asymptotically Work-Conserving Disciplines in Communication Systems

Evsey Morozov and Lyubov Potakhina[✉]

Institute of Applied Mathematical Research, Karelian Research Centre
and Petrozavodsk State University, Petrozavodsk, Russia
emorozov@karelia.ru, lpotahina@gmail.com

Abstract. In this paper, we define a class of asymptotically work – conserving service disciplines which, in particular, are used in some existing wireless systems. Then we describe a class of state-dependent queueing systems with focus on the workload-dependent and queue-dependent systems with asymptotically work-conserving discipline. We outline the regenerative approach to stability analysis of state-dependent systems and formulate corresponding sufficient stability conditions. In the last section these conditions are verified by simulation a system with workload-dependent and queue-dependent input rates and a wireless system with Markov-modulated transmission rates.

Keywords: Work-conserving service discipline · Workload-dependent system · Queue-dependent system · Stability conditions · Regeneration · Markov-modulated transmission rate · Simulation

1 Introduction

In this paper, we introduce and discuss *asymptotically work-conserving* (AWC) service discipline, and it is an important new contribution of this research. This discipline appears in state-dependent systems and allows to control flexibly both service and input rate depending on the current state of a basic process (workload or queue size). This control mechanism opens an opportunity to optimize the system utilization. An important modern area of application of such models is green computing. To the best of our knowledge, the AWC discipline has been mentioned for the first time in paper [2], where stability analysis of a multi-server system with classical retrial policy has been developed. (More results on state-dependent systems can be found, for instance, in [1].) Then we describe in brief regenerative stability analysis method and outline the proof of sufficient stability conditions of a wide class of workload-dependent and queue-dependent systems with AWC discipline. Roughly speaking, this discipline means that server uses full its capacity (maximal service rate) when system is heavy-loaded, while server capacity utilization below a threshold can be smaller (or equals zero). The main advantage of the regenerative approach is that it covers a wide class of non-Markov processes, and only requires the existence of regenerations with finite

© Springer International Publishing Switzerland 2015
P. Gaj et al. (Eds.): CN 2015, CCIS 522, pp. 326–335, 2015.
DOI: 10.1007/978-3-319-19419-6_31

mean regeneration cycle length. A key feature of regenerative stability analysis is that it focuses on the drift of the basic process at *high levels* while its behavior at *low levels* can be freely chosen, and it does not affect the stability of the system. The latter may be used to improve design of the communication/computer system, for instance, to repair the system while it is low-loaded (or idle), to switch off the server to save energy consumption, etc. In this research we use this possibility to demonstrate various state-depending disciplines when system is not overloaded, and this is another contribution of this paper. More exactly, we simulate systems with workload-dependent and queue-dependent input rates and also a wireless system in which transmission (service) rates are Markov-modulated.

The paper is organized as follows. In Sect. 2 we first discuss the work-conserving discipline, and then introduce and discuss AWC discipline. Then, in Sect. 3.1, we present in brief regenerative method of stability analysis and outline the proofs of sufficient stability conditions of the workload-dependent and queue-dependent communication systems. In Sect. 3.2 we describe a wireless state-dependent system, where the transmission rates are governing by the irreducible finite Markov chains. This system models behaviour of the schedulers in heavy-loaded wireless networks. We remark that all considered systems, being state-dependent, use AWC discipline. Finally, in Sect. 4, simulation results of particular state-dependent models verifying stability conditions are presented.

2   Asymptotically Work-Conserving Discipline

Consider a classic FIFO queueing system $GI/G/1$, denoted by Σ, with arrival instants $\{t_n\}$, the i.i.d. inter-arrival times $\tau_n := t_{n+1} - t_n$, the i.i.d. service times $\{S_n, n \geq 0\}$, where customer 0 arrives in an empty system at instant $t_0 = 0$. (To denote a generic element of an i.i.d. sequence, we will suppress serial index.) We assume that expectations $\lambda := 1/\mathsf{E}\tau \in (0, \infty)$, $\mathsf{E}S \in (0, \infty)$. Also, at instant t, denote by $\nu(t)$ the queue size (number of customers in the system) and by $W(t)$ the workload (remaining work in the system). Define the aggregated idle time $I(t)$ of the system in interval $[0, t]$, so $B(t) := t - I(t)$ is the busy time in the same interval. Denote by $D(t)$ the work departed system in interval $[0, t]$. It is assumed in classic models that server operates at full service rate (which is assumed to be equal to 1) whenever the system is non-empty, implying the equality $B(t) = D(t)$, $t \geq 0$. This property is called *work-conserving*. Equivalently, if $W(t) > 0$, then $dI(t) = dt$ and $dB(t) = 0$, while if $W(t) > 0$, then $dB(t) = dt$, $dI(t) = 0$. Thus, work-conserving property can be also expressed in the following form,

$$\int_0^t \mathbf{1}(W(u) > 0)dI(u) = 0 \, ; \tag{1}$$

$$\int_0^t \mathbf{1}(W(u) > 0)dB(u) = D(t) \, , \tag{2}$$

where $1(\cdot)$ is indicator function. (In the expressions above we equally could use the basic process $\{\nu(t)\}$ instead of $\{W(t)\}$.) Relation (1) allows to call work-conserving discipline *non-idling*. Introduce, for arbitrary $x \geq 0$, the process

$$B_x(t) = \int_0^t 1(W(u) > x)du, \tag{3}$$

which is the time the process $\{W(t)\}$ exceeds the threshold x in interval $[0, t]$. Note that for the system Σ, $D(B_x(t)) = B_x(t)$ for all $x > 0$. Unlike classic systems, state-dependent systems can flexibly change service (and input) rate depending on the state of a basic process. In this case we assume that the *maximal* service rate is equal to 1. In many cases server uses full its capacity when the process exceeds some (finite) threshold x, however in some other systems the maximal rate is achieved only in the limit as $x \to \infty$, see [2]. Thus, for a formal definition of AWC discipline we require that, for each $x \geq 0$,

$$D(B_x(t)) \leq B_x(t), \ t \geq 0, \tag{4}$$

and

$$\lim_{x \to \infty} \lim_{t \to \infty} \frac{D(B_x(t))}{B_x(t)} = 1. \tag{5}$$

While inequality (4) shows that server may do not use full its capacity if the workload exceeds arbitrary fixed (finite) threshold x, equality (5) reflects an equivalence between departed workload and busy time as $x \to \infty$. One also can give the following alternative formulation of the AWC discipline: if the workload $W(t) \Rightarrow \infty$ (in probability), then

$$\lim_{t \to \infty} \frac{\mathsf{E}D(t)}{t} = 1. \tag{6}$$

Because in this case $\mathsf{P}(W(t) > x) \to 1$ for each x, and since $D(t) = D(t)\Big(1(W(t) > x) + 1(W(t) \leq x)\Big) \leq t$, then (6) is equivalent to the convergence of the conditional expectation: for each x,

$$\lim_{t \to \infty} \frac{\mathsf{E}(D(t) \mid W(t) > x)}{t} = 1. \tag{7}$$

Note that convergence in mean in (6) is preferable in the regenerative stability analysis, in which *saturated regime* of the system is assumed a priori. Then a pre-determined *negative drift condition* shows that the saturated regime is indeed not achieved, implying stability [2].

Remark 1. The AWC discipline in the retrial system in [2] can be also called *asymptotically non-idling* because the idle time of the server, *being positive for any orbit size*, converges (in average) to 0 as the orbit size increases.

3 Regenerative Stability Analysis

3.1 Workload-Dependent and Queue-Dependent Systems

In this section, we briefly discuss regenerative stability analysis method and its application to workload-dependent and queue-dependent systems with AWC discipline.

First we consider a single-server state-dependent FIFO system with a control mechanism allowing to change input and service rates depending on the workload (waiting time) W_n of customer n which he meets upon arrival, $n \geq 0$. To motivate our interest to this model, we mention a feedback which is used in high-performance clusters where users are required to provide a runtime estimate for the submitted job [3]. In such systems the workload is completely determined at the arrival instants, and this information can be implemented in the corresponding control mechanism. Keeping main notation from Sect. 2, we define the corresponding processes $\nu_n = \nu(t_n^-)$ and $W_n = W(t_n^-)$ embedded at the arrival instants. Assume that, if $W_n = x$, then service time S_n of customer n and the next inter-arrival time τ_n are distributed as random variables $S(x)$ and $\tau(x)$, respectively, with given (conditional) distributions depending on x only. The sequence $\{W_n\}$ satisfies *Lindley's recursion*

$$W_{n+1} = (W_n + S_n - \tau_n)^+ \quad ((x)^+ := \max(0, x)), \quad n \geq 0, \tag{8}$$

and constitutes a Markov chain with general state space. Moreover, the regenerations of the basic queueing processes are defined as follows: $\theta_0 = 0$,

$$\theta_{n+1} = \min_k \{t_k > \theta_n : \nu_k = W_k = 0\}, \quad n \geq 0, \tag{9}$$

and form a *renewal process*. Denote by θ generic regeneration period, then the renewal process $\{\theta_n\}$ is called *positive recurrent* if $\mathsf{E}\theta < \infty$. Positive recurrence is the crucial step to establish stability of the system by regenerative method [4]. Define the forward regeneration time at instant t_n as

$$\theta(n) = \min_k \{\theta_k - t_n : \theta_k - t_n > 0\}, \quad n \geq 0.$$

It is known [5] that, as $n \to \infty$,

$$\theta(n) \Rightarrow \infty \quad \text{if and only if} \quad \mathsf{E}\theta = \infty. \tag{10}$$

($\Rightarrow$ stands for convergence in probability.) Thus, the key idea of the regenerative stability analysis method is to establish that, under predetermined assumptions, $\theta(n) \not\Rightarrow \infty$, implying $\mathsf{E}\theta < \infty$.

Now, for the workload-dependent system, we outline the regenerative proof of the following intuitively clear (*negative drift*) sufficient stability condition,

$$\limsup_{x \to \infty} \mathsf{E}(S(x) - \tau(x)) < 0, \tag{11}$$

which, together with the uniform integrability of the family $\{S(x) - \tau(x)\}$ and with the assumption that for each $C > 0$

$$\inf_{x \le C} \mathsf{P}(\tau(x) > S(x) + \delta) > 0 \quad \text{(for some } \delta > 0), \tag{12}$$

imply $\mathsf{E}\theta < \infty$. The detailed proof of this result (and (13) below) is based on the arguments from [6] and will be given in a future paper, while here we outline the proof. First, we define the increment $\Delta_n := W_{n+1} - W_n$ and evaluate conditional expectation $\mathsf{E}\Delta_n$ on the event $\{W_n = x\}$ as $x \to \infty$. Then condition (11) implies that $\mathsf{E}\Delta_n$ becomes negative for all n large. This shows that the sequence $\{W_n\}$ is indeed bounded. Then, based on assumption (12), we show that, within a finite interval of time, a customer arrives which meets an empty system. Because it means regeneration, then $\theta(n) \not\to \infty$, and thus $\mathsf{E}\theta < \infty$. An important difference with the stability analysis of classic system $GI/G/1$, under well-known condition $\rho := \mathsf{E}S/\mathsf{E}\tau < 1$, is that in the latter case negative drift holds whenever system is busy, while in the state-dependent system it happens only in the limit as $W_n \Rightarrow \infty$. We note that while negative drift assumption (11) prevents unbounded increasing of the workload, condition (12) implies the existence of the infinite sequence of regenerations. (Condition (11) solely does not guarantee that the workload process, being bounded, achieves zero state with a positive probability, see [4].)

Now, keeping main notation, we consider a less studied queue-dependent system with renewal input $\{t_n\}$ and with M thresholds,

$$0 = x_0 < x_1 < \cdots < x_M < x_{M+1} := \infty,$$

such that, if the queue size $\nu_n \in [x_i, x_{i+1})$ then service time of customer n is selected from an i.i.d. sequence $\{S_n^{(i)}, n \ge 0\}$ (with generic element $S^{(i)}$) with $\mathsf{E}S^{(i)} < \infty$, $i = 0, \ldots, M$. Regenerations of this system are defined by (9) as well. Based on regenerative method we prove that this system is positive recurrent if the following negative drift assumption holds:

$$\lambda \mathsf{E}S^{(M)} < 1. \tag{13}$$

We only remark that, unlike workload-dependent system, for queue-dependent system recursion (8) no longer defines a Markov chain, and it makes stability analysis more difficult.

3.2   Wireless System with Markov-Modulated Transmission Rates

The main challenge of scheduling in wireless networks is that the available transmission rate of each job varies in time due to fading and user mobility. Because base station can estimate the achievable transmission rates with a high precision, it is desirable to identify good channel-aware schedulers. It has been detected that the base station's resources should be allocated to the jobs with the highest transmission rate, or the *best rate (BR) transmission*, see Max-Rate scheduler [7], the Proportional Fair scheduler [8,9], Relatively Best scheduler [10]. Such situation naturally leads to the following state-dependent formulation of a wireless time-slotted M-server system with K classes of jobs. It is assumed that the number of class-k jobs $A_k(t)$ arriving to the system at each instant t form an i.i.d. input sequence with generic element A_k and the mean $\lambda_k := \mathsf{E}A_k \in (0, \infty)$. The input sequences are assumed to be mutually independent. The (integer-valued) *job size* (service requirement) b_k of class-k job is measured in *bits* and has a general distribution with mean $\mathsf{E}b_k < \infty$, $k = 1, \ldots, K$. For each k-class job, the channel condition is changing from slot to slot (independently of all other jobs), and these transitions are described by a finite irreducible Markov chain with an $N_k \times N_k$-dimensional transition probabilities matrix $\boldsymbol{Q}^{(k)}$. The decision is taken at instant $t \in \{0, 1, \ldots\}$, and is applied during the slot $[t, t+1)$. When a class-k job is in channel condition n, it receives data at transmission rate $r_n^{(k)}$ bits per slot. Assume that $0 \le r_1^{(k)} < r_2^{(k)} < \cdots < r_{N_k}^{(k)}$, and define $B_k = \left\lceil b_k / r_{N_k}^{(k)} \right\rceil$, the number of slots in the BR channel condition needed to complete transmission of the job. At that, the channel uses its full capacity, so the departed work (during a slot), measured in the BR, is equal to 1. An important assumption is that the jobs in the BR states have *absolute* priority over the non-BR jobs, while the tie-breaking rule for the non-BR jobs is not specified. This system is indeed queue-dependent, however, unlike previously considered systems, the control mechanism is *indirect*: the increasing of the queue size yields the growth of the number of the BR jobs, and thus discipline becomes AWC in saturated regime. Define the BR traffic intensity of class-k job and the total traffic intensity, respectively, as

$$\rho_k = \lambda_k \mathsf{E}B_k, \quad \rho = \sum_{k=1}^{K} \rho_k. \tag{14}$$

Note that regenerations of this system are again defined by (9). The following statement for a wide class of job size distributions is in [11].

Theorem 1. *If*

$$\rho < M \quad and \quad \min_{1 \le k \le K} \mathsf{P}(A_k = 0) > 0, \tag{15}$$

then the system is positive recurrent, that is $\mathsf{E}\theta < \infty$.

We remark that heavy-loaded servers operate most of time with the BR state jobs only, and it explains the appearance of condition $\rho < M$ in the statement of Theorem 1.

4   Simulation Results

In this section, we present simulation results for the workload-dependent and queue-dependent systems with AWC service disciplines, and also for the wireless system described above. The main aim of simulation is to verify stability conditions (We remark that simulation has been carried by means of the system R [12].)

First we consider an $M/G/1$-type system with the i.i.d service times and either (1) with workload-dependent input rate, or (2) with queue-dependent input rate. Service times follow either (i) exponential distribution with parameter $\mu = 1$, or (ii) Pareto distribution $F(x) = 1 - x^{-2}$, $x \geq 1$ ($F(x) = 0$, $x \leq 1$). In each experiment the total number of arrivals are $50\,000$.

In case (i) (see Figs. 1a and 2a), in the state-dependent system (denoted by (1)), input rate λ is changed from $\lambda = \rho = 1.03$ to $\lambda = \rho = 0.97$, when the workload W_n crosses (from below) the threshold 300. This is a particular case of the workload-dependent system from Sect. 3.1, because now service times remain i.i.d. In parallel, we simulate an $M/G/1$ state-independent system (denoted by (2)) with the same service times and with input rate $\lambda = \rho = 1.03$. Thus system $M/G/1$ is unstable, while in the state-dependent system stability condition (11) is violated, if $W_n \leq 300$, and is satisfied if $W_n > 300$, implying stabilization. In case (ii) (see Figs. 1b and 2b), in the state-dependent system (1), provided the queue size $\nu_n \leq 300$, input rate is $\lambda = 0.515$ (then $\rho = 1.03$), while $\lambda = 0.485$ (then $\rho = 0.97$), if $\nu_n > 300$. Because in all cases $\rho > 1$ in the corresponding state-independent system (2), then it is unstable, as we see on Figs. 1 and 2.

Thus simulation confirms stabilization of the state-dependent systems with AWC discipline which satisfies negative drift conditions (11) (or (13)), when the basic process is above a threshold, allowing the system to be unstable (using only a part of capacity) when the process is below this threshold. It is in contrast to the state-independent system which remains unstable in all scenarios.

Now we consider a two-server system with Markov-modulated transmission rates and two classes of jobs, where channel condition states are governed by 3-state aperiodic, irreducible Markov chains with the following transition matrices

$$Q^{(1)} = \begin{pmatrix} 1/2 & 1/2 & 0 \\ 4/9 & 4/9 & 1/9 \\ 0 & 2/9 & 7/9 \end{pmatrix}, \quad Q^{(2)} = \begin{pmatrix} 1/3 & 1/3 & 1/3 \\ 4/9 & 4/9 & 1/9 \\ 0 & 1/9 & 8/9 \end{pmatrix},$$

for class-1 and class-2 jobs, respectively. We assume that, for each newly arriving job, channel is initially in condition 1, and that conditions (randomly) change at the end of each slot. The number of class-k jobs arriving in each slot follows a Poisson distribution with parameter λ_k, $k = 1, 2$. The following job size distributions are considered: i) Pareto distribution $F(x) = 1 - x_m x^{-\alpha}$, $x \geq x_m > 0$, $\alpha > 0$; and ii) Weibull distribution $F(x) = 1 - e^{-(x/s)^i}$ with parameters $s > 0$, $i > 0$ and $x \geq 0$. In all experiments simulation time is equal to 5000 slots. Transmission rates $(r_1^{(k)}, r_2^{(k)}, r_3^{(k)})$ are $(1, 3, 5)$ for class-1 jobs, and are $(1, 5, 7)$ for class-2 jobs. Recall that the BR-jobs have absolute priority, and, if no BR jobs exist, a non-BR jobs is selected randomly.

Figure 3 shows dynamics of the workload process for different job size distributions in 2-server system for two cases: when $\rho < 2$ (denoted by (1)) and when $\rho > 2$ (denoted by (1)). Figure 3a presents the workload dynamics for Weibull job size with the same scale parameter $s = 7$ for both classes and

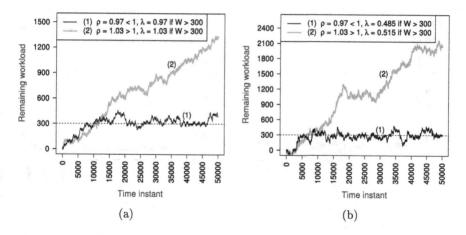

Fig. 1. Workload dynamics: (a) Exponential service time, (b) Pareto service time

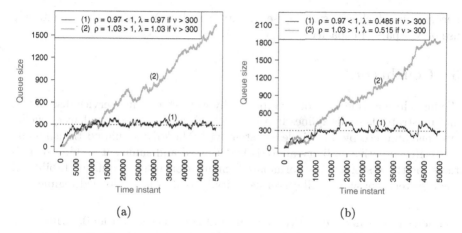

Fig. 2. Queue size dynamics: (a) Exponential service time, (b) Pareto service time

with different parameters $i_1 = 2$, $i_2 = 1.1$ for classes 1,2 respectively. Then, for $\lambda_1 = 0.65$, $\lambda_2 = 0.6$, we obtain $\rho < 2$, while for $\lambda_1 = 0.8$, $\lambda_2 = 0.5$, it follows that $\rho > 2$. Finally, Fig. 3b demonstrates the workload dynamics for Pareto job size with the same parameter $x_m = 4$ for both classes and with different parameters $\alpha_1 = 3.5$, $\alpha_2 = 5$ for classes 1,2 respectively. Then, for input rates $\lambda_1 = 0.7$, $\lambda_2 = 0.9$, we obtain $\rho < 2$, while for input rates $\lambda_1 = 0.9$, $\lambda_2 = 0.9$, it follows that $\rho > 2$. Thus, as Fig. 3 shows, the workload process remains stable when $\rho < 2$, and is unstable if $\rho > 2$. It indicates that condition $\rho < M$ in Theorem 1 is indeed stability criteria of this system. Now we explain one more time why the system with Markov-modulated transmission rates and priority of the BR users is queue-dependent with indirect control mechanism. Indeed, in this system the more jobs present, the larger the probability that a BR-job exists

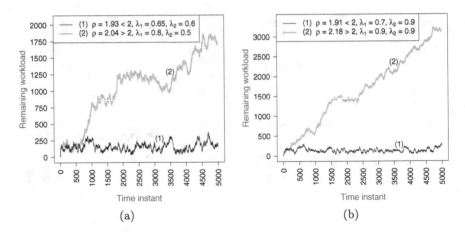

Fig. 3. Workload dynamics: (a) Weibull job size, (b) Pareto job size

becomes, in which case system uses full its capacity. Otherwise, the system may use only a part of capacity.

5   Conclusion

We introduce a wide class of asymptotically work-conserving service disciplines, and illustrate these disciplines by the workload- and queue-dependent queueing systems, and also by a a wireless system with Markov-modulated transmission rates. Regenerative stability analysis is discussed and sufficient stability conditions for these systems are formulated. Simulation results confirm stability of these systems, provided full server capacity is used in heavy-loaded regime.

Acknowledgements. Research is supported by Russian Foundation for Basic Research, projects 15-07-02341 A, 15-07-02354 A, 15-07-02360 A and by the Program of strategy development of Petrozavodsk State University.

References

1. Bekker, R., Borst, S.C., Boxma, O.J., Kella, O.: Queues with workload-dependent arrival and service rates. Queueing Syst. **46**, 537–556 (2004)
2. Morozov, E.: A multiserver retrial queue: regenerative stability analysis. Queueing Syst. **56**, 157–168 (2007)
3. Tsafrir, D., Etsion, Y., Feitelson, D.G.: Modeling user runtime estimates. In: Feitelson, D.G., Frachtenberg, E., Rudolph, L., Schwiegelshohn, U. (eds.) JSSPP 2005. LNCS, vol. 3834, pp. 1–35. Springer, Heidelberg (2005)
4. Morozov, E., Delgado, R.: Stability analysis of regenerative queueing systems. Autom. Remote Control **70**(12), 1977–1991 (2009)
5. Feller, W.: An Introduction to Probability Theory and Its Applications, vol. II. Wiley, New York (1971)

6. Morozov, E.: Stability analysis of a general state-dependent multiserver queue. J. Math. Sci. **200**(4), 462–472 (2014)
7. Knopp, R., Humblet, P.: Information capacity and power control in single-cell multiuser communications. In: Proceedings of IEEE International Conference on Communications, pp. 331–335 (1995)
8. Kushner, H., Whiting, P.: Convergence of proportional-fair sharing algorithms under general conditions. IEEE Trans. Wireless Commun. **3**, 1250–1259 (2004)
9. Bender, P., Black, P., Grob, M., Padovani, R., Sindhushayana, N., Viterbi, A.: CDMA/HDR: a bandwidth-efficient high-speed wireless data service for nomadic users. IEEE Commun. Mag. **38**(7), 70–77 (2000)
10. Borst, S.: User-level performance of channel-aware scheduling algorithms in wireless data networks. IEEE/ACM Trans. Networking **13**(3), 636–647 (2005)
11. Jacko, P., Morozov, E., Potakhina, L., Verloop, I.M.: Maximal flow-level stability of best-rate schedulers in heterogeneous wireless systems. Transactions on Emerging Telecommunications Technologies (2015) (accepted)
12. R Foundation for Statistical Computing, Vienna, Austria. http://www.R-project.org/

Unreliable Queueing System with Cold Redundancy

Valentina Klimenok[1]([✉]) and Vladimir Vishnevsky[2]

[1] Department of Applied Mathematics and Computer Science,
Belarusian State University, 220030 Minsk, Belarus
{klimenok,dudin@}bsu.by
[2] Institute of Control Sciences of Russian Academy of Sciences and Closed
Corporation "Information and Networking Technologies", Moscow, Russia
vishn@inbox.ru

Abstract. In this paper, we analyze a queueing system with so called "cold" redundancy. The system consists of an infinite buffer, the main unreliable server (server 1) and the absolutely reliable reserve server (server 2). The input flow is a $BMAP$ (Batch Markovian Arrival Process). Breakdowns arrive to the server 1 according to a MAP (Markovian Arrival Process). If the server 1 is fault-free, it serves a customer, if any. After breakdown occurrence the server 1 fails and the repair period starts immediately. The customer, whose service is interrupted by a breakdown, goes to the second server, where its service is restarted. When the repair period ends, the customer whose service on the server 2 has not yet completed goes back to the server 1 and its service begins anew. We assume that the switching from one server to another takes time. Switching times as well as service times and repair time have PH (Phase type) distribution. The queue under consideration can be applied for modeling of a hybrid communication system consisting of the FSO – Free Space Optics channel (server 1) and the radio-wave channel (server 2). We derive a condition for stable operation of the system, calculate its stationary distribution and base performance measures and derive an expression for the Laplace-Stieltjes transform of the sojourn time distribution.

Keywords: Unreliable queueing system · Batch markovian arrival process · Phase-type distribution · Stationary state distribution · Sojourn time distribution

1 Introduction

In recent years, the FSO – Free Space Optics technologies have become widespread due to their undoubted advantages. The main advantages of atmospheric optical (laser) communication link are high capacity and quality of communication. However, optical communication systems have also disadvantages, the main of which is the dependence of the communication channel on the weather condition. The unfavorable weather conditions such as rain, snow, fog, aerosols, smog

© Springer International Publishing Switzerland 2015
P. Gaj et al. (Eds.): CN 2015, CCIS 522, pp. 336–346, 2015.
DOI: 10.1007/978-3-319-19419-6_32

can significantly reduce visibility and thus significantly reduce the effectiveness of atmospheric optical communication link.

As it is mentioned in [1], one of the main directions of creating the ultra-high speed (up to 10 Gbit/s) and reliable wireless means of communication is the development of hybrid communication systems based on laser and radio-wave technologies. Unlike the FSO channel, radio-wave IEEE802.11n channel is not sensitive to weather conditions and can be considered as absolutely reliable. However, it has a lower transmission speed compared with the FSO-channel. In hybrid communication system consisting of the FSO channel and the radio-wave IEEE802.11n channel the latter can be considered as as a backup communication channel. Because of the high practical need for hybrid communication systems, a considerable amount of studies of this class of systems have appeared recently. Some results of these studies are presented in [2–4].

Papers from [2] are mainly focused on the study of stationary reliability characteristics, methods and algorithms for optimal channel switching in hybrid systems by means of simulation. The paper [3] deals with hybrid communication channel with so called "hot" redundancy, where the backup IEEE 802.11n channel continuously transmits data along with the FSO channel, but, unlike the latter, at low speed. In the paper [4], the hybrid communication system with "cold" redundancy is considered, where the radio-wave link is assumed to be absolutely reliable and backs up the atmospheric optical communication link only in cases when the latter interrupts its functioning because of the unfavorable weather conditions. The paper [1] is devoted to the study of a hybrid communication system where the millimeter-wave radio channel is used as a backup one. To model this system, the authors consider two-channel queueing system with unreliable heterogeneous servers which fail alternately.

In the present paper, we consider queueing system suitable to model a hybrid communication channel with "cold" reserve under more general, in comparison with the papers cited above, assumptions about the pattern of arrival processes of customers and breakdowns, distributions of service and repair times. Besides, we assume that the switching from one server to another takes time. Switching times as well as service times and repair time have PH (Phase type) distributions. The queue under consideration can be applied for modeling of hybrid communication system with "cold" redundancy where the radio-wave link is assumed to be absolutely reliable (its work does not depend on the weather conditions) and backs up the atmospheric optical communication link only in cases when the latter interrupts its functioning because of the unfavorable weather conditions. Upon the occurrence of favorable weather conditions the data packets begin to be transmitted over the FSO channel.

2   Mathematical Model

We consider a queueing system consisting of two heterogeneous servers and infinite waiting room. One of the servers (main server, server 1) is unreliable and the other one (reserve server, server 2) is absolutely reliable. The latter is in the so-called cold standby and connects to the service of a customer only in the case when the main server is under repair.

Customers arrive into the system in accordance with a Batch Markovian Arrival Process ($BMAP$). The $BMAP$ is defined by the underlying process ν_t, $t \geq 0$, which is an irreducible continuous-time Markov chain with the finite state space $\{0, \ldots, W\}$, and the matrix generating function $D(z) = \sum_{k=0}^{\infty} D_k z^k$, $|z| \leq 1$. The batches of customers enter the system only at the epochs of the chain ν_t, $t \geq 0$, transitions. The $(W+1) \times (W+1)$ matrices D_k, $k \geq 1$, (non-diagonal entries of the matrix D_0) define the intensities of the process ν_t, $t \geq 0$, transitions which are accompanied by generating the k-size batch of customers. The matrix $D(1)$ is an infinitesimal generator of the process ν_t, $t \geq 0$. The intensity (fundamental rate) of the $BMAP$ is defined as $\lambda = \boldsymbol{\theta} D'(1)\mathbf{e}$ where $\boldsymbol{\theta}$ is the unique solution of the system $\boldsymbol{\theta} D(1) = \mathbf{0}$, $\boldsymbol{\theta}\mathbf{e} = 1$, and the intensity of batch arrivals is defined as $\lambda_b = \boldsymbol{\theta}(-D_0)\mathbf{e}$. Here and in the sequel $\mathbf{e}(\mathbf{0})$ is a column (row) vector of appropriate size consisting of 1's (0's). For more information about the $BMAP$ see, e.g., [5].

If there is no customers in the system and the server 1 is fault-free at an arrival epoch, it immediately starts the service of an arriving customer. If the server 1 is under repair or serves a customer at an arrival epoch, an arriving customer is placed at the end of the queue in the buffer and is picked-up for service later on, according the FIFO discipline.

If a breakdown arrive to the server 1 during service of a customer, the repair of the server 1 and the switching to the server 2 begin immediately. After the switching time has expired, the customer goes to the server 2 where it starts its service anew. However, if during the switching time the server 1 becomes fault-free, the customer restarts its service on this server.

If the repair period on the server 1 ended but the service of a customer by the server 2 does not complete, then the switching to the server 1 begins. It is assumed that during the switching time the server 1 can not serve customers from the queue. After the switching time has expired, the customer goes to the server 1 where it starts its service anew. If at the end of the switching time the server 1 is under repair, the customer restarts its service on the server 2.

Breakdowns arrive to the server 1 according to a MAP which is defined by the $(V+1) \times (V+1)$ matrices H_0 and H_1. The breakdowns fundamental rate is calculated as $h = \boldsymbol{\gamma} H_1 \mathbf{e}$ where the row vector $\boldsymbol{\gamma}$ is the unique solution of the system $\boldsymbol{\gamma}(H_0 + H_1) = \mathbf{0}$, $\boldsymbol{\gamma}\mathbf{e} = 1$.

The service time of a customer by the k-th server, $k = 1, 2$, has PH type distribution with an irreducible representation $(\boldsymbol{\beta}^{(k)}, S^{(k)})$. The service process on the k-th server is directed by the Markov chain $m_t^{(k)}$, $t \geq 0$, with the state space $\{1, \ldots, M^{(k)}, M^{(k)}+1\}$ where $M^{(k)}+1$ is an absorbing state. The intensities of transitions into the absorbing state are defined by the vector $\mathbf{S}_0^{(k)} = -S^{(k)}\mathbf{e}$. The service rates are calculated as $\mu^{(k)} = -[\boldsymbol{\beta}^{(k)}(S^{(k)})^{-1}\mathbf{e}]^{-1}$, $k = 1, 2$.

The switching time from the server 2 to the server 1 has PH type distribution with an irreducible representation $(\boldsymbol{\alpha}^{(1)}, A^{(1)})$. The switching time from the server 1 to the server 2 has PH type distribution with an irreducible representation $(\boldsymbol{\alpha}^{(2)}, A^{(2)})$. The switching process on the k-th server is directed by the Markov chain $l_t^{(k)}$, $t \geq 0$, with the state space $\{1, \ldots, L^{(k)}, L^{(k)} + 1\}$ where $L^{(k)} + 1$ is an absorbing state, $k = 1, 2$. The intensities of transitions into the absorbing state are defined by the vector $\mathbf{A}_0^{(k)} = -A^{(k)}\mathbf{e}$. The switching rates are defined as $\kappa_k = -[\boldsymbol{\alpha}^{(k)}(A^{(k)})^{-1}\mathbf{e}]^{-1}$, $k = 1, 2$.

The repair period has PH type distribution with an irreducible representation $(\boldsymbol{\tau}, T)$. The repair process is directed by the Markov chain ϑ_t, $t \geq 0$, with the state space $\{1, \ldots, R, R+1\}$ where $R+1$ is an absorbing state. The intensities of transitions into the absorbing state are defined by the vector $\mathbf{T}_0 = -T\mathbf{e}$. The repair rate is $\phi = -(\boldsymbol{\tau}T^{-1}\mathbf{e})^{-1}$.

3   Process of the System States

Let at the moment t:

- i_t be the number of customers in the system, $i_t \geq 0$;
- $n_t = 0$ if the server 1 is fault-free and $n_t = 1$ if the server 1 is under repair;
- $r_t = 0$ if one of the servers serves a customer and $r_t = 1$ if the switching period takes place;
- $m_t^{(k)}$ be the state of the directing process of the service at the k-th busy server, $m_t^{(k)} = \overline{1, M^{(k)}}$, $k = 1, 2$;
- $l_t^{(1)}$ be the state of the directing process of the switching time from the server 2 to the server 1, $l_t^{(1)} = \overline{1, L^{(1)}}$;
- $l_t^{(2)}$ be the state of the directing process of the switching time from the server 1 to the server 2, $l_t^{(2)} = \overline{1, L^{(2)}}$;
- ϑ_t be the state of the directing process of the repair time at the server 1, $\vartheta_t = \overline{1, R}$;
- ν_t and η_t be the states of the directing process of the $BMAP$ and the MAP correspondingly, $\nu_t = \overline{0, W}$, $\eta_t = \overline{0, V}$.

The process of the system states is described by the regular irreducible continuous time Markov chain, $\xi_t, t \geq 0$, with the state space

$$X = \{(i, n, \nu, \eta),\ i = 0, n = 0, \nu = \overline{0, W}, \eta = \overline{0, V}\} \bigcup$$

$$\{(i, n, \nu, \eta, \vartheta),\ i = 0, n = 1, \nu = \overline{0, W}, \eta = \overline{0, V}, \vartheta = \overline{0, R}\} \bigcup$$

$$\{(i, n, r, \nu, \eta, m^{(1)}),\ i > 0, n = 0, r = 0, \nu = \overline{0, W}, \eta = \overline{0, V}, m^{(1)} = \overline{1, M^{(1)}}\} \bigcup$$

$$\{(i, n, r, \nu, \eta, l^{(1)}),\ i > 0, n = 0, r = 1, \nu = \overline{0, W}, \eta = \overline{0, V}, l^{(1)} = \overline{1, L^{(1)}}\} \bigcup$$

$$\{(i, n, r, \nu, \eta, m^{(2)}, \vartheta),\ i > 0, n = 1, r = 0, \nu = \overline{0, W}, \eta = \overline{0, V}, m^{(2)} = \overline{1, M^{(2)}},$$

$$\vartheta = \overline{1, R}\} \bigcup \{(i, n, r, \nu, \eta, \vartheta, l^{(2)}),\ i > 0, n = 1, r = 1, \nu = \overline{0, W},$$

$$\eta = \overline{0,V}, \vartheta = \overline{1,R}, l^{(2)} = \overline{1,L^{(2)}}\}.$$

In the following, we will assume that the states of the chain $\xi_t, t \geq 0$, are ordered as follows. Within the indicated above subsets of the set X the states of the chain are enumerated in the lexicographic order. Denote the obtained ranked sets as $X(0,0), X(0,1), X(i,n,r), i \geq 1, n, r = 1,2, r = 1,2$, and arrange these sets in the lexicographic order. Let $Q_{ij}, i,j \geq 0$, be the matrices formed by intensities of the chain transition from the state corresponding to the value i of the component i_n to the state corresponding to the value j of this component and $Q = (Q_{ij})_{i,j\geq0}$, be the generator of the chain. For further use in the sequel, we also introduce the following notation:

- $\otimes, \oplus$ are the symbols of Kronecker's product and sum of matrices;
- $\bar{W} = W + 1, \bar{V} = V + 1, a = \bar{W}\bar{V}$;
- $\text{diag}\{B_i, i = \overline{1,n}\}$ is the n-size diagonal matrix with diagonal blocks B_i.

Lemma 1. *Infinitesimal generator Q of the Markov chain $\xi_t, t \geq 0$, has the following block structure*

$$Q = \begin{pmatrix} \tilde{Q}_0 & \tilde{Q}_1 & \tilde{Q}_2 & \tilde{Q}_3 & \cdots \\ \hat{Q}_0 & Q_1 & Q_2 & Q_3 & \cdots \\ O & Q_0 & Q_1 & Q_2 & \cdots \\ O & O & Q_0 & Q_1 & \cdots \\ \vdots & \vdots & \vdots & \vdots & \ddots \end{pmatrix},$$

where the non-zero blocks are of the form

$$\tilde{Q}_0 = \begin{pmatrix} D_0 \oplus H_0 & I_{\bar{W}} \otimes H_1 \otimes \boldsymbol{\tau} \\ I_a \otimes T_0 & D_0 \oplus H \oplus T \end{pmatrix}, \quad \hat{Q}_0 = \begin{pmatrix} I_a \otimes \boldsymbol{S}_0^{(1)} & O \\ O_{aL^{(1)}\times a} & O \\ O & I_{aR} \otimes \boldsymbol{S}_0^{(2)} \\ O_{aRL^{(2)}\times a} & O \end{pmatrix},$$

$$\tilde{Q}_k = \begin{pmatrix} D_k \otimes I_{\bar{V}} \otimes \boldsymbol{\beta}^{(1)} & O_{a\times aL^{(1)}} & O & O_{a\times aRL^{(2)}} \\ O & O & D_k \otimes I_{\bar{V}} \otimes T \otimes \boldsymbol{\beta}^{(2)} & O \end{pmatrix}, k \geq 1,$$

$$Q_0 = \begin{pmatrix} I_a \otimes \boldsymbol{S}_0^{(1)}\boldsymbol{\beta}^{(1)} & O & O & O \\ O & O_{aL^{(1)}} & O & O \\ O & O & I_{aR} \otimes \boldsymbol{S}_0^{(2)}\boldsymbol{\beta}^{(2)} & O \\ O & O & O & O_{aRL^{(2)}} \end{pmatrix},$$

$$Q_k = \text{diag}\{D_{k-1} \otimes I_{\bar{V}} \otimes I_{M^{(1)}}, D_{k-1} \otimes I_{\bar{V}} \otimes I_{L^{(1)}},$$
$$D_{k-1} \otimes I_{\bar{V}} \otimes I_R \otimes I_{M^{(2)}}, D_{k-1} \otimes I_{\bar{V}} \otimes I_R \otimes I_{L^{(2)}}\}, k \geq 2,$$

$$Q_1 = \begin{pmatrix} D_0 \oplus \mathcal{Q}_{1,1} & O & O & I_{\bar{W}} \otimes \mathcal{Q}_{1,4} \\ I_{\bar{W}} \otimes \mathcal{Q}_{2,1} & D_0 \oplus \mathcal{Q}_{2,2} & I_{\bar{W}} \otimes \mathcal{Q}_{2,3} & O \\ O & I_{\bar{W}} \otimes \mathcal{Q}_{3,2} & D_0 \oplus \mathcal{Q}_{3,3} & O \\ I_{\bar{W}} \otimes \mathcal{Q}_{4,1} & O & I_{\bar{W}} \otimes \mathcal{Q}_{4,3} & D_0 \oplus \mathcal{Q}_{4,4} \end{pmatrix},$$

where

$$\mathcal{Q}_{1,1} = H_0 \oplus S^{(1)}, \ \mathcal{Q}_{1,4} = H_1 \otimes \mathbf{e}_{M^{(1)}} \otimes \boldsymbol{\tau} \otimes \boldsymbol{\alpha}^{(2)},$$

$$\mathcal{Q}_{2,1} = I_{\bar{V}} \otimes A_0^{(1)} \otimes \boldsymbol{\beta}^{(1)}, \ \mathcal{Q}_{2,2} = D_0 \oplus H_0 \oplus A^{(1)}, \ \mathcal{Q}_{2,3} = H_1 \otimes \mathbf{e}_{L^{(1)}} \otimes \boldsymbol{\tau} \otimes \boldsymbol{\beta}^{(2)},$$

$$\mathcal{Q}_{3,2} = I_{\bar{V}} \otimes T_0 \otimes \mathbf{e}_{M^{(2)}} \otimes \boldsymbol{\alpha}^{(1)}, \ \mathcal{Q}_{3,3} = (H_0 + H_1) \oplus T \oplus S^{(2)},$$

$$\mathcal{Q}_{4,1} = I_{\bar{V}} \otimes T_0 \otimes \mathbf{e}_{L^{(2)}} \otimes \boldsymbol{\beta}^{(1)}, \ \mathcal{Q}_{4,3} = I_{\bar{V}} \otimes I_R \otimes A_0^{(2)} \otimes \boldsymbol{\beta}^{(2)}, \ \mathcal{Q}_{4,4} = H \oplus T \oplus A^{(2)}.$$

Corollary 1. *The Markov chain $\xi_t, t \geq 0$, belongs to the class of continuous time quasi-Toeplitz Markov chains (QTMC), see [6].*

The proof follows from the form of the generator given by Lemma 1 and the definition of $QTMC$ given in [6].

In the following it will be useful to have expressions for the generating functions $\tilde{Q}(z) = \sum_{k=1}^{\infty} \tilde{Q}_k z^k$, $Q(z) = \sum_{k=0}^{\infty} Q_k z^k$, $|z| \leq 1$.

Corollary 2. *The matrix generating functions $\tilde{Q}(z)$, $Q(z)$ are of the form*

$$\tilde{Q}(z) = \begin{pmatrix} (D(z) - D_0) \otimes I_{\bar{V}} \otimes \boldsymbol{\beta}^{(1)} \ O & O & O \\ O & O \ (D(z) - D_0) \otimes I_{\bar{V}} \otimes T \otimes \boldsymbol{\beta}^{(2)} \ O \end{pmatrix}, \ (1)$$

$$Q(z) = Q_0 + \mathcal{Q}z + z \, \mathrm{diag}\,\{D(z) \otimes I_{\bar{V}} \otimes I_{M^{(1)}}, D(z) \otimes I_{\bar{V}} \otimes I_{L^{(1)}}, \quad (2)$$
$$D(z) \otimes I_{\bar{V}} \otimes I_R \otimes I_{M^{(2)}}, D(z) \otimes I_{\bar{V}} \otimes I_R \otimes I_{L^{(2)}}\},$$

where the matrix $\mathcal{Q}$ has the following block form:

$$\mathcal{Q} = \begin{pmatrix} \mathcal{Q}_{1,1} & O & O & \mathcal{Q}_{1,4} \\ \mathcal{Q}_{2,1} & \mathcal{Q}_{2,2} & \mathcal{Q}_{2,3} & O \\ O & \mathcal{Q}_{3,2} & \mathcal{Q}_{3,3} & O \\ \mathcal{Q}_{4,1} & O & \mathcal{Q}_{4,3} & \mathcal{Q}_{4,4} \end{pmatrix}, \quad (3)$$

and blocks $\mathcal{Q}_{i,j}$ are defined in the Lemma 1.

4   Stationary Distribution Performance Measures

Theorem 1. *The necessary and sufficient condition for existence of the stationary distribution of the Markov chain ξ_t, $t \geq 0$, is the fulfillment of the inequality*

$$\lambda < q_1 S_0^{(1)} + q_3 S_0^{(2)}, \quad (4)$$

where the vectors q_1, q_3 are calculated as $q_1 = x_1(\mathbf{e}_{\bar{V}} \otimes I_{M^{(1)}})$, $q_3 = x_3(\mathbf{e}_{\bar{V}R} \otimes I_{M^{(2)}})$, and the vectors x_1, x_3 are sub-vectors of the vector $\mathbf{x} = (x_1, x_2, x_3, x_4)$, which is the unique solution of the system of linear algebraic equations

$$\mathbf{x} \left[\mathcal{Q} + \mathrm{diag}\,\left\{ I_a \otimes S_0^{(1)} \boldsymbol{\beta}^{(1)}, O_{aL^{(1)}}, I_{aR} \otimes S_0^{(2)} \boldsymbol{\beta}^{(2)}, O_{aRL^{(2)}} \right\} \right] = 0, \quad \mathbf{x}\mathbf{e} = 1. \ (5)$$

Here the vectors x_1, x_2, x_3, x_4 have the dimensions $a, aL^{(1)}, aRM^{(2)}, aRL^{(2)}$ correspondingly.

Proof. It follows from [6], that a necessary and sufficient condition for existence of the stationary distribution of the chain ξ_t, $t \geq 0$, can be formulated in terms of the matrix generating function $Q(z)$ and has the form of the inequality

$$\mathbf{y}Q'(1)\mathbf{e} < 0, \tag{6}$$

where the vector $\mathbf{y}$ is the unique solution of the system of linear algebraic equations

$$\mathbf{y}Q(1) = \mathbf{0}, \tag{7}$$

$$\mathbf{y}\mathbf{e} = 1.$$

Let $\mathbf{x}$ be a stochastic vector separated into parts as $\mathbf{x} = (\mathbf{x}_1, \mathbf{x}_2, \mathbf{x}_3, \mathbf{x}_4)$. Represent the vector $\mathbf{y}$ in the form $\mathbf{y} = (\boldsymbol{\theta} \otimes \mathbf{x}_1, \boldsymbol{\theta} \otimes \mathbf{x}_2, \boldsymbol{\theta} \otimes \mathbf{x}_3, \boldsymbol{\theta} \otimes \mathbf{x}_4)$.

Then, using (2) and taking into account that $\boldsymbol{\theta} \sum\limits_{k=0}^{\infty} D_k = 0$, system (7) is reduced to the form

$$\mathbf{x}\left[\operatorname{diag}\left\{I_{\bar{V}} \otimes \boldsymbol{S}_0^{(1)}\boldsymbol{\beta}^{(1)}, O_{\bar{V}L^{(1)}}, I_{\bar{V}R} \otimes \boldsymbol{S}_0^{(2)}\boldsymbol{\beta}^{(2)}, O_{\bar{V}RL^{(2)}}\right\} + \mathcal{Q}\right] = 0. \tag{8}$$

Adding to system (8) the normalization condition, we obtain system (5).

Similarly, we conclude that inequality (4) is reduced to the following inequality:

$$\lambda + \mathbf{x}\mathcal{Q}\mathbf{e} < 0. \tag{9}$$

Taking into account the structure of the matrix $\mathcal{Q}$ given by (3), we reduce inequality (9) to the form

$$\lambda < \mathbf{x}_1\left(\mathbf{e}_{\bar{V}} \otimes S_0^{(1)}\right) + \mathbf{x}_3\left(\mathbf{e}_{\bar{V}R} \otimes S_0^{(2)}\right). \tag{10}$$

Applying mixed product rule, we derive from (10) ergodicity condition (4).

Remark 1. Intuitive explanation of stability condition (4) is as follows. The left hand side of inequality (4) is the rate of customers arriving into the system. The right hand side of the inequality is a rate of customers leaving the system after service under overload condition. It is obvious that in steady state the former rate must be less that the latter one.

Corollary 3. *In the case of stationary Poisson flow of breakdowns and exponential distribution of service and repair times, ergodicity condition (4)–(5) is reduced to the following inequality:*

$$\lambda < q_1\mu_1 + q_3\mu_2,$$

where

$$q_1 = \frac{\phi + \kappa^{(2)}}{h}\left[\frac{\phi + \kappa^{(2)}}{h} + \frac{\kappa^{(2)}}{\kappa^{(1)}} + \frac{\kappa^{(2)}(\kappa^{(1)} + h)}{\kappa^{(1)}\phi} + 1\right]^{-1}, \quad q_3 = q_1\frac{h\kappa^{(2)}(\kappa^{(1)} + h)}{\kappa^{(1)}\phi(\phi + \kappa^{(2)})}.$$

In what follows we assume inequality (4) be fulfilled. Let us enumerate the steady state probabilities in accordance with the introduced above order of the states of the chain and form the row vectors p_i of steady state probabilities corresponding the value i of the first component of the Markov chain, $i \geq 0$.

To calculate the vectors p_i, $i \geq 0$, we use the numerically stable algorithm, see [6], which has been elaborated for calculating the stationary distribution of multi-dimensional continuous time quasi-Toeplitz Markov chains.

Having the stationary distribution p_i, $i \geq 0$, been calculated we can find a number of stationary performance measures of the system. When calculating the performance measures, the following result will be useful, especially in the case when the distribution p_i, $i \geq 0$, is heavy tailed.

Lemma 2. *The vector generating function* $P(z) = \sum\limits_{i=1}^{\infty} p_i z^i$, $|z| \leq 1$, *satisfies the following equation:*

$$P(z)Q(z) = z\left[p_1 Q_0 - p_0 \tilde{Q}(z)\right]. \tag{11}$$

In particular, formula (11) can be used to calculate the value of the generating function $P(z)$ and its derivatives at the point $z = 1$ without the calculation of infinite sums. Having these derivatives been calculated, we will able to find a number of performance measures of the system. The problem of calculating the value of the $P(z)$ and its derivatives at the point $z = 1$ from Eq. (11) is non-trivial because the matrix $Q(z)$ is singular at the point $z = 1$.

Let us denote $f^{(n)}(z)$ the n-th derivative of the function $f(z)$, $n \geq 1$, and $f^{(0)}(z) = f(z)$.

Corollary 4. *The m-th, $m \geq 0$, derivatives of the vector generating function* $P(z)$ *at the point* $z = 1$ *are recursively calculated from the system of linear algebraic equations*

$$\begin{cases} P^{(m)}(1)Q(1) = \Gamma^{(m)}(1) - \sum\limits_{l=0}^{m-1} C_m^l P^{(l)}(1)Q^{(m-l)}(1), \\ P^{(m)}(1)Q'(1)\mathbf{e} = \frac{1}{m+1}\left[\Gamma^{(m+1)}(1) - \sum\limits_{l=0}^{m-1} C_{m+1}^l P^{(l)}(1)Q^{(m+1-l)}(1)\right]\mathbf{e}. \end{cases} \tag{12}$$

where

$$\Gamma^{(m)}(1) = \begin{cases} p_1 Q_0 - p_0 \tilde{Q}(1), & m = 0, \\ p_1 Q_0 - p_0 \tilde{Q}(1) - p_0 \tilde{Q}'(1), & m = 1, \\ -p_0\left[m\tilde{Q}^{(m-1)}(1) + Q^{(m)}(1)\right], & m > 1, \end{cases}$$

and the derivatives $Q^{(m)}(1)$, $\tilde{Q}^{(m)}(1)$ are calculated using formulas (1)–(2).

The proof of the corollary is parallel to the one outlined in [7] and is omitted here.

Now we are able to calculate a number of performance measures of the system under consideration.

- Throughput of the system $\varrho = q_1 S_0^{(1)} + q_3 S_0^{(2)}$.
- Mean number of customers in the system $L = \boldsymbol{P}^{(1)}(1)\mathbf{e}$.
- Variance of the number of customers in the system $V = \boldsymbol{P}^{(2)}(1)\mathbf{e} + L - L^2$.
- Probability that a system is empty and the server 1 is fault-free (is faulty)

$$P_0^{(0)} = \boldsymbol{p}_0^{(0)}\mathbf{e}, \ P_0^{(1)} = \boldsymbol{p}_0^{(1)}\mathbf{e}.$$

- Probability that the server 1 serves a customer (the server 1 is faulty and the server 2 serves a customer)

$$P_0^{(0,0)} = \boldsymbol{P}(1)\operatorname{diag}\left\{I_{aM^{(1)}}, 0_{L^{(1)}+aR(M^{(2)}+L^{(2)})}\right\}\mathbf{e}.$$

$$P_0^{(1,0)} = \boldsymbol{P}(1)\operatorname{diag}\left\{0_{a(M^{(1)}+L^{(1)})}, I_{aRM^{(2)}}, 0_{aRL^{(2)}}\right\}\mathbf{e}.$$

- Probability that the switching period from the server 2 to the server 1 (from the server 1 to the server 2) takes place

$$P_0^{(0,1)} = \boldsymbol{P}(1)\operatorname{diag}\left\{0_{aM^{(1)}}, I_{aL^{(1)}}, 0_{aR(M^{(2)}+L^{(2)})}\right\}\mathbf{e}.$$

$$P_0^{(1,1)} = \boldsymbol{P}(1)\operatorname{diag}\left\{0_{a(M^{(1)}+L^{(1)}+RM^{(2)})}, I_{aRL^{(2)}}\right\}\mathbf{e}.$$

5   Sojourn Time Distribution

Let $V(x)$ be the stationary distribution function of the sojourn time of an arbitrary customer in the system, $v(s) = \int_0^\infty e^{-sx}dV(x)$, $Re(s) \geq 0$, be the Laplace-Stieltjes transform of this function.

Theorem 2. *The Laplace-Stieltjes transform of the sojourn time stationary distribution is calculated as*

$$v(s) = \lambda^{-1}\left\{\boldsymbol{p}_0\sum_{k=1}^\infty \tilde{Q}_k\sum_{l=1}^k \varPhi^l(s) + \sum_{i=1}^\infty \boldsymbol{p}_i\sum_{k=2}^\infty Q_k\sum_{l=1}^{k-1}\varPhi^{i+l}(s)\right\}\mathbf{e} \qquad (13)$$

where
$$\varPhi(s) = (sI - \bar{Q})^{-1}Q_0, \ \bar{Q} = Q(1) - Q_0.$$

Proof. The proof is based on the probabilistic interpretation of the Laplace-Stieltjes transform. We assume that, independently on the system operation, the stationary Poisson input of so called catastrophes arrives. Let s, $s > 0$, be the rate of this flow. Then, the Laplace-Stieltjes transform $v(s)$ is interpreted as the probability of no catastrophe arrival during the sojourn time of a customer. This allows to derive the expression for $v(s)$ by means of probabilistic reasonings.

Let us assume that at the moment of the beginning of a customer service the initial phases of service time at the servers are already determined. Then the matrix of probabilities of no catastrophes arrival during the service time of the customer and corresponding transitions of the finite components of the Markov chain ξ_t, $t \geq 0$ is calculated as $\hat{\Phi}(s) = \int_0^\infty e^{(-sI+Q)t}\hat{Q}_0 dt = (sI - \bar{Q})^{-1}\hat{Q}_0$, if at the departure epoch there are no customers in the queue, and $\Phi(s) = \int_0^\infty e^{(-sI+\bar{Q})t}Q_0 dt = (sI - \bar{Q})^{-1}Q_0$, if at the departure epoch there are customers in the queue. Note, that $\hat{\Phi}(s)\mathbf{e} = \Phi(s)\mathbf{e}$ because $\hat{Q}_0\mathbf{e} = Q_0\mathbf{e}$.

Assuming that an arbitrary customer arriving in a group of size k is placed on the j-th position with probability $1/k$ and using the law of total probability, we obtain the following expression

$$v(s) = \mathbf{p}_0 \sum_{k=1}^\infty \frac{k}{\lambda}\tilde{Q}_k \sum_{l=1}^k \frac{1}{k}\Phi^l(s)\mathbf{e} + \sum_{i=1}^\infty \mathbf{p}_i \sum_{k=2}^\infty \frac{k-1}{\lambda}Q_k \sum_{l=1}^{k-1} \frac{1}{k-1}\Phi^{i+l}(s)\mathbf{e}. \quad (14)$$

Formula (13) immediately follows from formula (14). The theorem is proved.

Corollary 5. *Mean sojourn time, $\bar{v}$, of an arbitrary customer in the system is calculated as*

$$\bar{v} = -\lambda^{-1}\left[\mathbf{p}_0 \sum_{k=1}^\infty \tilde{Q}_k \sum_{l=1}^k \sum_{m=0}^{l-1}\Phi^m(0) + \sum_{i=1}^\infty \mathbf{p}_i \sum_{k=2}^\infty Q_k \sum_{l=1}^{k-1}\sum_{m=0}^{i+l-1}\Phi^m(0)\right]\Phi'(0)\mathbf{e} \quad (15)$$

where $\Phi'(0) = -(\bar{Q})^{-2}Q_0$.

Proof. To obtain formula (15), we used the relation $\bar{v} = -v'(0)$ and the fact that the matrix $\Phi(0)$ is a stochastic one.

6   Conclusion

In this paper, we study a single-server queueing system with $BMAP$ input and cold redundancy. The queue under consideration can be applied for modeling of a hybrid communication system consisting of the FSO – Free Space Optics channel and the radio wave channel. The system is studied in steady state using matrix-analytic methods. We derive a condition for stable operation of the system, calculate its stationary distribution and key performance measures and derive an expression for the Laplace-Stieltjes transform of the sojourn time distribution. This research was carried out in the framework of the applied project. Further studies suggest the computer realization of proposed algorithms and the implementation of numerical experiments to investigate the qualitative nature of the system under study.

Acknowledgments. The research is supported by the Russian Foundation for Basic Research (grant No. 14-07-90015) and the Belarusian Republican Foundation for Fundamental Research (grant No. F14R-126).

References

1. Vishnevsky, V., Kozyrev, D., Semenova, O.V.: Redundant queueing system with unreliable servers. In: Proceedings of the 6th International Congress on Ultra Modern Telecommunications and Control Systems and Workshops (ICUMT), pp. 383–386, Moscow (2014)
2. Arnon, S., Barry, J., Karagiannidis, G., Schober, R., Uysal, M. (eds.): Advanced Optical Wireless Communication Systems. Cambridge University Press, New York (2012)
3. Vishnevsky, V.M., Semenova, O.V., Sharov, SYu.: Modeling and analysis of a hybrid communication channel based on free-space optical and radio-frequency technologies. Autom. Remote Control **72**, 345–352 (2013)
4. Sharov, S.Yu., Semenova, O.V.: Simulation model of wireless channel based on FSO and RF technologies. In: Distributed Computer and Communication Networks, Theory and Applications (DCCN-2010), pp. 368–374 (2010)
5. Lucantoni, D.M.: New results on the single server queue with a batch Markovian arrival process. Commun. Stat.-Stoch. Models **7**, 1–46 (1991)
6. Klimenok, V.I., Dudin, A.N.: Multi-dimensional asymptotically quasi-Toeplitz Markov chains and their application in queueing theory. Queueing Syst. **54**, 245–259 (2006)
7. Dudin, A., Klimenok, V., Lee, M.H.: Recursive formulas for the moments of queue length in the $BMAP/G/1$ queue. IEEE Commun. Lett. **13**, 351–353 (2009)

Egalitarian Processor Sharing System with Demands of Random Space Requirement

Oleg Tikhonenko[✉]

Institute of Mathematics, Czestochowa University of Technology,
Al. Armii Krajowej 21, 42-200 Czestochowa, Poland
oleg.tikhonenko@gmail.com

Abstract. We investigate processor sharing systems with demands having some random space requirements (volumes) and demands lengths depending on their volumes. For such systems we determine non-stationary total demands capacity distribution in terms of Laplace and Laplace-Stieltjes transforms.

Keywords: Queueing system · Egalitarian processor sharing system · Demand space requirement · Total demands capacity · Laplace transform · Laplace-Stieltjes transform

1 Introduction

Egalitarian processor sharing (EPS) queueing systems are being often used in modeling data transmission in computer networks (see e.g. [1–4]). Indeed, they are an important tool in the flow level modeling of the Internet, where the TCP shares the resources of the network equally between all the flows being in progress (e.g. transfers of web pages). Of course, a processor sharing system is a slightly idealized model, since, in general, the bandwidth cannot be shared continuously (in real-valued parts). However, it is a good approximation e.g. for the file or document transfer, when it is divided into small-sized packets which are being served successively or which transmission rates from the sources have been equalized.

In the EPS service discipline, when k demands present in the system, all of them receive the service simultaneously with a service rate C/k, where C denotes the total capacity (bandwidth) of the server. In consequence, the sojourn time of a tagged demand is determined not only by number of demands (and their remaining works) found at the arrival epoch, but also by later arrivals. In fact, in the EPS system the arriving demands do not have to wait in the queue at all: the service process begins immediately after the arrival.

The literature devoted to processor sharing queueing models is very rich. In [5] the representation for the Laplace transform of the delay distribution of a tagged demand, conditioned on the required service time and the number of packets present in the system upon the arrival, was found for the $M/M/1 - EPS$ queue. An asymptotic approximation of the stationary sojourn time distribution

© Springer International Publishing Switzerland 2015
P. Gaj et al. (Eds.): CN 2015, CCIS 522, pp. 347–356, 2015.
DOI: 10.1007/978-3-319-19419-6_33

in the finite-buffer $M/M/1/K - EPS$ model was derived in [6]. A similar result for the sojourn time distribution was obtained in [7]. In [8–10], independently, the formulae for the Laplace-Stieltjes transform of the sojourn time distribution was derived for the $M/G/1 - EPS$ model. An interesting approach to the same characteristic can be found in [11], where the idea of the feedback queue was used.

The most important theoretical results for EPS-systems were obtained by S.F. Yashkov (see e.g. [10,12]). He also compiled the exhaustive reviews of the mathematical models to study EPS systems (see [13]).

We consider EPS systems of $M/G/1 - EPS$-type with demands of random space requirement. It means that each demand is characterized by some non-negative random indication named the demand space requirement or demand volume ζ. We also assume in general that demand length ξ and it's space requirement ζ are dependent.

Note that by demand length we mean the amount of work required to serve it, that is, the time of demand sojourn in the system at hand, provided that there are no other demands in the system during this time [10]. By residual length of the demand we mean the amount of work required to complete its service after some time instant, that is, the residual time of demand sojourn, provided that there are no other demands in the system during this time.

The joint distribution of ζ and ξ random variables we characterize by the joint distribution function $F(x,t) = \mathsf{P}\{\zeta < x, \xi < t\}$. Note that in this case the distribution functions of the random variables ξ and ζ take the form $B(t) = \mathsf{P}\{\xi < t\} = F(\infty, t)$ and $L(x) = \mathsf{P}\{\zeta < x\} = F(x, \infty)$ consequently. The buffer space is occupied by the demand at the epoch it arrives and is released entirely at the epoch it completes service. Denote by $\eta(t)$ the number of demands present in the system at the time instant t. We also denote by $\sigma(t)$ the total sum of space requirements of these demands. The random process $\sigma(t)$ is called the total (demands) capacity. It is clear that its characteristics can be used for buffer space determination in the nodes of computer and communication networks. Note that in the paper [14] the stationary characteristics of the process $\sigma(t)$ were obtained for the case of bounded system capacity (when there exists such $V > 0$ that $\sigma(t) < V$).

Our aim is to obtain non-stationary characteristics of the process $\sigma(t)$ under assumption that demands number and their total capacity in the system are unbounded. Our results are the generalization of ones obtained in [15]. We also present some relations for special cases of systems under consideration in steady state.

This work is organized as follows. In Sect. 2, we give the necessary notation and introduce the Markov random process describing the system behavior. Then, in Sect. 3, we obtain the non-stationary demands number distribution under zero initial condition in terms of Laplace transforms. In Sect. 4, we derive the non-stationary distribution of the total demands capacity in terms of Laplace-Stieltjes and Laplace transforms under the same initial condition. In Sect. 5, we derive the Laplace-Stieltjes transform of stationary total demands capacity distribution function and analyze some special cases when this function can be expressed by explicit form. Section 6 contains concluding remarks.

2 Random Process and Notation

We shall use the following notation. Let a be the rate of demands arrival process, $\varphi_i = \int_0^\infty x^i dL(x)$ and $\beta_i = \int_0^\infty x^i dL(x)$ be the i-th moment of the random variable ζ and ξ consequently, $i = 1, 2, \ldots$. Denote by $D(x, t) = \mathsf{P}\{\sigma(t) < x\}$ the distribution function of the total capacity of demands present in the system at time instant t. Let $\hat{\delta}(s, t)$ be the Laplace-Stieltjes transform (LST) of the function $D(x, t)$ with respect to x:

$$\hat{\delta}(s, t) = \mathsf{E}e^{-s\sigma(t)} = \int_0^\infty e^{-sx} d_x D(x, t).$$

We also introduce the Laplace transform (LT) $\overline{\delta}(s, q)$ of the function $\hat{\delta}(s, t)$ with respect to t:

$$\overline{\delta}(s, q) = \int_0^\infty e^{-qt} \mathsf{E}e^{-s\sigma(t)} dt = \int_0^\infty e^{-qt} \hat{\delta}(s, t) dt.$$

Denote by $\alpha(s, q)$ the double LST of the function $F(x, t)$:

$$\alpha(s, q) = \int_0^\infty \int_0^\infty e^{-sx-qt} dF(x, t).$$

Then, the proper LST $\varphi(s)$ and $\beta(q)$ of the distribution functions $L(x)$ and $B(t)$ take the form $\varphi(s) = \alpha(s, 0)$ and $\beta(q) = \alpha(0, q)$. The mixed $(i+j)$-th moment (if it exists) of the random vector (ζ, ξ) can be calculated as

$$\alpha_{ij} = \frac{\partial^{i+j} \alpha(s, q)}{\partial s^i \partial q^j} \bigg|_{s=0, q=0}, \quad i, j = 1, 2, \ldots.$$

Let us also introduce the following notation for vectors:

$$Y_k = (y_1, \ldots, y_k), Y_k^j = (y_1, \ldots, y_{j-1}, y_{j+1}, \ldots, y_k), \quad (Y_k, u) = (y_1, \ldots, y_k, u).$$

It is known [10] that the behavior of the system under consideration can be described by the following Markov random process:

$$\left(\eta(t), \xi_1^*(t), \ldots, \xi_{\eta(t)}^*(t) \right), \tag{1}$$

where $\xi_j^*(t)$ is the residual length of j-th demand present in the system at time instant t, $j = \overline{1, \eta(t)}$. Note that, in the case of $\eta(t) = 0$, the components $\xi_j^*(t)$ are absent in (1).

We assume that the system is empty at time instant $t = 0$, i.e. $\eta(0) = 0$ and, consequently, $\sigma(0) = 0$ (zero initial condition).

3   Non-stationary Distribution of Demands Number

The process (1) is characterized by the functions having the following probability sense:

$$P_k(t) = \mathsf{P}\{\eta(t) = k\}, \quad k = 0, 1, \dots; \tag{2}$$

$$\Theta_k(t, Y_k) = \mathsf{P}\left\{\eta(t) = k, \xi_j^*(t) < y_j, j = \overline{1, k}\right\}, \quad k = 1, 2, \dots. \tag{3}$$

It is clear that, for $k \geq 1$, we have $P_k(t) = \Theta_k(t, \infty_k)$, where $\infty_k = \underbrace{(\infty, \dots, \infty)}_{k}$.

Note that the functions $\Theta_k(t, Y_k)$ are symmetrical about permutations of components of the vector Y_k (see [10,13], where this fact was established for densities $\frac{\partial \Theta_k(t, Y_k)}{\partial y_1, \dots, \partial y_k}$).

Denote by $\hat{\theta}_k(q, Y_k)$ the LT of the function $\Theta_k(t, Y_k)$ with respect to t. Then for $k \geq 1$ we have:

$$\hat{p}_k(q) = \int_0^\infty e^{-qt} P_k(t)\mathrm{d}t = \hat{\theta}_k(q, \infty_k),$$

where $\hat{p}_k(q)$ is the LT of the function $P_k(t)$.

Let $\hat{p}_0(q)$ be the LT of the function $P_0(t)$. It's known that, for zero initial condition, the following relation holds [10]:

$$\hat{p}_0(q) = [q + a - a\pi(q)]^{-1}, \tag{4}$$

where $\pi(q)$ is the LST of busy period of the system.

Lemma 1. *For zero initial condition, the functions $\hat{\theta}_k(q, Y_k)$ can be expressed by relations:*

$$\hat{\theta}_k(q, Y_k) = \frac{(q+a)^k}{q + a - a\pi(q)} \prod_{j=1}^{k} \int_0^{y_j} \left(1 - \frac{a}{q+a}B(t)\right) \mathrm{d}t, \quad k = 1, 2, \dots. \tag{5}$$

Proof. It can be easy shown, taking into consideration the aforementioned symmetry, that the functions (3) satisfy the following equations (see e.g. [10], where these equations are presented in other notation):

$$\frac{\partial \Theta_1(t, y)}{\partial t} - \frac{\partial \Theta_1(t, y)}{\partial y} + \frac{\partial \Theta_1(t, y)}{\partial y}\bigg|_{y=0}$$
$$= aP_0(t)B(y) - a\Theta_1(t, y) + \frac{\partial \Theta_2(t, (y, u))}{\partial u}\bigg|_{u=0}; \tag{6}$$

$$\frac{\partial \Theta_k(t, Y_k)}{\partial t} - \frac{1}{k}\sum_{j=1}^{k}\left[\frac{\partial \Theta_k(t, Y_k)}{\partial y_j} - \frac{\partial \Theta_k(t, Y_k)}{\partial y_j}\bigg|_{y_j=0}\right]$$
$$= \frac{a}{k}\sum_{j=1}^{k}\Theta_{k-1}(t, Y_k^j)B(y_j) - a\Theta_k(t, Y_k) + \frac{\partial \Theta_{k+1}(t, (Y_k, u))}{\partial y}\bigg|_{u=0}, \quad k = 2, 3, \dots. \tag{7}$$

If we apply the LT to Eqs. (6) and (7), we obtain the following ones for the functions $\hat{\theta}_k(q, Y_k)$:

$$-\frac{\partial \hat{\theta}_1(q,y)}{\partial y} + \frac{\partial \hat{\theta}_1(q,y)}{\partial y}\Big|_{y=0} = a\hat{p}_0(q)B(y) - (q+a)\hat{\theta}_1(q,y) + \frac{\partial \hat{\theta}_2(q,(y,u))}{\partial u}\Big|_{u=0}; \quad (8)$$

$$-\frac{1}{k}\sum_{j=1}^{k}\left[\frac{\partial \hat{\theta}_k(q,Y_k)}{\partial y_j} - \frac{\partial \hat{\theta}_k(q,Y_k)}{\partial y_j}\Big|_{y_j=0}\right] = \frac{a}{k}\sum_{j=1}^{k}\hat{\theta}_{k-1}(q,Y_k^j)B(y_j)$$

$$-(q+a)\hat{\theta}_k(q,Y_k) + \frac{\partial \hat{\theta}_{k+1}(q,(Y_k,u))}{\partial u}\Big|_{u=0}, \quad k = 2,3,\ldots. \quad (9)$$

We shall find a solution of the Eqs. (8) and (9) in the form

$$\hat{\theta}_k(q,Y_k) = C(q)(q+a)^k \prod_{j=1}^{k}\int_0^{y_j}\left(1 - \frac{a}{q+a}B(t)\right)dt, \quad k = 1,2,\ldots, \quad (10)$$

where $C(q)$ is some function. From direct substitution, we can see that the functions (10) satisfy Eq. (9), and, if we substitute the functions (10) to the Eq. (8), we have $C(q) = \hat{p}_0(q)$. So, we obtain the statement of the lemma taking into account the relation (4).

Note that $\pi(q)$ is a unique solution of the functional equation $\pi(q) = \beta(q + a - a\pi(q))$, such that $|\pi(q)| \leq 1$ (see e.g. [10]).

Corollary 1. *In the case of $\rho = a\beta_1 < 1$, independent of initial conditions limits $\theta_k(Y_k) = \lim_{t\to\infty}\Theta_k(t,Y_k)$, $k = 1,2,\ldots$, exist and can be calculated by the relation*

$$\theta_k(Y_k) = (1-\rho)a^k \prod_{j=1}^{k}\int_0^{y_j}[1 - B(t)]dt. \quad (11)$$

Proof. If $\rho < 1$, the process (1) is regenerative, and its points of regeneration coincide with busy period termination epochs. It follows from the regeneration theory [16] and (5) that the limit $\theta_k(Y_k) = \lim_{t\to\infty}\Theta_k(t,Y_k)$ exists, and

$$\theta_k(Y_k) = \lim_{q\to 0+}q\hat{\theta}_k(q,Y_k) = \lim_{q\to 0+}\frac{q(q+a)^k}{q+a-a\pi(q)}\prod_{j=1}^{k}\int_0^{y_j}\left(1 - \frac{a}{q+a}B(t)\right)dt$$

$$= (1-\rho)a^k \prod_{j=1}^{k}\int_0^{y_j}[1 - B(t)]dt,$$

where, for $\rho < 1$, we have $\lim_{q\to 0+}\frac{q}{q+a-a\pi(q)} = 1 - \rho$ (see e.g. [17]).

Note that formulae for densities

$$f_k(Y_k) = \frac{\partial \theta_k(Y_k)}{\partial y_1 \ldots \partial y_k} = (1-\rho)a^k \prod_{j=1}^{k}[1 - B(y_j)], \quad k = 1,2,\ldots,$$

were obtained in [10].

Corollary 2. *Let $\hat{p}_k(q)$ be the LT of the function $P_k(t)$, $k = 0, 1, \ldots$, under zero initial condition. Then we have:*

$$\hat{p}_k(q) = \frac{a^k(1 - \pi(q))^k}{(q + a - a\pi(q))^{k+1}}. \tag{12}$$

Proof. It is clear that

$$\hat{p}_k(q) = \hat{\theta}_k(q, \infty_k). \tag{13}$$

Let us prove the identity

$$\int_0^\infty \left(1 - \frac{a}{q+a}B(t)\right) dt = \frac{a(1 - \pi(q))}{(q + a)(q + a - a\pi(q))}. \tag{14}$$

Indeed, we have from the normalization condition that $\hat{p}_0(q) + \sum_{k=1}^\infty \hat{\theta}_k(q, \infty_k) = 1/q$, whereas, taking into consideration (4) and Lemma 1, we obtain

$$1 + \sum_{k=1}^\infty (q + a)^k \left[\int_0^\infty \left(1 - \frac{a}{q+a}B(t)\right) dt\right]^k = \frac{q + a - a\pi(q)}{q}.$$

The identity (14) follows now from the last relation. So, the statement of the corollary follows from (14), if we take into account (4) and Lemma 1.

From (12) or from (11) we can easily obtain the known [10] result for steady-state distribution of demands number in the system (when $\rho < 1$):

$$p_k = \lim_{t \to \infty} P_k(t) = (1 - \rho)\rho^k, \quad k = 0, 1, \ldots.$$

4 Total Demands Capacity Distribution Under Zero Initial Condition

Let $\chi(t)$ be a space requirement of a demand present in the system at time instant t, and $\xi^*(t)$ be its residual length. Introduce the function $E_y(x) = \mathsf{P}\{\chi(t) < x | \xi^*(t) = y\}$. The LST $e_y(s)$ of this function with respect to x is expressed by the following relation [18]:

$$e_y(s) = [1 - B(y)]^{-1} \int_{x=0}^\infty e^{-sx} \int_{u=y}^\infty dF(x, u). \tag{15}$$

Theorem 1. *For zero initial condition, the function $\overline{\delta}(s, q)$ can be represented as*

$$\overline{\delta}(s, q) = \{[q + a - a\pi(q)][1 - I(s, q)]\}^{-1},$$

where

$$I(s, q) = (q + a) \int_0^\infty \left(1 - \frac{a}{q+a}B(y)\right) e_y(s) dy,$$

and $e_y(s)$ is determined by the relation (15).

Proof. The distribution function $D(x,t)$ can be represented in the following form:

$$D(x,t) = P_0(t) + \sum_{k=1}^{\infty} \int_0^{\infty} \cdots \int_0^{\infty} D_k(x,t|Y_k) d_{Y_k} \Theta_k(t,Y_k), \tag{16}$$

where $D_k(x,t|Y_k) = \mathsf{P}\{\sigma(t) < x | \eta(t) = k, \xi_j^*(t) = y_j, j = \overline{1,k}\}$ is the conditional distribution function of $\sigma(t)$, under condition that there are k demands in the system and residual demands lengths are equal to $y_1, \ldots, y_k$ at time instant t, $d_{Y_k} \Theta_k(t,Y_k) = \frac{\partial^k \Theta_k(t,Y_k)}{\partial y_1 \ldots \partial y_k} dy_1 \ldots dy_k$.

It's clear that the function $D_k(x,t|Y_k)$ can be expressed by Stieltjes convolution:

$$D_k(x,t|Y_k) = E_{y_1} * \cdots * E_{y_k}(x). \tag{17}$$

Passing to LST with respect to x in (16) and taking into consideration (17), we obtain:

$$\hat{\delta}(s,t) = P_0(t) + \sum_{k=1}^{\infty} \int_0^{\infty} \cdots \int_0^{\infty} \prod_{i=1}^{k} e_{y_i}(s) d_{Y_k} \Theta_k(t,Y_k). \tag{18}$$

Passing in (18) to LT with respect to t, we have:

$$\overline{\delta}(s,q) = \hat{p}_0(q) + \sum_{k=1}^{\infty} \int_0^{\infty} \cdots \int_0^{\infty} \prod_{i=1}^{k} e_{y_i}(s) d_{Y_k} \hat{\theta}_k(q,Y_k).$$

From the last relation and Lemma 1 we obtain:

$$\overline{\delta}(s,q) = [q + a - a\pi(q)]^{-1} \left\{ 1 + \sum_{k=1}^{\infty} \left[(q+a) \int_0^{\infty} \left(1 - \frac{a}{q+a} B(y) \right) e_y(s) dy \right]^k \right\}.$$

This relation is equivalent to the statement of the theorem.

Corollary 3. *Let $\delta_1(t)$ be the first moment of the process $\sigma(t)$ under zero initial condition, and $\hat{\delta}_1(q) = \int_0^{\infty} e^{-qt}\delta_1(t)dt$ be the LT of $\delta_1(t)$. Then, we have:*

$$\hat{\delta}_1(q) = \frac{(q+a)(q+a-a\pi(q))}{q^2} \int_0^{\infty} \left(1 - \frac{a}{q+a} B(y) \right) e(y)dy,$$

where $e(y) = [1 - B(y)]^{-1} \int_{x=0}^{\infty} \int_{u=y}^{\infty} x dF(x,u).$

Proof. The corollary follows from the relation $\hat{\delta}_1(q) = -\frac{\partial \overline{\delta}(s,q)}{\partial s}\big|_{s=0}$ (see [17]).

Corollary 4. *If the random variables ζ and ξ are independent, we have:*

$$\overline{\delta}(s,q) = [q + a(1 - \varphi(s))(1 - \pi(q))]^{-1}. \tag{19}$$

Proof. In this case, we obtain, taking into consideration (14), (15) and the fact that $F(x,t) = L(x)B(t)$,

$$I(s,q) = (q+a) \int_0^\infty \left(1 - \frac{a}{q+a}B(y)\right) \left[(1 - B(y))^{-1}\varphi(s) \int_{u=y}^\infty dB(u)\right] dy$$

$$= \varphi(s)(q+a) \int_0^\infty \left(1 - \frac{a}{q+a}B(y)\right) dy = \frac{a\varphi(s)(1 - \pi(q))}{q + a - a\pi(q)}.$$

Now, the formula (19) follows from the statement of the theorem.

Corollary 5. *For zero initial condition, LT $g(q,z)$ of the generation function $P(z,t)$ of the demands number in the system at time instant t has the form:*

$$g(q,z) = [q + a(1 - z)(1 - \pi(q))]^{-1}. \tag{20}$$

Proof. Follows from Corollary 4 and the fact that, in this case, we have $\zeta \equiv 1$.

Note that formula (20) was first obtained by S.F. Yashkov (see e.g. [13]). In the paper, we obtain it as a special case of the theorem and Corollary 4.

5   Steady-State Total Demands Capacity Distribution

Corollary 6. *Let $\rho = a\beta_1 < 1$. Then stationary mode exists for the system under consideration, and LST $\delta(s)$ of steady-state demands total capacity has the form:*

$$\delta(s) = \frac{1 - \rho}{1 + a\alpha_q'(s,q)|_{q=0}}. \tag{21}$$

Proof. The existence of the limit $\delta(s) = \lim_{t\to\infty} \hat\delta(s,t)$ follows from the theory of regenerative processes [16]. From this theory we also have:

$$\delta(s) = \lim_{q\to 0^+} q\overline\delta(s,q) = (1 - \rho) \lim_{q\to 0^+} [1 - I(s,q)]^{-1},$$

where

$$\lim_{q\to 0^+} I(s,q) = a \int_0^\infty [1 - B(y)]e_y(s)dy$$

$$= a \int_{x=0}^\infty \int_{u=0}^\infty ue^{-sx}dF(x,u) = -a\alpha_q'(s,q)|_{q=0},$$

whereas the statement of the corollary follows.

Note that formula (21) was first presented in [15]. From this relation, we can obtain the following formulae for the first and second moment of the steady-state total demands volume:

$$\delta_1 = \frac{a\alpha_{11}}{1 - \rho}, \quad \delta_2 = 2\delta_1^2 + \frac{a\alpha_{21}}{1 - \rho}.$$

From formula (21), we can (in some cases) determine the relation for the distribution function $D(x)$ of the stationary total demands capacity σ.

For example, consider the case when the demand capacity ζ and its length ξ are connected by the relation $\xi = c\zeta + \xi_1, c \geq 0$, where the random variables ζ and ξ_1 are independent (such dependence between demand capacity and its length is typical for many real information systems). Denote by $\kappa_1 = \mathsf{E}\xi_1$ the first moment of the random variable ξ_1. In this case we have [17]: $\alpha(s, q) = \varphi(s + cq)\kappa(s)$, where $\kappa(s)$ is the LST of the distribution function of the random variable ξ_1. Then, the relation (21) takes the following form:

$$\delta(s) = \frac{1 - \rho}{1 + a[c\varphi'(s) - \kappa_1\varphi(s)]}. \tag{22}$$

Now, let us assume that demand capacity has an exponential distribution with parameter f. In this case, from the relation (22) we obtain

$$\delta(s) = \frac{(1 - \rho)(s + f)^2}{(s + f)^2 - \rho_1 f^2 - \rho_2 f(s + f)}, \tag{23}$$

where $\rho_1 = ac/f$, $\rho_2 = a\kappa_1$, so that $\rho = a\beta_1 = \rho_1 + \rho_2$.

So, we can determine the original of the LT $\delta(s)/s$, where $\delta(s)$ is defined by formula (23), and obtain the stationary distribution function $D(x)$ in the following form:

$$D(x) = 1 - \frac{(1 - \rho)e^{-fx}}{2b}\left[\frac{(\rho_2 + b)^2 e^{(\rho_2 + b)fx/2}}{2 - \rho_2 - b} - \frac{(\rho_2 - b)^2 e^{(\rho_2 - b)fx/2}}{2 - \rho_2 + b}\right], \tag{24}$$

where $b = \sqrt{\rho_2^2 + 4\rho_1}$.

If demand length does not depend on its capacity ($c = 0$, $\rho = \rho_2$), we have from the relation (24) that

$$D(x) = 1 - \rho e^{-(1-\rho)fx}. \tag{25}$$

If demand length is proportional to its capacity ($c > 0$, $\kappa_1 = 0$), we have from (24) that

$$D(x) = 1 - \frac{\sqrt{\rho}}{2}\left[(1 + \sqrt{\rho})e^{-(1-\sqrt{\rho})fx} - (1 - \sqrt{\rho})e^{-(1+\sqrt{\rho})fx}\right], \tag{26}$$

where, in this case, $\rho = \rho_1$.

6 Conclusions

In the paper, we investigated egalitarian processor sharing (EPS) system with demands of random capacity and demand length depending on its capacity. The non-stationary demands total volume distribution was obtained in terms of Laplace and Laplace-Stieltjes transforms under zero initial condition. The non-stationary demands number distribution under the same initial condition was obtained as a special case.

We also calculated the steady-state distribution function of the total demands capacity for some special cases.

The results obtained in the paper can be used for estimating of total demands capacity characteristics for systems with EPS discipline. Such estimation is applicable to buffer space determination in the nodes of computer and communication networks.

References

1. Berger, A.W., Kogan, Y.: Dimensioning bandwidth for elastic traffic in high-speed data networks. IEEE/ACM Trans. Netw. **8**(5), 643–654 (2000)
2. Bonald, T., Massouli, L.: Impact of fairness on Internet performance. In: Proceeedings of SIGMETRICS 2001, pp. 82–91 (2001)
3. Bonald, T., May, M., Bolot, J.-C.: Analytic evaluation of RED performance. In: Proceedings of the Nineteenth Annual Joint Conference of the IEEE Computer and Communications Societies, vol. 3, pp. 1415–1424 (2000)
4. Chen, N., Jordan, S.: Throughput in processor-sharing queues. IEEE Trans. Autom. Contr. **52**(2), 299–305 (2007)
5. Coffman, J.E.G., Muntz, R.R., Trotter, H.: Waiting time distributions for processor-sharing systems. J. ACM **17**(1), 123–130 (1970)
6. Morrison, J.A.: Asymptotic analysis of the waiting-time distribution for a large closed processor-sharing system. SIAM J. Appl. Math. **46**, 140–170 (1986)
7. Knessl, C.: On the sojourn time distribution in a finite capacity processor shared queue. J. ACM **40**(5), 1238–1301 (1993)
8. Ott, T.J.: The sojourn-time distribution in the $M/G/1$ queue with processor sharing. J. Appl. Prob. **21**, 360–378 (1984)
9. Schassberger, R.: A new approach to the $M/G/1$ processor-sharing queue. Adv. Appl. Prob. **16**, 202–213 (1984)
10. Yashkov, S.F.: Analysis of Computer Queues. Radio i Svyaz, Moscow (1989). [in Russian]
11. Van den Berg, J.L., Boxma, O.: The $M/G/1$ queue with processor sharing and its relation to a feedback queue. Queueing Syst. **9**, 365–402 (1991)
12. Yashkov, S.F.: A derivation of response time distribution for a $M/G/1$ processor sharing queue. Probl. Contr. Info. Theor. **12**, 133–148 (1983)
13. Yashkov, S.F., Yashkova, A.S.: Processor sharing: a survey of the mathematical theory. Autom. Remote Control. **68**(9), 1662–1731 (2007)
14. Tikhonenko, O.M.: Queuing systems with processor sharing and limited resources. Autom. Remote Control. **71**(5), 803–815 (2010)
15. Sengupta, B.: The spatial requirement of M/G/1 queue or: how to design for buffer space. In: Baccelli, F., Fayolle, G. (eds.) Modeling and Performance Evaluation Methodology. LNCIS, vol. 60, pp. 545–562. Springer, Berlin (1984)
16. Lakatos, L., Szeidl, L., Telek, M.: Introduction to Queueing Systems with Telecommunication Applications. Springer, New York (2010)
17. Tikhonenko, O.M.: Queueing Models in Information Systems. Universitetskoe, Minsk (1990). [in Russian]
18. Tikhonenko, O.: Computer Systems Probability Analysis. Akademicka Oficyna Wydawnicza EXIT, Warsaw (2006). [in Polish]

Stochastic Bounds for Markov Chains
with the Use of GPU

Jarosław Bylina[1]([✉]), Jean-Michel Fourneau[2], Marek Karwacki[1],
Nihal Pekergin[3], and Franck Quessette[2]

[1] Institute of Mathematics, Marie Curie-Skłodowska University, Lublin, Poland
jaroslaw.bylina@umcs.pl
[2] PRiSM, CNRS and Univ. Versailles St Quentin, Versailles, France
[3] LACL, UPEC, Créteil, France

Abstract. The authors present a new approach to find stochastic bounds
for a Markov chain – namely with the use of the GPU for computing the
bounds. A known algorithm [1,2] is used and it is rewritten to suit the
GPU architecture with the cooperation of the CPU. The authors do
some experiments with matrices from various models as well as some
random matrices. The tests are analyzed and some future considerations
are given.

Keywords: Markov chains · GPU · Stochastic bounds · Sparse matrices · Heterogeneous algorithms

1 Introduction

Modeling real systems – network systems, in particular – with the use of Markov
chains is a well known and recognized method [3]. It gives very good results.
However, among many advantages, it has some drawbacks.

The main problem is that to get a reasonable precision of the model (and here
we are not talking about the numerical precision, but rather about the needed
accuracy of the representation of reality in the model – especially when we are
to express the system "memory" in the "memory-less" Markov process [3]), we
have to prepare a quite big model. That means that the model have a lot of
states, hence it demands quite a lot of time and space to get the solution [4].

However, there are some ways to decrease the number of the investigated
states or change the structure of the matrix to more convenient for computations.
Some of the methods [5–9] could use the Vincent's algorithm [1,2].

The main idea of this work is to adapt the Vincent's algorithm to machines
with GPUs [10].

GPUs – that is Graphics Processing Units – are specialized electronic chips
with many cores used to process rapidly a lot of similar data (as encountered

---

This research is partly supported by a PHC Polonium grant. The French teams are
supported by grant ANR-12-MONU-00019.

P. Gaj et al. (Eds.): CN 2015, CCIS 522, pp. 357–370, 2015.
DOI: 10.1007/978-3-319-19419-6_34

in computer graphics). In the form of General Purpose Graphics Processing Units (GPGPUs) they are used more and more often as stream processors – not necessarily devoted solely to graphical operations.

The main advantage of GPUs is its ability to work in parallel with large amount of data – like big matrices which represent Markov chains. However, a good work distribution and proper data formats are needed to fully utilize the power of the GPU.

The article is put as follows. Section 2 presents some mathematical background and Sect. 3 describes the original Vincent's algorithm (with its parallel version). Section 4 is dedicated to chosen formats for storing sparse matrices. Section 5 describes GPU implementation of the Vincent's algorithm and its hybrid (GPU-CPU) form. Section 6 shows some experimental results and Sect. 7 discusses the achieved improvements and issues related to the devices used. The last one, Sect. 8 gives some conclusion and future possibilities tied to the algorithm.

2   Theoretical Background

The method used in the article is based on stochastic ordering [11–14].

2.1   Stochastic Ordering

Briefly, we can define a stochastic ordering ($\leq_{st}$) of two probability vectors as follows:

$$\mathbf{p} \leq_{st} \mathbf{q} \quad \text{iff} \quad \forall_{k \in \{1,\dots,N\}} \sum_{i=k}^{N} p_i \leq \sum_{i=k}^{N} q_i, \tag{1}$$

where:

- $\mathbf{p}$ and $\mathbf{q}$ are probability (row) vectors of size N each;
- $\mathbf{p} = [p_1, \dots, p_N]$;
- $\mathbf{q} = [q_1, \dots, q_N]$.

Further, we can define the stochastic ordering ($\leq_{st}$) of two random variables as follows:

$$X \leq_{st} Y \quad \text{iff} \quad \mathbf{p} \leq_{st} \mathbf{q}, \tag{2}$$

where:

- X and Y are both random variables with values from $\{1, \dots, N\}$;
- $\mathbf{p}$ is the probability vector (of size N) describing the probability mass function of X;
- $\mathbf{q}$ is the probability vector (of size N) describing the probability mass function of Y.

It could be easily proven that:

$$X \leq_{st} Y \quad \text{iff} \quad \forall_{\text{non-decreasing function } f} \quad E[f(X)] \leq E[f(Y)]. \tag{3}$$

The statement (3) is very important because a lot of interesting characteristics of probability models are non-decreasing functions. That means, that if we are interested in estimating an index given by a non-decreasing function f of a random variable X and if we find a random variable Y (or Z) which restricts the variable X in the sense of the stochastic ordering from the top (bottom, respectively) we have also an upper (lower, respectively) limitation of $E[f(X)]$ given as $E[f(Y)]$ $(E[f(Z)]$, respectively).

The point is to efficiently find Y (or Z) which could be easier to investigate than X.

2.2   Comparison of Discrete Time Markov Chains

We can also compare stochastic processes analogously. For discrete time Markov chains (DTMCs) we can define the stochastic ordering ($\leq_{st}$) as follows:

$$\{X(n)\} \leq_{st} \{Y(n)\} \quad \text{iff} \quad \forall_{n>0}\big[X(0) \leq_{st} Y(0) \Rightarrow X(n) \leq_{st} Y(n)\big], \quad (4)$$

where $\{X(n)\}$ and $\{Y(n)\}$ are DTMCs with $n \in \mathbb{N}$.

Thus, taking $n \to \infty$ we have also:

$$\{X(n)\} \leq_{st} \{Y(n)\} \Rightarrow \mathbf{\Pi}^X \leq_{st} \mathbf{\Pi}^Y, \quad (5)$$

where $\mathbf{\Pi}^X$ and $\mathbf{\Pi}^Y$ are steady-state distribution of $\{X(n)\}$ and $\{Y(n)\}$, respectively.

A time-homogeneous DTMC $\{X(n)\}$ is defined by its initial distribution $X(0)$ and its stochastic matrices of transition probabilities $\mathbf{P}$. We can define the stochastic ordering also for stochastic matrices:

$$\mathbf{R} \leq_{st} \mathbf{S} \quad \text{iff} \quad \forall_{i \in \{1,\dots,N\}} \ \mathbf{R}_{i,*} \leq_{st} \mathbf{S}_{i,*}, \quad (6)$$

where:

– $\mathbf{R}$ and $\mathbf{S}$ are stochastic matrices of size $N \times N$;
– $\mathbf{M}_{k,*}$ denotes the k-th row of the matrix $\mathbf{M}$.

One more definition of some stochastic matrices properties is needed – that is stochastic monotonicity (st-monotonicity) of a stochastic matrix $\mathbf{P}$ of a size $N \times N$:

$$\mathbf{P} \quad \text{is } st\text{-monotone} \quad \text{iff} \quad \forall_{i \in \{2,\dots,N\}} \ \mathbf{P}_{i-1,*} \leq_{st} \mathbf{P}_{i,*}. \quad (7)$$

(This definition is not the most general definition of st-monotonicity, but quite sufficient for our considerations.)

The main corollary necessary for the next subsection is following: given two time-homogeneous DTMCs $\{X(n)\}$ and $\{Y(n)\}$ with transition probabilities matrices $\mathbf{P}^X$ and $\mathbf{P}^Y$, respectively, if:

– $X(0) \leq_{st} Y(0)$,
    and

- $\mathbf{P}^X$ or $\mathbf{P}^Y$ (or both) is *st*-monotone,
  and
- $\mathbf{P}^X \leq_{st} \mathbf{P}^Y$,

it holds that:

$$\{X(n)\} \leq_{st} \{Y(n)\}. \tag{8}$$

3   Vincent's Algorithm

3.1   Sequential Vincent's Algorithm

We can define an operator $\max_{st}$ which takes two vectors $\mathbf{p}$, $\mathbf{q}$ of the same size N and returns another vector of the same size which is its arguments' upper bound in the sense of $\leq_{st}$:

$$\mathbf{r} = \max{}_{st}(\mathbf{p}, \mathbf{q}) \quad \text{iff} \quad \forall_{i \in \{1,\dots,N\}} \ \sum_{j=i}^{N} r_j = \max\left(\sum_{j=i}^{N} p_j, \sum_{j=i}^{N} q_j \right), \tag{9}$$

where:

- $\mathbf{p} = [p_1, \dots, p_N]$;
- $\mathbf{q} = [q_1, \dots, q_N]$;
- $\mathbf{r} = [r_1, \dots, r_N]$.

This bound is optimal, that is:

$$\forall_{\mathbf{s}}\left(\mathbf{p} \leq_{st} \mathbf{s} \wedge \mathbf{q} \leq_{st} \mathbf{s}\right) \Rightarrow \max{}_{st}(\mathbf{p}, \mathbf{q}) \leq_{st} \mathbf{s}. \tag{10}$$

Given a time-homogeneous DTMC with a transition probabilities matrix $\mathbf{P}$, we can use formulas (1), (6), (7) and (9) to obtain an algorithm (known as Vincent's Algorithm [1,2] – shown in Fig. 1) which can produce a stochastic matrix $\mathbf{V}$ which is the optimal and unique *st*-monotone upper stochastic bound for the matrix $\mathbf{P}$. That is, $\mathbf{V}$ cannot be decreased in the sense of $\leq_{st}$, conserving the *st*-monotonicity at the same time.

Input: $\mathbf{P}$
Output: $\mathbf{V}$

$\mathbf{V}_{1,*} \leftarrow \mathbf{P}_{1,*}$
for $i = 2, \dots, N$:
    $\mathbf{V}_{i,*} \leftarrow \max{}_{st}\left(\mathbf{P}_{i,*}, \mathbf{V}_{i-1,*}\right)$

Fig. 1. Vincent's Algorithm for the upper bound ($\mathbf{P}$ and $\mathbf{V}$ are of size $N \times N$, while $\mathbf{M}_{k,*}$ denotes the k-th row of the matrix $\mathbf{M}$ and $\max_{st}$ is defined in (9))

$$\begin{bmatrix} .1 & .7 & .0 & .2 \\ .3 & .3 & .1 & .3 \\ .2 & .4 & .4 & .0 \\ .5 & .0 & .1 & .4 \end{bmatrix} \rightarrow \begin{bmatrix} .1 & .7 & .0 & .2 \\ .1 & .5 & .1 & .3 \\ .1 & .5 & .1 & .3 \\ .1 & .4 & .1 & .4 \end{bmatrix}$$

Fig. 2. A stochastic matrix (left) and its upper bound (right) obtained with the use of Vincent's Algorithm

In other words:

$$\forall_{st\text{-monotone stochastic matrix } \mathbf{S}} \quad \mathbf{P} \leq_{st} \mathbf{S} \Rightarrow \mathbf{V} \leq_{st} \mathbf{S}. \tag{11}$$

An example of the upper bound computed by Vincent's Algorithm is in Fig. 2.

There is quite an analogous version of Vincent's Algorithm for finding the optimal and unique st-monotone lower stochastic bound for the given matrix – but they behave in the same way, considering their technical and computational properties, so we focus on the former one (for the upper bound).

3.2 Parallel Vincent's Algorithm

Vincent's Algorithm can be quite easily parallelized if we split this algorithm in three steps and define following three operations for every step (see also Figs. 3, 4 and 5 which present the operations as algorithms performed in-place):

$$\mathbf{B} = S(\mathbf{A}) \quad \text{iff} \quad \forall_{i,j \in \{1,\ldots,N\}} \; \mathbf{B}_{i,j} = \sum_{k=j}^{N} \mathbf{A}_{i,k}, \tag{12}$$

$$\mathbf{B} = M(\mathbf{A}) \quad \text{iff} \quad \forall_{i,j \in \{1,\ldots,N\}} \; \mathbf{B}_{i,j} = \max_{k \in \{1,i\}} \mathbf{A}_{k,j}, \tag{13}$$

$$\mathbf{B} = D(\mathbf{A}) \quad \text{iff} \quad \forall_{i \in \{1,\ldots,N\}} \; \mathbf{B}_{i,N} = \mathbf{A}_{i,N}$$

$$\wedge \quad \forall_{i \in \{1,\ldots,N\}, j \in \{1,\ldots,N-1\}} \; \mathbf{B}_{i,j} = \mathbf{A}_{i,j} - \mathbf{A}_{i,j+1}, \tag{14}$$

where:

- $\mathbf{A}$ and $\mathbf{B}$ are matrices of size $N \times N$;
- $\mathbf{M}_{i,j}$ denotes an element in the i-th row and j-th column of the matrix $\mathbf{M}$;
- $\mathbf{M}_{k,*}$ denotes the k-th row of the matrix $\mathbf{M}$;
- $\mathbf{M}_{*,k}$ denotes the k-th column of the matrix $\mathbf{M}$.

Now, we can represent the result of Vincent's Algorithm as a composition of these three functions:

$$V(\mathbf{P}) = D(M(S(\mathbf{P}))), \tag{15}$$

where $V(\mathbf{P})$ denotes the matrix obtained from Vincent's Algorithm.

Each of these functions can be easily parallelized row-wise (D and S) or column-wise (M), because each one works on rows (D and S) or columns (M) independently.

An example of these three steps performed on the matrix from Fig. 2 is shown in Fig. 6.

```
Input:    A
Output:    A (in-place)

parallel for i = 1, . . . , N:
    for j = (N − 1), . . . , 1:
        A_{i,j} ← A_{i,j} + A_{i,j+1}
```

Fig. 3. The first step (12) of parallelized Vincent's Algorithm (15) ($\mathbf{A}$ is a matrix of size $N \times N$, while $\mathbf{A}_{i,j}$ denotes an element in the i-th row and j-th column of the matrix $\mathbf{A}$

```
Input:    A
Output:    A (in-place)

parallel for j = 1, . . . , N:
    for i = 2, . . . , N:
        A_{i,j} ← max(A_{i,j}, A_{i−1,j})
```

Fig. 4. The second step (13) of parallelized Vincent's Algorithm (15) ($\mathbf{A}$ is a matrix of size $N \times N$, while $\mathbf{A}_{i,j}$ denotes an element in the i-th row and j-th column of the matrix $\mathbf{A}$

```
Input:    A
Output:    A (in-place)

parallel for i = 1, . . . , N:
    for j = (N − 1), . . . , 1:
        A_{i,j} ← A_{i,j} − A_{i,j+1}
```

Fig. 5. The third step (14) of parallelized Vincent's Algorithm (15) ($\mathbf{A}$ is a matrix of size $N \times N$, while $\mathbf{A}_{i,j}$ denotes an element in the i-th row and j-th column of the matrix $\mathbf{A}$

4   Sparse Matrices on GPU

Probabilistic matrices representing Markov chains tend to be very sparse. There are a few formats to store sparse matrices [15,16]. We chose CSR (Compressed Sparse Rows), because there is a lot of row-wise operations (S and D) and CSR is efficient for such processing. Unfortunately, M is column-wise and for such an operation CSC (Compressed Sparse Columns) would be better. However, this operation is a minority.

$$\begin{bmatrix} .1 & .7 & .0 & .2 \\ .3 & .3 & .1 & .3 \\ .2 & .4 & .4 & .0 \\ .5 & .0 & .1 & .4 \end{bmatrix} \xrightarrow{S} \begin{bmatrix} 1.0 & .9 & .2 & .2 \\ 1.0 & .7 & .4 & .3 \\ 1.0 & .8 & .4 & .0 \\ 1.0 & .5 & .5 & .4 \end{bmatrix}$$

$$M \swarrow$$

$$\begin{bmatrix} 1.0 & .9 & .2 & .2 \\ 1.0 & .9 & .4 & .3 \\ 1.0 & .9 & .4 & .3 \\ 1.0 & .9 & .5 & .4 \end{bmatrix} \xrightarrow{D} \begin{bmatrix} .1 & .7 & .0 & .2 \\ .1 & .5 & .1 & .3 \\ .1 & .5 & .1 & .3 \\ .1 & .4 & .1 & .4 \end{bmatrix}$$

Fig. 6. An example of three steps of parallelized Vincent's Algorithm

The CSR format stores only non-zero entries of the matrix (an array `val`, sorted left-to-right-then-top-to-bottom by their position in the original matrix), their corresponding column indices (an array `col_ind`) and the indices of the start of every row in the first two arrays (an array `row_ptr` – with an additional index pointing to the first non-existent entry). Figure 7 shows a stochastic matrix and its CSR form.

The CSR format is good for the input matrix P which is presumably sparse and will be used for the output matrix $D(M(S(P)))$ (which, unfortunately, can be quite dense, but will be naturally constructed row-wise, and there are some chance that it is sparse anyway).

On the other hand, matrices $S(P)$ and $M(S(P))$ are dense matrices. However, we can easily see that P (and $D(M(S(P)))$ respectively) has a zero element in a position if and only if $S(P)$ (and $M(S(P))$ respectively – because $S(D(A)) = A$) has a zero elements in the same position or has an element equal to its right neighbor (see also Fig. 6).

We can take advantage of this fact to develop a modified CSR format (called here VCSR) adapted to store intermediate matrices from Fig. 6 (namely, $S(P)$ and $M(S(P))$). Here, we store the matrix $S(A)$ in the same CSR structure as A – however, absent elements are not taken to be zero, but are taken to be equal to their right neighbor (unless they do not have such a neighbor – in which case

$$A = \begin{bmatrix} .0 & .8 & .0 & .2 \\ .1 & .2 & .3 & .4 \\ .0 & 1.0 & .0 & .0 \\ .5 & .0 & .0 & .5 \end{bmatrix}$$

```
val     = [ .8 .2 .1 .2 .3 .4 1.0 .5 .5 ]
col_ind = [ 1 3 0 1 2 3   1 0 3 ]
row_ptr = [ 0 2 6 7 9 ]
```

Fig. 7. A stochastic matrix (top) and its CSR form (bottom); rows and columns indexed from 0

$$S(\mathbf{A}) = \begin{bmatrix} 1.0 & 1.0 & .2 & .2 \\ 1.0 & .9 & .7 & .4 \\ 1.0 & 1.0 & .0 & .0 \\ 1.0 & .5 & .5 & .5 \end{bmatrix}$$

```
val     = [ 1.0 .2 1.0 .9 .7 .4 1.0 1.0 .5 ]
col_ind = [  1  3  0  1  2  3  1  0  3 ]
row_ptr = [  0  2   6  7  9 ]
```

Fig. 8. The result of the operation S performed on the matrix from Fig. 7 (top) and its VCSR form (bottom); rows and columns indexed from 0; note that col_ind and row_ptr are the same as in Fig. 7

Table 1. Performance time [s] of CPU-CSR and GPU-CSR (and its steps) for selected matrices

NZ	CPU-CSR				GPU-CSR			
	S	M	D	Total	S	M	D	Total
20 813	0.0001	0.0010	0.0010	0.0021	0.0001	0.0651	0.0002	0.0654
37 353	0.0002	0.0022	0.0012	0.0036	0.0001	0.0707	0.0002	0.0710
35 080	0.0002	0.0021	0.0017	0.0040	0.0001	0.0959	0.0002	0.0962
62 084	0.0004	0.0040	0.0033	0.0077	0.0002	0.1341	0.0004	0.1347
49 102	0.0003	0.0030	0.0023	0.0056	0.0002	0.1535	0.0005	0.1542
87 491	0.0005	0.0050	0.0030	0.0085	0.0003	0.1708	0.0005	0.1716
63 163	0.0007	0.0067	0.0036	0.0110	0.0004	0.2146	0.0006	0.2156
112 798	0.0004	0.0036	0.0031	0.0071	0.0002	0.1567	0.0005	0.1574
11 447 597	0.0723	0.4097	0.3437	0.8257	0.0004	17.2520	0.0005	17.2529

they are taken to be zero). Figure 8 shows the format VCSR for the matrix $S(\mathbf{A})$ (where $\mathbf{A}$ and its CSR form is shown in Fig. 7).

Thus, if we store our input matrix $\mathbf{P}$ in CSR format and the first intermediate matrix $S(\mathbf{P})$ in VCSR, then the operation S does not change the structure of the matrix. Moreover it does not need any auxiliary data structures to compute elements of $S(\mathbf{P})$. Similarly, storing the second intermediate matrix $M(S(\mathbf{P}))$ in VCSR and the output matrix $D(M(S(\mathbf{P})))$ in CSR allows finding output without changing the structure and without any auxiliary data structures.

Unfortunately, the matrix structure changes between $M(S(\mathbf{P}))$ and $D(M(S(\mathbf{P})))$ (and the change is significant) so in this step the matrix must be rebuilt and this step cannot be performed in-place.

5   GPU Implementations of Vincent's Algorithm

Such a three-step algorithm was implemented in two versions – at first:

– CPU-CSR, where we used the CPU alone and the CSR and VCSR formats for storing the matrices;

Table 2. Performance time [s] of HYB-CSR (and its steps) for selected matrices (speedup given in relation to CPU-CSR)

NZ	S (on GPU)	From GPU to CPU	M (on CPU)	From CPU to GPU	D (on GPU)	Total	Speedup
20 813	0.0001	0.0001	0.0010	0.0001	0.0002	0.0015	1.40
37 353	0.0001	0.0001	0.0022	0.0002	0.0002	0.0028	1.29
35 080	0.0001	0.0002	0.0021	0.0002	0.0002	0.0028	1.43
62 084	0.0002	0.0002	0.0040	0.0004	0.0004	0.0052	1.48
49 102	0.0002	0.0002	0.0030	0.0003	0.0005	0.0042	1.33
87 491	0.0003	0.0004	0.0050	0.0005	0.0005	0.0067	1.27
63 163	0.0004	0.0004	0.0067	0.0006	0.0006	0.0087	1.26
112 798	0.0002	0.0002	0.0036	0.0004	0.0005	0.0049	1.45
11 447 597	0.0004	0.0289	0.4097	0.0004	0.0005	0.4399	1.88

Table 3. Hardware and software used in the experiments

CPU	$2 \times$ Intel Xeon X5650 2.67 GHz
Host memory	48 GB DDR3 1333 MHz
GPU	$2 \times$ Tesla M2050 (515 Gflops DP, 3 GB memory)
OS	Debian GNU Linux 6.0
Libraries	CUDA Toolkit 4.0, CUSP 0.2

– GPU-CSR, where we used the GPU (for all the main computations) and the CSR and VCSR formats for storing the matrices.

After some tests (see Sect. 6 below, Table 1) we spotted that the first (S) and the third (D) parts of the GPU-CSR implementation are faster than the same parts of the CPU-CSR implementation. However, the second part (M) is much faster in the CPU-CSR implementation.

It is caused by the distinct flow of the matrix manipulation. The matrix is stored by rows but it is processed by columns. Moreover, it must be rebuilt because its structure changes.

That is why we prepared and tested another algorithm – a hybrid one (HYB-CSR). Now, the first and the third steps are performed on the GPU and the second step – on the CPU. Clearly, it requires the data to move from the GPU to the CPU (before M) and back (after M). Such communication can take quite a long time – but, fortunately, it didn't (Table 2). Moreover, the HYB-CSR algorithm turned out to be the fastest, despite the communication between CPU and GPU.

Table 4. Performance time [s] of CPU-2D and GPU-2D for selected matrices

NZ	CPU-2D	GPU-2D
20 813	0.04	0.0505
37 353	0.04	0.0500
35 080	0.11	0.1003
62 084	0.13	0.1002
49 102	0.22	0.1916
87 491	0.23	0.1922
63 163	0.37	0.2835
112 798	0.36	0.2850

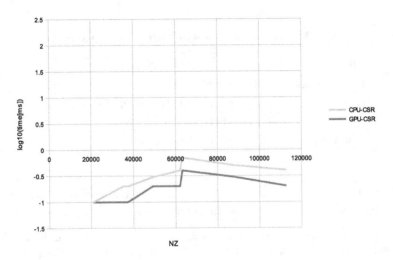

Fig. 9. Performance time of the step S on CPU and GPU (lower is better)

6   Experimental Results

Our algorithms were tested on a node of a computational cluster – the node characteristics is given in Table 3.

We used various stochastic matrices of no exceptional structure, coming from various models.

All the tables and figures show only the number of non-zeros (NZ) because there was no dependency on the size of the matrix. That is because the real size of the problem is the number of non-zeros if we store the sparse matrix in a packed manner (like CSR/VCSR).

We also tested – for comparison – two algorithms processing matrices as two-dimensional unpacked arrays, namely CPU-2D and GPU-2D (running on the

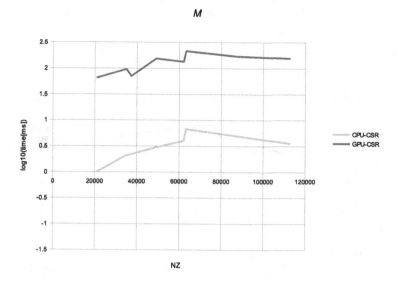

Fig. 10. Performance time of the step M on CPU and GPU (lower is better)

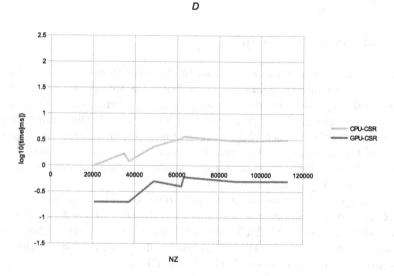

Fig. 11. Performance time of the step D on CPU and GPU (lower is better)

CPU and the GPU, respectively) – which gave very poor performance (Table 4), due to storage (and processing) of a lot of unnecessary (zeros or repeating) values.

We can see that the HYB-CSR algorithm is the fastest – almost twice as fast as the one performed on the CPU (CPU-CSR) – see also Figs. 9, 10, 11 and 12.

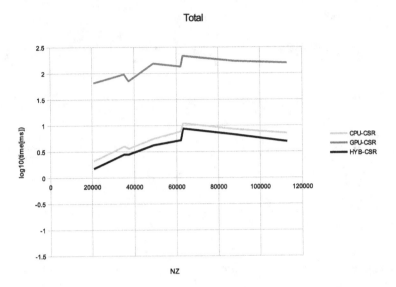

Fig. 12. Performance time of the algorithms CPU-CSR, GPU-CSR and HYB-CSR (lower is better)

7   Analysis of the Results

We can see that there is some potential in the algorithm – the steps S and D could be even a couple of hundreds times faster (as we can see for the biggest matrix – Table 1). However, the middle step, M, is much slower on the GPU.

This is because in the steps S and D a lot of threads (hundreds, literally) are working together – each thread processing a row of the matrix. Moreover, because of the proximity of the data in the GPU memory, there are plenty of opportunities that threads of the same warp can utilize the coalescent access (which is a very important feature of the GPU).

In the step M the situation is virtually reversed. The formats CSR/VCSR are row-wise so the column-wise algorithm (as the step M itself) cannot utilize the coalescent access and a lot of threads have to wait for the others. In the worst case, actually only one thread can work, efficiently. And single threads on GPUs are much slower than single threads on CPUs.

8   Conclusion and Future Works

We can see that using GPU for the described problem can speed up the performance of the algorithm – although not much. The gain is the biggest for the large matrices.

It seems that the crucial thing to work on is a suitable distribution of computations between GPU and CPU to utilize both types of computing units the best. It could be done by dividing the work in a less traditional way [17–19] – that is to exchange fragments of the matrix (especially, a big one) between GPU and CPU and perform on them operations M (on CPU) and D/S (on GPU) in parallel. In that way, we could prevent downtimes of CPU while GPU is working and vice versa.

Moreover, it could improve the performance further because a good distribution of work allows us to process larger matrices (we could see that the speedup increases with the size of the matrix) – since not the whole matrix must be in the (very limited) memory of the GPU at the same time.

References

1. Abu-Amsha, O., Vincent, J.-M.: An algorithm to bound functionals on Markov chains with large state space. In: 4th INFORMS Conference on Telecommunications, Boca Raton, Floride, E.U. INFORMS (1998)
2. Fourneau, J.-M., Pekergin, N.: An algorithmic approach to stochastic bounds. In: Calzarossa, M.C., Tucci, S. (eds.) Performance 2002. LNCS, vol. 2459, pp. 64–88. Springer, Heidelberg (2002)
3. Stewart, W.J.: Introduction to the numerical Solution of Markov Chains. Princeton University Press, New Jersey (1995)
4. Bylina, J., Bylina, B., Karwacki, M.: A Markovian model of a network of two wireless devices. In: Kwiecień, A., Gaj, P., Stera, P. (eds.) CN 2012. CCIS, vol. 291, pp. 411–420. Springer, Heidelberg (2012)
5. Mamoun, M.B., Busic, A., Pekergin, N.: Generalized class C Markov chains and computation of closed-form bounding distributions. Probab. Eng. Inf. Sci. **21**, 235–260 (2007)
6. Busic, A., Fourneau, J.-M.: A matrix pattern compliant strong stochastic bound. In: 2005 IEEE/IPSJ International Symposium on Applications and the Internet Workshops (SAINT 2005 Workshops), Trento, Italy, pp. 260–263. IEEE Computer Society (2005)
7. Dayar, T., Pekergin, N., Younès, S.: Conditional steady-state bounds for a subset of states in Markov chains. In: Structured Markov Chain (SMCTools) Workshop in the 1st International Conference on Performance Evaluation Methodolgies and Tools, VALUETOOLS 2006, Pisa, Italy, ACM (2006)
8. Fourneau, J.-M., Le Coz, M., Pekergin, N., Quessette, F.: An open tool to compute stochastic bounds on steady-state distributions and rewards. In: 11th International Workshop on Modeling, Analysis, and Simulation of Computer and Telecommunication Systems (MASCOTS 2003), Orlando, FL, IEEE Computer Society (2003)
9. Fourneau, J.-M., Le Coz, M., Quessette, F.: Algorithms for an irreducible and lumpable strong stochastic bound. Linear Algebra Appl. **386**, 167–185 (2004)
10. Thompson, C.J., Hahn, S., Oskin, M.: Using modern graphics architectures for general-purpose computing: a framework and analysis. In: Proceedings of the 35th Annual ACM/IEEE International Symposium on Microarchitecture, pp. 306–317. IEEE Computer Society Press, Los Alamitos (2002)
11. Kijima, M.: Markov Processes for Stochastic Modeling. Chapman & Hall, London (1997)
12. Muller, A., Stoyan, D.: Comparison Methods for Stochastic Models and Risks. Wiley, New York (2002)
13. Shaked, M., Shantikumar, J.G.: Stochastic Orders and Their Applications. Academic Press, San Diego (1994)

14. Stoyan, D.: Comparison Methods for Queues and Other Stochastic Models. Wiley, Berlin (1983)
15. Bylina, B., Bylina, J., Karwacki, M.: Computational aspects of GPU-accelerated sparse matrix-vector multiplication for solving Markov models. Theor. Appl. Inform. **23**(2), 127–145 (2011)
16. Bylina, J., Bylina, B., Karwacki, M.: An efficient representation on GPU for transition rate matrices for Markov chains. In: Wyrzykowski, R., Dongarra, J., Karczewski, K., Waśniewski, J. (eds.) PPAM 2013, Part I. LNCS, vol. 8384, pp. 663–672. Springer, Heidelberg (2014)
17. Bylina, B., Karwacki, M., Bylina, J.: A CPU-GPU hybrid approach to the uniformization method for solving Markovian models – a case study of a wireless network. In: Kwiecień, A., Gaj, P., Stera, P. (eds.) CN 2012. CCIS, vol. 291, pp. 401–410. Springer, Heidelberg (2012)
18. Lee, J., Samadi, M., Park, Y., Mahlke, S.: Transparent CPU-GPU collaboration for data-parallel kernels on heterogeneous systems. In: Proceedings of the 22nd International Conference on Parallel Architectures and Compilation Techniques (PACT), September 2013
19. Ohshima, S., Kise, K., Katagiri, T., Yuba, T.: Parallel processing of matrix multiplication in a CPU and GPU heterogeneous environment. In: Daydé, M., Palma, J.M.L.M., Coutinho, A.L.G.A., Pacitti, E., Lopes, J.C. (eds.) VECPAR 2006. LNCS, vol. 4395, pp. 305–318. Springer, Heidelberg (2007)

An Analysis of the Extracted Parts of Opte Internet Topology

Monika Nycz[1], Tomasz Nycz[1], and Tadeusz Czachórski[2(✉)]

[1] Institute of Informatics, Silesian University of Technology,
Akademicka 16, 44-100 Gliwice, Poland
{monika.nycz,tomasz.nycz}@polsl.pl
[2] Institute of Theoretical and Applied Informatics,
Polish Academy of Sciences, Bałtycka 5, 44-100 Gliwice, Poland
tadek@iitis.pl

Abstract. The paper defines a method to extract a part of a graph, which corresponding to a network (Internet) topology covering a certain area. The method refers to data gathered within Opte Project and describing the Internet topology. The extracted parts of graphs have been analyzed in terms of the number of neighbours, longest path length and existence of cycles. Then the resulting topologies are used to model transient behaviour of wide area networks with the use of fluid-flow approximation.

Keywords: Internet topology · Fluid-flow approximation · Computer networks · Opte project

1 Introduction

Queueing theory is a useful tool in modelling and performance evaluation of computer networks. The literature concerning this subject is abundant. However, traditional queueing models do not fit well to the problems arising in contemporary networks. There are two problems we address here: the need of transient analysis and the need of modelling very large topologies. The traffic intensity in networks is time-dependent: we observe its perpetual changes due to the nature of users, sending variable quantities of data, and also due to the performance of traffic control algorithms, which are trying to avoid congestion in networks, e.g. the algorithm of congestion window used in TCP protocol adapting the rate of the sent traffic to the observed losses or transmission delays. The complexity of network topologies is also increasing rapidly and realistic models should reflect their real structure. In computer applications a mathematical model is useful only when it furnishes quantitative results. Therefore practical issues related to numerical side of models are of importance. We have studied [1] three approaches possible in transient analysis of queueing models: numerical solution of Markov chains [2], diffusion approximation [3] and fluid-flow approximation [4,5]. Each method has the advantages and drawbacks, but the fluid-flow analysis is the simplest one. Therefore it allows us to model the largest topologies and at the same

© Springer International Publishing Switzerland 2015
P. Gaj et al. (Eds.): CN 2015, CCIS 522, pp. 371–381, 2015.
DOI: 10.1007/978-3-319-19419-6_35

time it gives us means to reflect the performance of TCP congestion control algorithm. The model assumptions simplifies some aspects of the network, offering high performance of computation, which is a big benefit in analyzing the topologies with millions of nodes and flows. The Markovian models may be adapted to flows and packet sizes observed in real networks, but the number of states to be solved numerically grows rapidly. If we include in the model also transmission rules, soon we attain the reasonable computational limits. The diffusion approximation calculations are more time-consuming then those of fluid-flow model and has also precision limits in its numerical computations, that makes impossible to analyse this way networks with large buffers at nodes. We compared the two approximation approaches in [6].

The real Internet topologies form dependent structures with varying degrees of complexity. The research on Internet topology mapping has begun relatively recently and is still very popular, especially because of the utility for simulating the designed network protocols or model analyzes. There are several network algorithms, projects and tools used for network mapping, [7–12], including Skitter [13], Rocketfuel [14], Scriptroute [15], nec [16].

This study aims to analyze selected Internet network topologies with a large number of nodes and flows with the use of fluid-flow approximation. As a base topology serves here the Internet map developed within The Opte Project, [17], but we are planning to focus on CAIDA topologies in our following research.

The Opte project was appointed to gather data from the global network and create an image of the Internet. During the initial phases of the project, data were collected using the traceroutes, but currently the BGP dumps are used for these purpose. The collected data on routing are stored in the database, from where they are exported to files in the Large Graph Layout (LGL) format. Text files are used as the data source for the visualization tools, i.e. LGL, Gephi, to create the final image. The project is still ongoing and, as its author wrote on his website, [17], the current works concern a new version the LGL program, enabling the animation the 'life' of the Internet, e.g. to show how the individual nodes are being changed, which nodes disappear, etc. The final result will give a map of the Internet 2014. The Opte Project provides currently two images of the Internet: from 2003 and from 2014. However, the source data are available to public only for the 2003 Internet at the time of writing this publication and they were used in this study. Figure 1 presents the global network classified by territory, completed by the IP address and marked by specific color:

- green – Europe, Middle East, Central Asia and Africa,
- red – Asia and Pacific,
- blue – North America,
- yellow – Latin America and Caribbean,
- cyan – RFC1918 IP Addresses,
- white – Unknown.

The image of 2010 Internet, Fig. 2, is based on the neighborhood. The color brightness represents the size of the connection – the highest temperatures (in terms of colors of light) are placed where connection points are most numerous. The data points come from the BGP dumps.

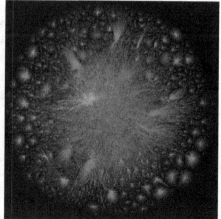

Fig. 1. The map of the Internet from 2003 generated by The Opte Project [18]

Fig. 2. The map of the Internet from 2010 generated by The Opte Project [19]

2 Topology Extraction

The project provides three types of source files for the Internet 2003 topology:

- LGL (Large Graph Layout), including undirected graph, with nodes and all its neighbors;
- 2D Coords, containing nodes together with the coordinates in the 2D space;
- Edge Colors, consisting of edges with assigned territory colours.

We may extract a part of network topology in two ways. The first approach is based on the particular area. From the edge file, we divide edges by their assigned RGB color and extract all row for a particular color using regular expressions. Thus, we obtain the list of edges in each territory. Then, we built the list of unique nodes in each of the territory, sorted by IP address, and assign an identifier to each node. As a last phase, we built the list of nodes with all their neighbours, according to Pajek NET Format [20], which is a supported graph format in Gephi (The Open Graph Viz Platform) tool [21]. The second method involves the selection of the primary node in the network and then studies its neighbourhood to a certain depth. We select one IP address as a base and from the edge file we determine all his neighbours (depth level 1). Then we search for the neighbours of those neighbours (depth level 2), and so on, until we reach the assumed level of depth.

In this paper we focus on the first strategy, which resulted with one graph per territory and consisted of following number of nodes and edges:

- T1: Europe, Middle East, Central Asia and Africa;
- T2: Asia and Pacific;
- T3: North America;

Territory	Number of nodes	Number of edges	Average degree	Maximum degree	Average path length
T1	20615	13692	1.322	10	2.561
T2	18200	12185	1.331	16	2.386
T3	26900	17432	1.281	8	2.392
T4	2989	1965	1.301	7	2.539
T5	13199	11775	–	–	–

– T4: Latin America and Caribbean;
– T5: RFC1918 IP Addresses.

During our research we wanted to check the complexity of the regional subnets, depending on the color, for further use in fluid-flow analysis. Unfortunately, the edge file does not contain the white links.

Each graph was transformed into picture using Gephi. In our scheme the size and color of the points determine the number of neighbour nodes – the darker the color of a node, the more inputs/outputs the node has. Due to the use of the real structure, for each network the information about IP addresses is provided.

Unfortunately, all topologies turned out to be a vast collection of disjoint clusters of few-node networks, Fig. 3, with the single exception in the form of some larger network (for up to 30 nodes), Fig. 4, which could be expected in case of T1 (Fig. 5a) area, but not in the case of T2 (Fig. 5b), T3 (Fig. 5c), T4 (Fig. 5d). Only in one territory graph, we detected a subnetwork with a cycle, Fig. 4a. As a result we observed that the method caused the significant loss of extremely important points from the global network point of view.

3   Fluid-Flow Approximation

The fluid approximation [4,5] uses first-order ordinary linear differential equations to determine the average values of node queues and the dynamics of TCP

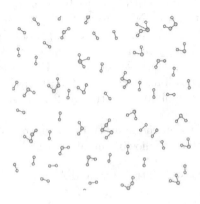

Fig. 3. Enlarged representative part of extracted topologies

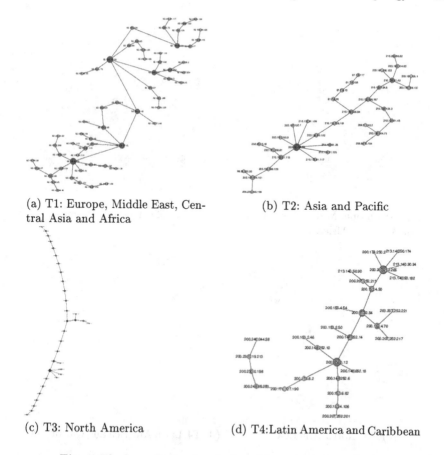

(a) T1: Europe, Middle East, Central Asia and Africa

(b) T2: Asia and Pacific

(c) T3: North America

(d) T4:Latin America and Caribbean

Fig. 4. The largest subnetworks found in extracted topologies

congestion windows in a modelled network. The changes of queue length in Eq. (1), are defined as the intensity of the input stream, i.e. all flows traversing that particular node, reduced by the number of served packets in time unit.

$$\frac{dq_v(t)}{dt} = \sum_{i=1}^{K} \frac{W_i(t)}{R_i(q(t))} - \mathbf{1}(q_v(t) > 0) \cdot C_v. \tag{1}$$

Each flow i $(i = 1, \ldots, N)$ is determined by a time varying congestion window size W_i kept by sender of i-th flow. A router receives traffic from K TCP flows $(K \leqslant N)$. The window size, Eq. (2), increases by one at each RTT (Round Trip Time) denoted here by R_i in the absence of packet loss and decreases by half after each packet loss occurring in nodes on the flow path. The amount of losses for each entire TCP connection is computed as the flow of losses equal to the flow of packets W_i/R_i times the probability that a packet is lost at any node on the route, based on matrix B storing drop probabilities in each router for all flows in the network:

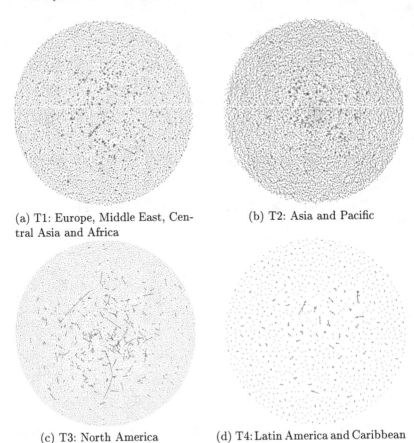

(a) T1: Europe, Middle East, Central Asia and Africa

(b) T2: Asia and Pacific

(c) T3: North America

(d) T4: Latin America and Caribbean

Fig. 5. Extracted topology based on territory division

$$\frac{dW_i(t)}{dt} = \frac{1}{R_i(\boldsymbol{q}(t))} - \frac{W_i(t)}{2} \cdot \frac{W_i(t-\tau)}{R_i(\boldsymbol{q}(t-\tau))} \cdot$$
$$\left(1 - \prod_{j \in V}(1 - B_{ij})\right), \tag{2}$$

$$R_i(\boldsymbol{q}(t)) = \sum_{j=1}^{K} \frac{q_j(t)}{C_j} + \sum_{j=1}^{K-1} Lp_j. \tag{3}$$

It is supposed in Eq. (2) that the delay after which the sender reacts on a packet loss by changing its congestion window is equal to the RTT time which is computed in Eq. (3) taking into account the queueing delay (first term) and propagation delay (second term).

The drop probability given by Eq. (4) in a single node is calculated according to RED mechanism, depending on the weighted average queue length $x_v(t)$ as the sum of instantaneous queue $q_v(t)$ with w weight parameter and previous average queue with $(1 - w)$ weight parameter.

$$p_v(x_v) = \begin{cases} 0, & 0 \leqslant x_v < t_{min_v} \\ \frac{x_v - t_{min_v}}{t_{max_v} - t_{min_v}} p_{max_v}, & t_{min_v} \leqslant x_v \leqslant t_{max_v} \\ 1, & t_{max_v} < x_v \end{cases} . \tag{4}$$

Based on the exploration of the graph nodes and the their connections, we could define the additional information used in subsequent steps to determine the model parameters, i.e.:

- the number of layers (based on neighborhood),
- the percentage of nodes in each layer,
- the average node degree (average number of neighbors in the layer).

The model assumes the flow paths and the parameters of flows and routers to be defined before the calculation phase. Therefore in preprocessing stage, the routing is determined based on the network structure and assigned links propagation delays – the shortest paths from the source to the destination are calculated. The flow route generation algorithm assumes: 1) marking and numbering the boundary nodes; 2) selecting a pair of boundary routers (source and destination) based on random number: $src = i \ / \ k$, $dest = i \ \% \ k$, where $i = (0, max^2)$, max is the number of boundary routers; 3) generating a propagation coefficients for all edges of the graph (weight) and computing on this basis the shortest paths in the graph using Dijkstra's algorithm.

4 Exemplary Network Scenario

As a test scenario, the structure of 25 routers was chosen and served as an input graph in fluid flow approximation model. In the studied structure, see Fig. 6, we determined 12 boundary nodes through which the algorithm led 7 classes of flows. In the presented case the network was small enough to be contained in a single AS, so it was sufficient to assume a static routing between nodes. Performed numerical calculations demonstrated that in the present network case, there is no queue overflow at nodes, Fig. 7. In most routers the queue length increases gradually. Only in some network points the RED mechanism has been activated, lowering queue occupancy.

In few flow classes (2, 5, 6, 7) the partial congestion window size, Fig. 9, reduction was observed but the total flow throughput was similar, Fig. 10.

The biggest losses, Fig. 8, were evident in the nodes 14 and 15, and later also in 24. Although the probability of losses is maintained at a relatively low level, these are nodes that can be suspected of possible overflow in the future. Nevertheless, in the illustrated period of time we do not observed a node, for which it would be necessary to completely reject the newly-arrived packet. Interestingly, the nodes 14, 15 and 24 are not the focal point network – these are edge routers. And as it turned out to be more congested than routers from the main path (through which most of the traffic traverse).

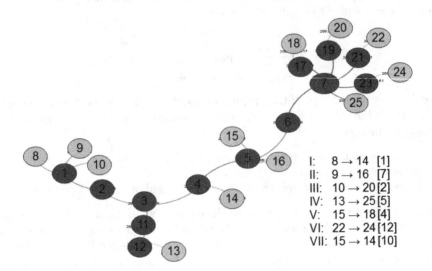

Fig. 6. Selected subnetwork from T4 area for the analysis

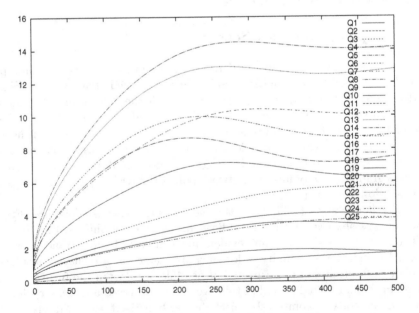

Fig. 7. Queue lengths of individual nodes

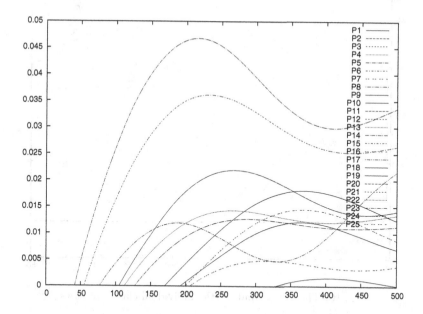

Fig. 8. A drop probabilities in individual nodes

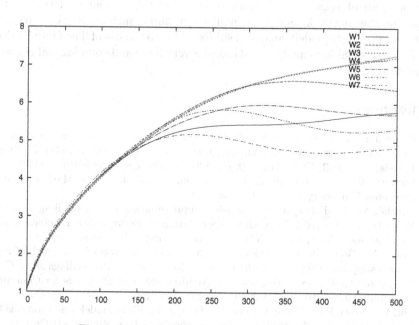

Fig. 9. Window sizes of individual classes of flows

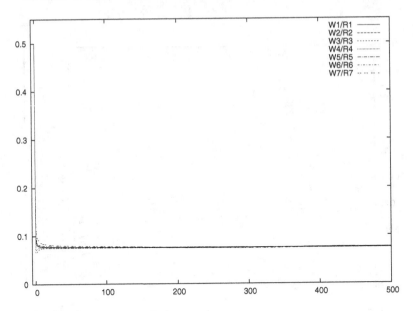

Fig. 10. The throughputs of the individual classes of flows

5 Conclusion

The problem of accurate and quick mapping of the current structure of the global Internet network, as well as smaller structures, such as ISPs or enterprise networks, is still the widely analyzed subject. Due to the use of The Opte Project topology we are able to analyze and model a very large and complex real network structures.

References

1. Czachórski, T., Nycz, M., Nycz, T., Pekergin, F.: Analytical and numerical means to model transient states in computer networks. In: Kwiecień, A., Gaj, P., Stera, P. (eds.) CN 2013. CCIS, vol. 370, pp. 426–435. Springer, Heidelberg (2013)
2. Stewart, W.J.: An Introduction to the Numerical Solution of Markov Chains. Princeton University Press, Princeton (1994)
3. Czachórski, T., Pekergin, F.: Diffusion approximation as a modelling tool. In: Kouvatsos, D.D. (ed.) Next Generation Internet: Performance Evaluation and Applications. LNCS, vol. 5233, pp. 447–476. Springer, Heidelberg (2011)
4. Misra, V., Gong, W.B., Towsley, D.: A fluid-based analysis of a network of aqm routers supporting tcp flows with an application to red. In: Proceedings of the Conference on Applications, Technologies, Architectures and Protocols for Computer Communication (SIGCOMM 2000), pp. 151–160 (2000)
5. Liu, Y., Presti, F.L., Misra, V., Towsley, D., Gu, Y.: Fluid models and solutions for large-scale ip networks. In: Proceedings of the 2003 ACM SIGMETRICS International Conference on Measurement and Modeling of Computer Systems SIGMETRICS 2003 (2003)

6. Nycz, T., Nycz, M., Czachórski, T.: A numerical comparison of diffusion and fluid-flow approximations used in modelling transient states of tcp/ip networks. In: Kwiecień, A., Gaj, P., Stera, P. (eds.) CN 2014. CCIS, vol. 431, pp. 213–222. Springer, Heidelberg (2014)

7. Pansiot, J.J., Grad, D.: On routes and multicast trees in the internet. ACM Comput. Commun. Rev. **28**(1), 41–50 (1998)

8. Rickard, J.: Mapping the internet with traceroute. Broadwatch Magazine, **10**(12) (1996)

9. Burch, H., Cheswick, B.: Mapping the internet. IEEE Comput. **32**(4), 97–98 (1999)

10. Tangmunarunkit, H., Govindan, R., Shenker, S., Estrin, D.: The impact of routing policy on internet paths. In: Proceedings of IEEE INFOCOM 2001 (2001)

11. Ramesh, G., Hongsuda, T.: Heuristics for internet map discovery. In: IEEE INFOCOM 2000, pp. 1371–1380 (2000)

12. Meyer, D.: University of oregon route views project. http://www.antc.uoregon.edu/route-views

13. Skitter. http://www.caida.org/tools/measurement/skitter/

14. Spring, N., Mahajan, R., Wetherall, D.: Measuring isp topologies with rocketfuel. In: Proceedings of ACM SIGCOMM (2002)

15. Spring, N., Wetherall, D., Anderson, T.: Scriptroute: a public internet measurement facility. In: Proceedings of the USENIX Symposium on Internet Technologies and Systems (USITS) (2003)

16. Magoni, D., Hoerdt, M.: Internet core topology mapping and analysis. Comput. Commun. **28**(5), 494–506 (2005)

17. Lyon, B.: The Opte Project. http://www.opte.org/

18. Lyon, B.: The Opte Project - Internet 2003. http://www.blyon.com/blyon-cdn/1069646562.2D.9000x9000.png

19. Lyon, B.: The Opte Project - Internet 2010. http://blyon.com/blyon-cdn/opte/maps/static/opte-2010.png

20. Pajek NET Format - a Gephi supported graph format. http://gephi.github.io/users/supported-graph-formats/pajek-net-format/

21. Gephi: The open graph viz platform. http://gephi.github.io/

Scheduling of Isochronous Data Transactions in Compliance with QoS Restrictions in the USB 3.0 Interface

Andrzej Kwiecień and Michał Sawicki[✉]

Institute of Informatics, Silesian University of Technology,
Akademicka 16, 44-100 Gliwice, Poland
{andrzej.kwiecien,michal.sawicki}@polsl.pl

Abstract. The newest versions of the most ubiquitous USB interface are 3.0/3.1. USB 3.0 improves utilization of bus throughput through the changing of system architecture and introduction a new scheduling algorithm. In some applications using USB ports (e.g. vision system), the quality of service (QoS) is required. This paper presents proposal of model of USB system based on theory of scheduling and definition of QoS for communication interfaces (in particular USB 3.0). Some experiments (using prepared USB 3.0 application) were performed to verify compliance with QoS and their results are shown in this article.

Keywords: Interface usb 3.0 · Isochronous transaction · Scheduling algorithm · QoS

1 Introduction

Developers of IT systems design their products to satisfy the needs and requirements of customers (end users), what means to assure quality of providing services (QoS). Many factors, which depend on some elements of the system, result in user satisfaction. Assurance of quality of service is based not only on well-known and optimized mechanisms of tasks scheduling on CPU and mechanisms of data transmission in computer network, but also on "efficient" communication (I/O subsystem [1]) between this system and peripheral devices (e.g. video camera).

In computer systems, peripheral devices communicate with a computer using different interfaces (e.g. wired USB and Thunderbolt, wireless Bluetooth). As an example, the vision system consists of a lot of video cameras, which are connected to the controller (computer) via USB 3.0 bus [2]. Therefore, the principal USB task is execution of many ordered data transfers, and previously creation of data exchange scenario, that it is possible to quasi-simultaneously execute of many various data transfers over single serial bus.

In the past, USB compliance with QoS and real-time requirements have been studied in [1,3,4], but those discussions only concerned USB 1.1/2.0. It is worth

© Springer International Publishing Switzerland 2015
P. Gaj et al. (Eds.): CN 2015, CCIS 522, pp. 382–388, 2015.
DOI: 10.1007/978-3-319-19419-6_36

mention, that IEEE 1394 (FireWire) compliance with QoS also has been studied in the past. Currently, it is hard to find articles concerning USB 3.0 compliance with QoS.

First of all, some theoretical aspects of transactions scheduling are shown in second section. Model of transactions scheduling process is proposed by using elements of theory of scheduling [5,6]. QoS is defined, and some metrics like variance of isochronous interval or idle time are given. Following section presents some experiments and their results. Experiments were performed to verify QoS compliance with specific USB 3.0 system.

2 Model of Data Transactions Scheduling Process

In USB communication system N various data transfers are quasi-simultaneously performed on a single serial bus. Therefore, each transfer[1] (e.g. ith transfer) is divided on M^i data transactions, which consist of many packets wherein the number of packets is identical for all transactions of the same transfer[2] and it depends on transfer types (control, interrupt, isochronous, bulk) and values (e.g. MaxPacketSize, BurstSize) contained within endpoint descriptor.

In the vision system (generally A/V system), a video transmission should be based on isochronous data transfer due to the nature of this transmission. It is worth mention, that first implementations of USB camera used bulk transfers, however the isochronous transfer has been designed just for multimedia applications. Further reasoning concerns of only isochronous transfer with assumption, that only read operation is performed. This is reflected in a real vision system, where host controller reads isochronous data from recording equipment, e.g. video camera.

In USB 3.0 specification, parallel execution of transactions is not permitted[3]. Next transaction cannot start before the end of previous transaction. Therefore, precedence constraints defined as an order relation $\prec$ is imposed on set of transactions. In Fig. 1, the model of USB communication system is depicted. There are two transmission channels: downstream and upstream. The N identical isochronous devices are connected to host controller via hub.

Basically, isochronous read transaction consists of following packets:

- TP – transaction packet, which initiates and parameterises transaction,
- DP – data packet, which transfers data.

Transaction packet (TP) is transmitted during t^i_{TP} from host controller to peripheral device, and this transmission is called "downstream". According to assumption, that only read operation is performed, data packet (DP) is transmitted from peripheral device to host controller during t^i_{DP}, and this transmission is

---

[1] The symbols associated with ith transfer are tagged by transfer number in superscript.

[2] This assumption simplifies further reasoning. In fact, the last transaction of a transfer may consist of less number of packets. Additionally, last transmitted packet may have smaller payload than previous packets.

[3] In the particular case, it is permitted for isochronous write operation.

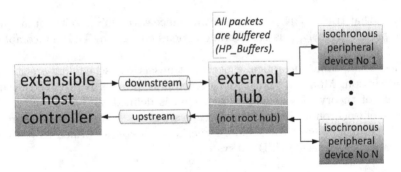

Fig. 1. Model of USB 3.0 communication system

called "upstream". Data packet consists of two separate sub-packets: data packet header (DPH) and data packet payload (DPP), which are transmitted back-to-back without any internal gap. In the newest version of USB, more than one data packets within transaction are allowed, and number of data packets B^i is called BurstSize (value contained within endpoint descriptor). Therefore, a transaction consists of $1 + B^i$ packets. This type of transaction is called burst. Each transaction (e.g. jth transaction) of ith transfer is parametrized by non-negative values:

- S_j^i – transaction start time (expressed as function $\pi(i, j)$ in further discussion),
- p^i – transaction performance time, which is identical for all transactions of the same transfer, and it is expressed as:

$$p^i = t_{\text{TP}}^i + t_{\text{DP}}^i B^i, \tag{1}$$

- C_j^i – transaction finishing time expressed as:

$$C_j^i = S_j^i + p^i. \tag{2}$$

Isochronous interval is an interval between starts of consecutive transactions of the same transfer, and it is defined below:

$$T_k^i = S_{k+1}^i - S_k^i \text{ where } k \in \{1, \dots, M^i - 1\}. \tag{3}$$

Idle time is defined as (based on [5]):

$$I = C_{\max} - \sum_{i=1}^{N} M^i p^i \tag{4}$$

where $C_{\max}$ is makespan and is defined as:

$$C_{\max} = \max_{\substack{i=1\dots N \\ j=1\dots M^i}} C_j^i. \tag{5}$$

If process of transactions scheduling is defined as discrete injection $\pi : X \times Y \to \mathbb{R}^+$, where X is a set of transfers, Y is a set of transactions, and π is a scenario of data exchange (schedule), than combinatorial optimization of transactions schedule can take the following definition:

Definition 1. *The transactions scheduling problem is to find function π_{opt} included in a set of feasible solutions Π (possible scenarios), that a given objective function is optimized.*

The scenario π is an injective function, because parallel execution of transaction is not permitted and it is true that:

$$\bigwedge_{i,j\in\{1,...,N\}} \quad \bigwedge_{\substack{k\in\{1,...,M^i\} \\ l\in\{1,...,M^j\}}} S_k^i = S_l^j \Rightarrow i = j \wedge k = l. \qquad (6)$$

Unfortunately, the transactions scheduling problem is not trivial, because quite often $\overline{\overline{\Pi}} \gg 1$, and this issue is similar to NP-hard knapsack problem as mentioned in [3].

3   Definition of QoS Restrictions

QoS in the communication systems, particularly in USB 3.0, is proposed as set of below requirements:

1. The regular supply of short lifetime data (e.q. multimedia).
2. The fastest execution of all ordered data transfers and the efficient utilization of bus bandwidth.

In ideal communication systems, where QoS is fulfilled, the following statement is true:

$$\bigwedge_{\substack{i\in\{1,...,N\} \\ T_{\mathrm{const}}^i \geqslant 125\,\mu s}} \quad \bigvee \quad \bigwedge_{k\in\{1,...,M^i-1\}} T_k^i = T_{\mathrm{const}}^i. \qquad (7)$$

Unfortunately, QoS is not completely fulfilled in USB 3.0 communication system in view of the algorithm of transactions scheduling, implementation of host controller and peripheral device drivers, as well as in view of following mechanisms:

- synchronization of isochronous devices, based on transmission of isochronous timestamps (ITP),
- smart isochronous transaction (PING/PING_RESPONSE),
- header packet flow control (LCRD),
- packets acknowledgement (LGOOD/LBAD) performed by link layer,
- power and link management (LDN/LUP),
- link calibration, using ordered sets (e.g. TSEQ).

It is also assumed, that USB 3.0 bus is constantly in U0 state, and transitions to another states of LTSSM are not permitted. For example, if "aggressive" energy conservation is turned on, the bus transitions to U1, U2, U3 states, and

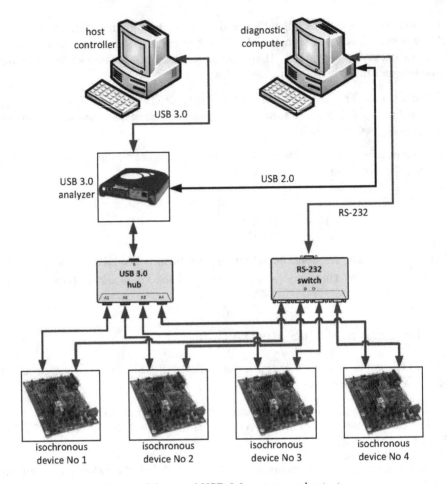

Fig. 2. Scheme of USB 3.0 system under test

return to U0 wastes time. Fulfilment of first QoS requirement may be equal to minimization of variance of isochronous intervals of one transfer:

$$V^i = \frac{1}{M^i - 1} \sum_{k=1}^{M^i - 1} \left(T_k^i - \frac{1}{M^i - 1} \sum_{k=1}^{M^i - 1} T_k^i \right)^2 \quad \text{for} \quad i \in \{1, \ldots, N\}. \quad (8)$$

Fulfilment of this requirement for all transfers, using basic property of variance $V^i \geqslant 0$, is based on minimization of sum of variances:

$$V = \sum_{i=1}^{N} V^i. \quad (9)$$

Lateness or earliness of one of transactions result in reordering other transactions. Therefore, this events are not stochastically independent and unfortunately it is not possible to simplify above sum of variances.

Fulfilment of second QoS restriction requires to minimize idle time (based on formulas (1), (4) and [5]):

$$I = \max_{\substack{i=1...N \\ j=1...M^i}} C_j^i - \sum_{i=1}^{N} M^i \left(t_{TP}^i + t_{DP}^i B^i \right) \tag{10}$$

and it results in the fastest execution of data transfers (minuend) or efficient utilization of the bus bandwidth (subtrahend).

4 Implementation of USB 3.0 System Under Test

Communication system under test was prepared (Fig. 2) and used to verify QoS guarantee of isochronous transfers in USB 3.0 port. Four evaluation boards (Cypress EZ-USB FX3) with ARM processor and USB 3.0 controller were used as isochronous peripheral devices, which were connected through USB to host controller via hub (LogiLink 4-port USB 3.0). USB protocol analyzer (Beagle 5000 SuperSpeed v2) was connected between hub and host controller, thus downstream and upstream transmissions capturing was possible. Real-time operating system ThreadX ran on this evaluation boards. Firmwares for USB controllers were responsible for one input isochronous endpoint service and used DMA mechanism to pass data from CPU to endpoint buffers. Device drivers ran on computer with USB 3.0 host controller (Intel Z77 xHC) and installed Microsoft Windows 7 Professional. The drivers generated quasi-simultaneously isochronous requests (one isochronous transfer per device). Experiment was controlled by diagnostic PC connected to USB analyzer through USB 2.0 and to evaluation boards through RS-232 port via special switch (MCOM 4-port RS). This connection simplified debugging and controlling experiment.

The system allowed to conduct some experiments for measure and evaluate variance of isochronous intervals and idle time in example USB system, where

Table 1. Results of performed experiments

		BurstSize (B^i)		
		1	2	3
Transfer number (i)	1	125 001 19	124 999 407	91 532 112 881
	2	125 002 908	125 030 1483	91 687 110 271
	3	125 000 1662	125 077 2390	91 846 110 363
	4	124 997 2446	124 998 1419	91 893 110 288
Objective functions	$\sum_{i=1}^{4} \sqrt{V^i}$ [ns]	378 287	5699	443 803
	$\frac{I}{C_{max}}$ [%]	93	87	81

four isochronous input transfer are quasi-simultaneously executed. The results are presented in Table 1. Each cell for isochronous transfers No 1–4 contains two values: mean of isochronous intervals and in the next line standard deviation of this intervals. Last row contains idle time expressed as a percentage of idle time. To verify obtained results experiments were repeated with another host controller (Renesas XHC). Second results are quite similar to previous values. However scheduling process may be depended on implementation of host controller driver.

5   Discussion and Summary

Conducted experiment shows that USB 3.0 is not completely compliance with defined QoS restrictions as expected earlier, because the statement (7) is not true for this experimental system. Although, the deviations of isochronous interval for $B^i \in \{1, 2\}$ are low and they range from a few microseconds. In addition, the means of isochronous intervals are quite similar to expected value ($125\,\mu s$). For third BurstSize, means are lower than expected values.

An increase of idle time is proportional to change of BurstSize because the subtrahend of formula (10) is a sum of linear functions, where BurstSize is variable. The experiment was repeated and similar results were obtained.

The conclusion is that the algorithm of transactions scheduling, implemented in host controller driver, has to be adapted to QoS restrictions. It will be the objective of further researches.

Acknowledgements. The work was performed using the infrastructure supported by POIG.02.03.01-24-099/13 grant: GeCONiI – Upper Silesian Center for Computational Science and Engineering.

References

1. Huang, C.-Y., Chang, L.-P., Kuo, T.-W.: A cyclic-executive-based QoS guarantee over USB. In: The 9th IEEE Real-Time and Embedded Technology and Applications Symposium (2003)
2. Anderson, D., Trodden, J.: USB 3.0 Technology. MindShare (2013)
3. Missimer, E., Li, Y., West, R.: Real-time USB communication in the quest operating system. In: The 19th IEEE Real-Time and Embedded Technology and Applications Symposium (2013)
4. Huang, C.-Y., Kuo, T.-W., Pang, A.-C.: QoS support for USB 2.0 periodic and sporadic device requests. In: The 25th IEEE International Real-Time Systems Symposium (2004)
5. Brucker, P.: Scheduling Algorithms. Springer, Berlin (2007)
6. Pinedo, M.: Scheduling Theory, Algorithms, and Systems. Springer, New York (2008)

Applications of Secure Data Exchange Method Using Social Media to Distribute Public Keys

Piotr Milczarski, Krzysztof Podlaski, and Artur Hłobaż[✉]

Faculty of Physics and Applied Informatics, University of Lodz, Lodz, Poland
{piotr.milczarski,podlaski,artur.hlobaz}@uni.lodz.pl

Abstract. The concern about the data security is still growing. The applications we use still lack basic security. The security can be obtained easily without heightening the awareness of the users by simply implementing methods that allow end-to-end security. In the paper we present the new method of public key distribution using Social Media Networks. We use groups or circles that users can create there and publish images in the gallery with proper public or private access. After establishing secure session user can switch into different medium of communication, e.g. different messenger or text-SMS to share their data in secure way. Then there are described several applications of the presented method Secure Data Exchange (*SDEx*), e.g. secure SMS texting, Social Media Networking and data sharing. We propose frame format of the published QRcode as a public key and data message frame format.

Keywords: QR codes · Data security · Secure transmission · Social media · Key distribution · End-to-end data security

1 Introduction

Security of the data and privacy of the communication is a vital topic since the beginning of the human history. In the age of the Internet it is more and more crucial and even normal Internet users are becoming more and more conscious of the problem [1–4]. In recent years there have been several major stories about how easy is to eavesdrop and breach our data or communication privacy like emails, e.g. secret documents leaked by former U.S. National Security Agency contractor Edward Snowden, WikiLeaks, HeartBleed OpenSSL, etc. The examples above show that not only the hackers are the problem to our data privacy but also the governments and their secret service agencies.

Majority of users is not conscious what encryption is and that applications they use does not support it. It will not help to change it by simply saying or advertising that we users should rise our security level. There are some research that show that your friends can raise your awareness and make you more willing to implement the security measures [5]. Of course we are being convinced that the communication is secure but to what extent nobody knows. The research done in Nov 2014 by Electronic Frontier Foundation (*EFS*) has shown that only

P. Gaj et al. (Eds.): CN 2015, CCIS 522, pp. 389–399, 2015.
DOI: 10.1007/978-3-319-19419-6_37

6 out of 39 popular messengers fulfills 7 criterions that may support end-to-end security properly [6].

The sophisticated online surveillance techniques used by the spy agencies are problem also. If the governments can do it maybe hackers can do that as well. The problem of different types of attacks is described in several publications, e.g. session hijacking [7–10].

More and more people switch into the Internet services like Skype or social networking services like Facebook, Twitter, etc. We are ensured that the communication is secure. In Facebook the messages exchanged are protected by Secure Sockets Layer, SSL encryption. The other problem is that even if the transfer seems to be secure in reality it is not. in the Facebook case the data is encrypted only between an end user and Facebook servers. Facebook can have access to the clear text of those conversations, which potentially could be surrendered to law enforcement under a court order. Facebook has said it could enable end-to-end encryption between users exchanging data, but said such technology is complicated and makes it harder for people to communicate [11]. We think that Facebook underestimates their users.

In the papers [12–14] we have shown several methods and techniques how to make end-to-end users communication more secure In the current paper we develop the idea presented in [12,15]. In the Sect. 2 we analyze the security used in popular communicators. In the next section we show some solutions that improves the data security over the Social Media Networking communication. The method Secure Data Exchange (*SDEx*) based on public key distribution using social media is described in details also. Apart from the method we discuss the QRcode use and its format as a public key and the data format as well. In the Sect. 4 we show several examples of the method implementation in the applications that allows secure data exchange using texting SMS, Social Network Messengers, etc. It is well known that the compromising of the public key in asymmetric encryption will not break the data transfer confidentiality, integrity and authentication [16].

2 Security in Social Media Networks and Known Messengers

The Internet services like Skype or social networking services like Facebook and Twitter are more and more popular. Most users are pretty sure that their data is secure. But even if we are ensured that the communication is secure in fact it is not.

In Facebook the messages exchanged are protected by Secure Sockets Layer, SSL encryption but even if the transfer seems to be secure it is not because in the Facebook case that only encrypts data between an end user and Facebook. Facebook can have access to the clear text of those conversations, which potentially could be surrendered. We share more and more data using such services not realizing the danger. How important is that problem shows Table 1. According to Animoto research 75 % US Internet users share their videos using Facebook [17].

Table 1. Ways in which US Internet users share videos in Aug 2014 [17]

Way used to share video	Users percentage
Facebook	75.6
Email	51.9
YouTube	49.7
SMS/MMS	34.1
Instagram	29.8
DVD	27.7
Google+	23.1
Twitter	19.7
Snapchat	16.1
Dropbox	15.7

In the recent research [6] (Nov 2014) done by Electronic Frontier Foundation, EFF 39 services was checked including popular tools from Apple, Google, Facebook, BlackBerry, Microsoft and Yahoo. According to the Electronic Frontier Foundation research only 6 applications pass the security test. The EFF was interested in 7 possible questions/features [6]: is data encrypted in transit; is it encrypted so the provider can't read it; can the service verify contacts' identities; are past communications secure if keys are stolen; is the code open to independent review; is security design properly documented; and has there been an independent security audit?

All 39 applications encrypt content in transit, but only six satisfied all of the EFF's security requirements and managed to fulfill all seven EFF requirements: ChatSecure + Orbot, Cryptocat, RedPhone, Silent Phone, Silent Text and TextSecure. Apple fulfills 5/7 requirements. It lost points for not verifying contacts' identities or opening its code to independent review. But most of other services only checked off two boxes (WhatsApp, Snapchat, Skype, Google Hangouts, Facebook chat) – usually encrypted in transit and having security audit [6]. AIM only satisfied the encrypted in transit bit. The EFF said that most of the tools that are easy for the general public to use don't rely on security best practices including end-to-end encryption and open source code.

One of the solutions used to improve security in the Internet and in Social Media Networking, SMN communication is presented by Cryptocat. Cryptocat can log a user into Facebook and pull his contact list in order to set up an end-to-end encrypted conversation. The Facebook users using Cryptocat solution have secure end-to-end communication.

Basing on the *EFS* criterions we prepared the outline of the method that supports end-to-end security. The idea of using the known methods of encryption to support end-to-end security is presented in our previous paper [15]. We show some new methods how to strengthen the security of data exchange.

In the *SDEx* method that general outline is presented in [15], we have sketched the idea of creating secure channel that supports end-to-end secure connectivity. That method also fulfills all of the *EFS* criterions supposing that we have the code open to independent review and there has been an independent security audit. It does not need the special architecture because it uses existing Social Medial Networking users profile.

Users can easily upload their public keys as an image into their gallery and share it within their group, circle or publicly. If the security is broken in some way the users can easily change passwords, groups/circles and users in them or switch into different account to distribute the public key. The method is described in detail in the next section.

3   Method Secure Data Exchange (*SDEx*)

3.1   The Method *SDEx* and Their Applications Prerequisites

The method *Secure Data Exchange (SDEx)* was presented generally in [15]. The detailed description of the method is given below. The *SDEx* method is designed so it can fulfill the first five requirements of the *EFS* research. The application prerequisites that uses the *SDEx* method are:

- the Internet access,
- Social Medias Network access, can upload and download files/ images from them,
- save data (keys) locally on the device,
- can generate QRcode,
- process QRcodes,
- can capture and send text and data from SMS, messengers,
- can capture voice telephony agent to work with the voice transmission,
- can encrypt and decrypt using well known methods.

The prerequisites of the application can be widened or shortened due to the application's functionality. Examples that uses text SMS, Social Media Networking and IPv6 are given in the next section. About QRcode you can find more in [18, 19]. To fulfill the *EFS* requirements described above the applications encrypt the private keys, passwords and the data that we want to save and keep it. That will prevent reading the data while i.e. the smartphone is compromised. Our application use also double encryption described in [15].

3.2   Frame Format of the Published QRcode as a Public Key

To facilitate filtering methods searching for the proper QRcodes, the public key stored in QRcode image should be tagged with a metadata header. The proposed in *SDEx* method QRcode frame format consisting of the metadata header and the key is given below:

| Frame metadata header of QRcode | Public key |

where Frame metadata header of QRcode contains set number of 16 bytes:

- unique set of bits – for filtering the gallery – 4 bytes,
- header length in bytes – 1 byte,
- timestamp of key generation in milliseconds – 8 bytes,
- version of the key – 1 byte,
- key length in bits – 2 bytes.

The QRcode is storing only the data without any tags. That data can be anything like names, phone numbers, web page links, etc. That is why apart from the key itself the QRcode must contain additional metadata.

3.3   Data Message Frame Format

The *SDEx* method can be used in different types of communication: simple data, SMS, voice calls, video, etc. In the paper we discuss the text data transmission. Below we present the example of the data message frame format that can be used in text messages communication, e.g. text SMS messages:

| Frame header of the message | message |

where Frame metadata header of the message contains set number of 14 bytes:

- application ID, e.g. XXS3 – 4 bytes, where XX stands for 2 bytes special unique code,
- message type and controls, e.g. data 01000000, new key request, key received, key acknowledged, unrecognized key, etc. – 1 byte,
- timestamp of key generation of the public key (user B) in seconds – 4 bytes; we assume that users will not be able to generate keys in less than 1 second,
- version of the key used – 1 byte, e.g. RSA – 01000001 (letter A),
- message length in bytes – 4 bytes.

The proposed header can vary depending on the medium used, e.g. phone calls. In the text messaging using SMS it can be simplified by omitting the message length part of it.

3.4   Description of Secure Data Exchange (*SDEx*) Method

In the method we use asymmetric cryptography, e.g. RSA when we discuss key pair. The way how the method *SDEx* works is given below and presented in 8 steps:

Step 1. After downloading the *SDEx* application from the store, e.g. Google Play, the application has to connect and login to the social media network user's account. If not user should be able to create his own profile. After this user can create his own group or use existing private or public one.

Step 2. The user application generate public and private key pair (asymmetric pair).

Step 3. The public key is saved in QRcode file. It can be JPEG or PNG file format, 480 x 480 px for example. The QRcode is locally checked whether the key can be read properly by the *SDEx* application. In the QRcode there is encoded the message metadata so as to the application could recognize proper application QRcode and the encryption key.

Step 4. The QRcode files are stored in users' Social Media Network gallery chosen by the users, see Fig. 1. The QRcode file is downloaded and checked whether the key can be read properly by the *SDEx* application. If not, the *SDEx* application compares the size of original file and the downloaded one and uses better condition. Then it starts from the step 3 again (e.g. 4 attempts).

Step5. At this step both users A and B (or more users) has successfully gone through Step 1–4, according to asymmetric cryptography presented in [15].

   5.a. To connect with user B the user A must have access to the user B social media network gallery to download images from it.

   5.b User A downloads public key *Pub_KeyB* published in the user B gallery, prepares encrypted message using *Pub_KeyB* and sends it to the user B.

   5.c The user B application checks whether it is proper key by examining regarded metadata written in the header of QRcode using proposed format of it. It is not a meta data of a file. It is our metadata. The user B decrypts message using *Priv_KeyB*.

   5.d If the user B wants to send message to the user A he does the similar steps from 5.a to 5.c.

   5.e In the case when the message is sent in offline mode and the client returns to online mode it can be chosen whether to use current recipient key or check and download the new one.

Step 6. Initial data exchange Fig. 2. At that step user can agree to exchange new keys using secure line at that moment. The methods of secure secret keys exchange are presented in our previous paper [15]. The session key is created from the exchanged secret keys. After the key of the session is set the users obtain secure end-to-end connectivity. Hence the users applications can start with the Step 8.

Step 7. New secret keys session exchange, Fig. 3. During the communication one of the users can request for the new keys generation (symmetrical or asymmetrical ones). Then the message with the corresponding message type in set on the new key request is sent (e.g. from user A.) The other user can acknowledge or reject the request. The first user can then take the proper action like aborting or continuing. During that session the users can also switch into different medium (SMS, messengers) for key exchanging as well as the completely new medium for subsequent data exchange. The whole key exchange session is encrypted.

Step 8. Applications can filter and capture the media channel, e.g. SMS, voice, video, etc. and then establish session and continues communication over that media Fig. 2.

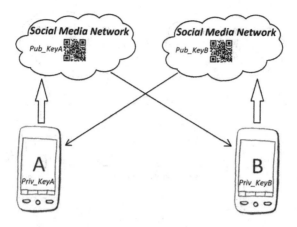

Fig. 1. Key sharing application using *SDEx* method

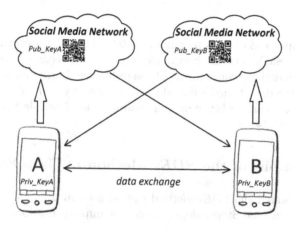

Fig. 2. Initial data exchange

Users at any moment can create a new group at a given social media network for certain users. The users can change and upload a new public key (QRcode image) to their corresponding gallery or group galleries.

3.5 New Key Publication

Users can publish the new public key. User A should store the old key or at least its timestamp. When it receives encrypted data with old key (timestamp of the key is given in the header) or unrecognized message it can send old key or unrecognized key control message. The user B must then (its application) download new key from the user A social media network gallery.

The applications periodically (in random time) check whether the public key/keys are not changed. In the case of compromising the user's social media

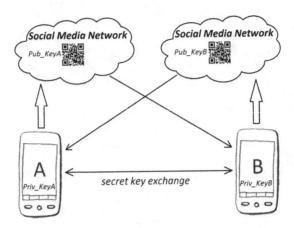

Fig. 3. Secret key exchange. They are independent from the published in social media network.

account the application warns user immediately. Then the user can, e.g. change password for his SMN account, inform the SMN admins, change the public key. It is well known that the compromising of the public key in asymmetric encryption will not break the data transfer confidentiality, integrity and authentication [16]. Users can quickly change key pair and group/circle of people that are allowed to access the key, also.

4 Applications of the *SDEx* Method in Data Exchange

As it was mentioned the *SDEx* method can be used in different types of applications that is why the Step 8 depends on that implementation.

4.1 SMS Communication

The steps shown above can be used in secure SMS messaging. The example is shown in Fig. 4. Of course the applications are installed on the users mobile phones and corresponding steps 1–5 described above are successfully finished.

At that point the application uses simplified message header. In the header at Fig. 4 the fields stand for:

- *From: xxxxxxxxx* – source phone number – provided by smartphone of user A,
- *To: yyyyyyyyy* – destination phone number – provided by smartphone of user B; both fields From and To are not contained in the data message frame format,
- Header of the data message: **XXS3@TËm|A** – frame header of the SMS message, where:
    - **XXS3** – application ID (for SMS capturing by the application),
    - **@** – type of the message – data message, (seen as a text),

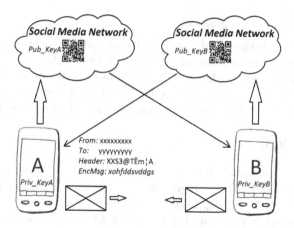

Fig. 4. Text exchange using SMS

- **TËm|** – timestamp as a string (1422618022 as an integer number),
- **A** – type of encryption method, A stands for RSA,
- *EncMsg: xohfddsvddgs* – encrypted message.

In the SMS messaging the header field message length is not necessary.

After receiving the message the user B application captures the SMS (because of the XXS3 field). The application has to correlate the phone number of the user A with the Social Media profile. The user B application checks other metadata in the header. If the process of the header filtering is successful it proceeds with the message decryption and answering the message if needed. If the process of the header filtering fails it can send control message, e.g. unrecognized key, or user B can abort the later communication with user A. It will depends on the user application setting. The advantage of such a solution is that the users can change the communication channel during the conversation.

4.2 Communication via Social Media Messengers

There are several popular messengers, mobile and desktop ones. As it was mentioned in the Sect. 2 only 6 out of 39 fulfill seven EFF requirements completely. In our method, after the steps 1–4 (from Sect. 3) are finished, users can communicate using, e.g. Social Media chats. The idea of application working is similar to SMS exchange, see Fig. 4, we can send more than 160 chars as text, but also pictures, documents, video, etc.

Second option of using the Social Media Networking, instead of staying at the same as key sharing, is to switch into different one, e.g. from Facebook to Twitter, Google+, Skype, Viber, etc. During their session users can also switch into different medium (SMS, messengers) for key exchanging as well as the completely new medium for subsequent data exchange.

4.3   Example of IPv6 Session

The other example is the use of global IPv6 addressing or peer-to-peer networks, e.g. Flooding networks or Distributed hash table networks.

After the session establishment (Steps 1–6 in Sect. 3) the users agree on way of communication and can send their IPv6 global addresses and set up direct p2p connection. Then the communication is as described in Sect. 3.

5   Conclusions

In the paper we presented the Secure Data Exchange (*SDEx*) method and its application that can fulfill the requirement of the *EFS* criterions for the secure end-to-end data exchange, e.g. in popular Internet and mobile messengers. Supposing that our applications are implementing *EFS* requirements and fulfill the last two requirements/criterions that only demand from the company to use proper documentation and software development process.

According to the *EFS* review presented in the Sect. 2 only 6 out of 39 applications fulfill all of the requirements. That research supports our view that our solution, *SDEx* method and its presented possible applications may be one of the answers to growing security concerns. The Internet services like Skype or social networking services like Facebook, Twitter etc. are more and more popular. Most users are pretty sure that their data is secure. But even if we are ensured that the communication is secure in fact it is not.

In the paper we presented the method that is easy to apply, does not need special architecture or additional services, because it uses existing Social Media networks. The method assures secure end-to-end communication and can fulfill the EFF requirements.

The Steps 1–8 presented in Sect. 3 can be implemented in mobile, web and desktop applications sharing data and the examples are in Sect. 4. Their advantages are:

- users can exchange keys as images using popular Social Media Networking;
- users can send data on the different media separately;
- during that session the users can also switch into different medium (SMS, messengers) for key exchanging as well as the completely new medium for subsequent data exchange;
- users does not need new infrastructure to share data or keys;
- application checks periodically in random time whether the public key is not changed;
- users can quickly change key pair and group or circle of people that are allowed to access the public key.

References

1. Nikiforakis, N., Meert, W., Younan, Y., Johns, M., Joosen, W.: Sessionshield: lightweight protection against session hijacking. In: Erlingsson, U., Wieringa, R., Zannone, N. (eds.) ESSoS 2011. LNCS, vol. 6542, pp. 87–100. Springer, Heidelberg (2011)

2. Adid, B.: Sessionlock: securing web sessions against eavesdropping. In: Proceedings of the 17th international conference on World Wide Web, pp. 517–524 (2008)
3. Lin, H.: The study and implementation of the wireless network data security model. In: Fifth International Conference on Machine Vision (ICMV 12). International Society for Optics and Photonics (2013)
4. Threats, I.: New challenges to corporate security. Res. J. Appl. Sci. **8**(3), 161–166 (2013)
5. Das, S., Kramer, A.D.I., Dabbish, L.A., Hong, J.I.: Increasing security sensitivity with social proof: a large-scale experimental confirmation. In: Proceedings of the 2014 ACM SIGSAC Conference on Computer and Communications Security (CCS '14), pp. 739–749. ACM, New York (2014)
6. Electronic Frontier Foundation. https://www.eff.org/secure-messaging-scorecard
7. Ahuja, M.S.: A review of security weaknesses in Bluetooth. Int. J. Comput. Distrib. Sys. **1**, 3 (2012)
8. Haataja, K., Hyppönen, K., Pasanen, S., Toivanen, P.: Reasons for Bluetooth network vulnerabilities. Bluetooth Security Attacks. Springer, Berlin Heidelberg (2013)
9. Sriram, S.V.S., Sahoo, G.: A mobile agent based architecture for securing WLANs. International Journal of Recent Trends Engineering **1**(1), 137–141 (2009)
10. Huang, J.: A study of the security technology and a new security model for WiFi network. In: Fifth International Conference on Digital Image Processing. International Society for Optics and Photonics (2013)
11. The Verge. http://www.theverge.com/2013/6/26/4468050/facebook-follows-google-with-tough-encryption-standard
12. Hłobaż, A., Podlaski, K., Milczarski, P.: Applications of QR codes in secure mobile data exchange. In: Kwiecień, A., Gaj, P., Stera, P. (eds.) CN 2014. CCIS, vol. 431, pp. 277–286. Springer, Heidelberg (2014)
13. Hłobaż, A.: ecurity of measurement data transmission - message encryption method with concurrent hash counting. Przeglad Telekomunikacyjny i Wiadomosci Telekomunikacyjne **80**(1), 13–15 (2007)
14. Hłobaż, A.: Security of Measurement Data Transmission - Modifications of the Message Encryption Method Along with Concurrent Hash Counting, pp. 39–42. Przeglad Wlokienniczy - Wlokno, Odziez (2008)
15. Podlaski, K., Hłobaż, A., Milczarski, P.: New Method for Public Key Distribution Based on Social Networks. arXiv:1503.03354 [cs.CR] (2015)
16. Katz, J., Lindell, Y.: Introduction to Modern Cryptography. Chapman & Hall/CRC Press, London (2007)
17. Animoto. http://www.adweek.com/socialtimes/social-video-sharing/613811
18. BS ISO/IEC 18004:2006. Information technology. Automatic identification and data capture techniques. QR Code 2005 bar code symbology specification (2006)
19. QRcode webpage. http://www.qrcode.com/en/codes/

Speaker Verification Performance Evaluation Based on Open Source Speech Processing Software and TIMIT Speech Corpus

Piotr Kłosowski[✉], Adam Dustor, and Jacek Izydorczyk

Silesian University of Technology, Institute of Electronics,
Akademicka Str. 16, 44-100 Gliwice, Poland
pklosowski@polsl.pl

Abstract. Creating of speaker recognition application requires advanced speech processing techniques realized by specialized speech processing software. It is very possible to improve the speaker recognition research by using speech processing platform based on open source software. The article presents the example of using open source speech processing software to perform speaker verification experiments designed to test various speaker recognition models based on different scenarios. Speaker verification efficiency was evaluated for each scenario using TIMIT speech corpus distributed by Linguistic Data Consortium. The experiment results allowed to compare and select the best scenario to build speaker model for speaker verification application.

Keywords: Speaker recognition · Speaker verification · Speech processing · Open source software · Speech corpus

1 Introduction

Department of Telecommunication, a part of the Institute of Electronics and Faculty of Automatic Control, Electronics and Computer Science Silesian University of Technology, for many years has been specializing in advanced fields of telecommunication engineering. One of them is speech signal processing. Main research areas on this field are: speech synthesis, speech recognition and speaker verification and identification.

Department of Telecommunication is working on the European project titled "Innovative speaker recognition methodology for communications network safety". The project is realized in cooperation with Samsung Research &Development Institute Poland. The main goal of this project is to increase the functionality and raise security level of mobile devices by using efficient speaker recognition algorithms [1,2]. The subject of the project is to create a research platform used for development and testing speaker identification algorithms based on voice and implement the best of them to mobile devises. Research platform allows to choose optimal solution under the criteria of distortion resistance, computational complexity and effectiveness of recognition, to be later implemented in a mobile device.

© Springer International Publishing Switzerland 2015
P. Gaj et al. (Eds.): CN 2015, CCIS 522, pp. 400–409, 2015.
DOI: 10.1007/978-3-319-19419-6_38

Research platform to evaluate speaker recognition algorithms require proper implementation. There are different ways of implementing speaker recognition research platform. As a criterion for choosing the appropriate method of implementation we can use implementation price or implementation time, or both of them. Creating of speaker recognition platform requires advanced speech processing techniques realized by specialized speech processing software [3]. Which software should be used for speaker recognition platform development? There are many possibilities. To create own speaker recognition application can be used:

1. programming languages such as C++, Java, Python, etc.,
2. high level commercial computing environment for implementation of speaker recognition algorithms such as: MATLAB with Signal Processing Toolbox,
3. open source speech processing software.

The interesting solution is to use of open source speech processing software [4]. This solution has the lowest price and the shortest implementation time. The article focuses on two open source applications that can be used to construct platform for speaker recognition research.

2   Speaker Recognition and Verification

Speaker recognition is the process of automatically recognizing who is speaking by analysis speaker-specific information included in spoken utterances. This process encompasses identification and verification. The purpose of speaker identification is to determine the identity of an individual from a sample of his or her voice and it can be divided into two main categories, i.e. closed-set and open-set. In a closed-set identification there is an assumption that only registered speakers have an access to the system which makes a decision 1 from N, where N is the number of previously registered speakers. In an open-set identification there is no such an assumption so the identification system has to determine whether the testing utterance comes from a registered speaker or not and if yes it should determine his or her identity. The purpose of speaker verification is to decide whether a speaker is whom he claims to be. Most of the applications in which voice is used to confirm the identity claim of a speaker are classified as speaker verification. Speaker recognition systems can also be divided into text-dependent and text-independent.

In text-dependent mode the speaker has to provide the same utterance for training and testing, whereas in text-independent systems there are no such constraints. The text-dependent systems are usually based on template matching techniques in which the time axes of an input speech sample and each reference template are aligned and the similarity between them is accumulated from the beginning to the end of the utterance. Because these systems can directly exploit voice individuality associated with each phoneme or syllable, they usually achieve higher recognition performance than text-independent systems [5].

Basic structure of speaker verification process is shown in Fig. 1. Speech signal is cut into short fragments, which usually last for 20–30 [ms] known as speech frames. Feature extraction is responsible for extracting from each frame a set of parameters known as feature vectors. Extracted sequence of vectors is then compared to speaker model (in verification) or speaker models (in identification) by pattern matching. The purpose of pattern matching is to measure similarity between test utterance and speaker model. In identification an unknown speaker is identified as the speaker whose model best matches the test utterance. In verification the similarity between input test sequence and claimed model must be good enough to accept the speaker as whom he claims to be. As a result, verification requires choosing decision threshold. If computed distance is less than this threshold, a decision can be made that the speaker is whom he claims to be. How to find an optimum value of this threshold still remains a problem for scientist [6]. Another very desired property of this threshold is its independence of a speaker, which means that there is one threshold for all speakers. Since these problems are not solved satisfactorily, they still remain very important research issues, apart from problem of finding the best set of speech parameters, which must be studied further to make an improvement in speaker recognition technology.

3   Speaker Recognition Research Platform Based on Open Source Software

The Table 1 presents the comparison of the most popular open source speech processing tools. This comparison can be helpful in choosing the right speech processing tool for a specific speaker and speech recognition application [4].

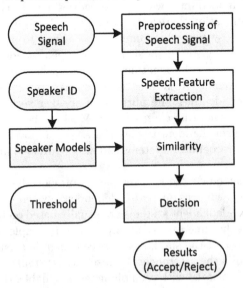

Fig. 1. Basic structure of speaker verification process

Table 1. The comparison of the most popular open source speech processing tools [4]

Name	Environment	Portability	Flexibility	Features
CMU Sphinx [7]	C++ Java	Unix (Linux) Embedded sys.	High	Model training, speech recognition framework
SPRACH [4]	C++ Tcl/Tk Perl	Unix (Linux)	High	Speech recognition framework
GMTK [8]	C++ Tcl/Tk	Unix (Linux)	Very high	Probabilistic modeling framework
SONIC [9]	C++ Tcl/Tk	UNIX (Linux, Solaris) Windows Mac OS	Very high	Continuous speech recognition
HTK [10]	ANSI C	Unix (Linux) Windows	High	Hidden Markov model toolkit
ALIZE [11]	C++	Unix (Linux) Windows Embedded sys	Medium	Speech processing and recognition framework
SPRO [4]	C++	Unix (Linux) Windows Embedded sys	Mdium	Seech features extraction

The most important elements of each speaker recognition system are speech features extraction and speaker classification. The paper presents examples of the use the open source speech processing software named SPRO for speech features extraction. Similarly, the open source software can be used to test and create efficient classifiers that are the basis for designing effective speaker recognition applications. For testing and construction of the various classifiers can be used e.g. the Hidden Markov Model Toolkit (HTK) [10] or open source platform for biometrics authentication called ALIZE [11]. Use of the open source speech processing software can significantly improve the construction and testing of modern speaker recognition application. Two applications: SPRO and ALIZE ware used to build speaker recognition platform for the European projekt. Additionally speech resources from TIMIT speech corpus was used. Below the three most important elements of the speaker recognition platform are presented.

3.1 SPRO Software

SPRO is an open source speech processing toolkit. It was designed for both: speaker and speech recognition. SPRO has been created for variable resolution spectral analysis, but it also supports classic mechanics used in speech processing [4]. It provides runtime commands, as well as standard C library for implementing new algorithms and applications. After compilation, the user is given access to tools used for following purposes: filter-bank based speech analysis, linear predictive speech analysis, comparing streams, extracting speech parameters. The library provides additional signal processing functions, which can be used in custom speech recognition applications, such as FFT, LPC analysis and feature processing, such as lifter, CMS and variance normalization. SPRO is distributed under GNU Public License agreement.

The speaker recognition platform uses MFCC (The Mel-Frequency Cepstrum Coefficient) as a speech signal features. MFCC is a representation of the short-term power spectrum of a sound, based on a linear cosine transform of a log power spectrum on a nonlinear mel scale of frequency. Mel-frequency cepstral coefficients (MFCCs) are coefficients that collectively make up an MFC. They are derived from a type of cepstral representation of the audio clip (a nonlinear "spectrumof-a-spectrum"). The difference between the cepstrum and the mel-frequency cepstrum is that in the MFC, the frequency bands are equally spaced on the mel scale, which approximates the human auditory system's response more closely than the linearly-spaced frequency bands used in the normal cepstrum. This frequency warping can allow for better representation of sound, for example, in audio compression. The European Telecommunications Standards Institute in the early 2000s defined a standardised MFCC algorithm to be used in mobile phones [12].

3.2   ALIZE Software

The second selected open source application is ALIZE. ALIZE is open source platform for biometrics authentication [11]. It provides single engine for face and voice recognition. ALIZE project's goal is to provide access to biometric technologies for industrial and academic usage. It consists of low-level API and high-level executables. ALIZE allows the user to quickly create speech recognition system, as well as provides tools for industrial voice processing applications. The project has been made open source, because it is believed that allowing broad scientific research on speech recognition algorithms results in quicker improvement of such systems, by making them more accurate and resistant to noise [13]. Project has been evaluated by The National Institute for Standards and Technology (NIST) and achieved very good results. It is written in C++ and allows multi-platform implementation, including a possible use in embedded devices.

ALIZE software was used to build models and preform speaker verification process. Both processes have used GMM (Gaussian Mixture Model). GMM is a parametric probability density function represented as a weighted sum of Gaussian component densities. GMMs are commonly used as a parametric model of the probability distribution of continuous measurements or features in a biometric system, such as vocal-tract related spectral features in a speaker recognition system. GMM parameters are estimated from training data using the iterative EM (Expectation Maximization) algorithm or MAP (Maximum A Posteriori) estimation from a well-trained prior model.

3.3   TIMIT Speech Corpus

Speech processing software is not sufficient to build speaker recognition platform. Creating advanced speech processing and speaker recognition systems requires dozens of real voice samples. Instead of creating numerous variations of different accents, researchers may use speech corpora, obtainable from multiple sources. Additionally, each possible voice sample set has been collected for specialized applications, focusing on different aspects of speech or speaker recognition. Available test results of other systems based on each corpus enables better evaluation of a system being in development.

TIMIT is a Acoustic-Phonetic Continuous Speech Corpus. The goal of TIMIT was to provide data for the acquisition of acoustic and phonetic information in order to evaluate and improve automatic speech recognition systems [14]. TIMIT was created by Massachusetts Institute of Technology (MIT), Stanford Research Institute (SRI), and Texas Instruments (TI), and was sponsored by Defense Advanced Research Projects Agency – Information Science and Technology Office (DARPA-ISTO) [15]. It contains 6300 sentences spoken by 630 speakers (438 male, 192 female). All 8 major dialects regions of the United States are covered. TIMIT also includes hand verified transcripts for all sentences [14]. TIMIT speech corpus has been adapted to support development of phoneme recognition systems. The recognition error rate of 24.6 % has been reached by using recurrent neural network, although it is described as one not being significantly different from other methods [16]. The recognition accuracy using this corpus has improved about 13 % over last 20 years. In 1990 the accuracy has reached around 66 % and then went up to 75 % in 1994. Since then, it has improved slowly and is currently estimated to be about 79 % [17].

4   Speaker Recognition Research Platform and Speaker Verification Experiment Results

The speaker recognition research platform was used to develop and testing speaker identification algorithms based on voice. It performs the tasks presented in the Table 2. The task include: speech files preprocessing, speech features extraction, building and training models, speaker recognition evaluation and results analysis. Speaker recognition platform works in two modes:

– training mode – to build world model and speaker recognition models,
– recognition mode – to evaluate speakers verification performance.

Speaker recognition models consists of the two types of models: world (non-client) model and speaker (client) models. The world model is common model of speakers who are not clients of the speaker recognition platform. To perform the verification, client and non-client (world) models are generally computed in a training phase. Purpose of these models is identify the differences between the client and impostors regarding an acoustic realization. The various scenarios of speaker verification experiment, presented in Table 3, were tested. The scenarios

Table 2. Speaker recognition research platform tasks

No.	Task	Implementation
1	Speech files preparation	Bash, Python
2	Speech features extraction	SPRO
3	Speech features normalization	ALIZE
4	Voice activity detection	ALIZE
5	Build world model	ALIZE
6	Build and train speakers models	ALIZE
7	Speaker recognition evaluation	ALIZE
8	Results analysis	Python

Table 3. The various scenarios of speaker verification experiment

No	World model		Speakers models		Speaker verification		Speakers sex	Results [EER]
	Speakers	Files	Speakers	Files	Speakers	Files		
1	130	1300	500	2500	500	2500	Men and women	0.0305
2	330	3300	300	1500	300	1500	Men and women	0.0855
3	30	300	600	3000	600	3000	Men and women	0.1742
4	130	1300	500	4000	500	1000	Men and women	0.0217
5	130	1300	500	4500	500	500	Men and women	0.0220
6	138	1380	300	1500	300	1500	Men only	0.0312
7	338	3380	100	500	100	500	Men only	0.0261
8	338	3380	100	800	100	200	Men only	0.0203

differ in the way of use and division of TIMIT speech corpus to build world model, to build speakers models and to perform speakers verification process. Speaker verification experiment can be classified as text-independent. The best results were obtained using scenario No. 8. Details of this scenario are as follows:

- 438 (338+100) make speakers,
- world model based on 338 speakers, 10 files for each speaker, 3380 files,
- speaker models based on 100 speakers, 8 files for each speaker, 800 files,
- speaker verification evaluation based on 100 speakers, 2 files for each speaker, 200 files.

The speaker verification experiment results of scenario No. 8 were presented on Fig. 2. This figure presents among others ROT (Relative Operating Characteristic) plot. The ROC plot is a visual characterization of the trade-off between the FAR (False Acceptance Rate) and the FRR (False Reject Rate). In general, the matching algorithm performs a decision based on a threshold which determines how close to a template the input needs to be for it to be considered a match.

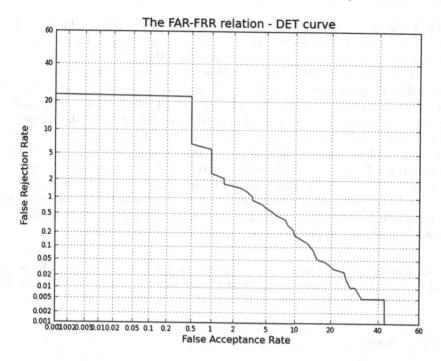

Fig. 2. Speaker verification experiment results – scenerio No. 8

If the threshold is reduced, there will be fewer false non-matches but more false accepts. Conversely, a higher threshold will reduce the FAR but increase the FRR. A common variation is the DET (Detection Error Trade-off) plot, which is obtained using normal deviation scales on both axes. The effectiveness of speech recognition for scenario No. 8 was presented below:

$$EER = 0.02034 \text{ width treshhold} = -0.46400. \tag{1}$$

The EER value is the rate at which both acceptance and rejection errors are equal. The EER is a quick way to compare the accuracy of speaker recognition applications with different ROC curves. In general, the speaker recognition application with the lowest EER is the most accurate.

5   Summary

The paper presents example of the use open source speech processing software to build speaker recognition research platform. The platform provides voice acquisition, preprocessing, feature extraction, voice activity detection, acoustic modeling, world model building, optimization decisions and performance evaluation score. The platform is a complete set of speaker recognition process modules. Each module integrates the most popular of today's most commonly used speaker recognition algorithms. The speaker recognition research platform uses

TIMIT speech corpus distributed by Linguistic Data Consortium. The platform enables to perform complex speaker recognition research experiments allowing testing new and modern speaker recognition algorithms.

The main goal of the speaker recognition experiments was to provide a method for dividing a set of TIMIT speech corpus speakers into two subsets. The fist speakers subset was used to build world model, the second to build client models. The both client and world (non-client) models are generally computed in a training phase. Purpose of these models is identify the differences between the client and impostors regarding an acoustic realization. Different scenarios of dividing a TIMIT corpus speakers were tested. The best speaker recognition results (EER = 0.02034) were obtained using scenario No. 8. In this scenario, data of 338 speakers were used to build world model and data of 100 speakers to build speakers model. The results of this experiment allow to construct effective speaker recognition application and implement it in the mobile device.

Acknowledgements. This work was supported by The National Centre for Research and Development (www.ncbir.gov.pl) under Grant number POIG.01.03.01-24-107/12 (Innovative speaker recognition methodology for communications network safety).

References

1. Dustor, A., Kłosowski, P., Izydorczyk, J.: Influence of feature dimensionality and model complexity on speaker verification performance. In: Kwiecień, A., Gaj, P., Stera, P. (eds.) CN 2014. CCIS, vol. 431, pp. 177–186. Springer, Heidelberg (2014)
2. Dustor, A., Kłosowski, P., Izydorczyk, J.: Speaker recognition system with good generalization properties. In: Proceedings of International Conference on Multimedia Computing and Systems 2014 p. 73, Marrakech, Morocco, IEEE (2014)
3. Rabiner, L.R., Schafer, R.W.: Introduction to digital speech processing. Found. Trends Sig. Process. 1(1–2), 1–194 (2007)
4. Kłosowski, P., Dustor, A., Izydorczyk, J., Kotas, J., Ślimok, J.: Speech recognition based on open source speech processing software. In: Kwiecień, A., Gaj, P., Stera, P. (eds.) CN 2014. CCIS, vol. 431, pp. 308–317. Springer, Heidelberg (2014)
5. Beigi, H.: Fundamentals of speaker recognition. Springer, New York (2011)
6. Togneri, R., Pullella, D.: An overview of speaker identification: accuracy and robustness issues. IEEE Circ. Sys. Mag. 11(2), 23–61 (2011)
7. Tsontzos, G., Orglmeister, R.: CMU Sphinx4 speech recognizer in a Service-oriented Computing style. In: IEEE International Conference on Service-Oriented Computing and Applications (SOCA), pp. 1–4 (2011)
8. Bilmes, J., Bartels, C.: Graphical model architectures for speech recognition. IEEE Sig Process. Mag. 22(5), 89–100 (2005)
9. Pellom, B., Hacioglu, K.: Recent improvements in the CU SONIC ASR system for noisy speech: the SPINE task. In: Proceedings of IEEE International Conference on Acoustics, Speech, and Signal Processing (ICASSP). Hong Kong (Apr 2003)
10. Young, S., Evermann, G., Hain, T., Kershaw, D., Moore, G., Odell, J., Ollason, D., Povey, D., Valtchev, V., Woodland, P.: The HTK Book. Cambridge University Engineering Department, Cambridge, UK (2002)

11. Bonastre, J.F., Wils, F., Meignier, S.: ALIZE, a free toolkit for speaker recognition. In: IEEE International Conference on Acoustics, Speech, and Signal Processing (ICASSP '05), vol. 1, pp. 737–740 (2005)
12. Speech Processing, Transmission and Quality Aspects (STQ); Distributed speech recognition; Front-end feature extraction algorithm; Compression algorithms. Technical standard ES 201 108, v1.1.3. European Telecommunications Standards Institute (2003)
13. Fauve, B.G.B., Matrouf, D., Scheffer, N., Bonastre, J.F., Mason, J.S.D.: State-of-the-art performance in text-independent speaker verification through open-source software. IEEE Trans. Audio, Speech, Lang. Process. 15(7), 1960–1968 (2007)
14. Garofolo, J.S., Lamel, L.F., Fisher, W.M., Fiscus, J.G., Pallett, D.S., Dahlgren, N.L., Zue, V.: TIMIT Acoustic-Phonetic Continuous Speech Corpus. Linguistic Data Consortium, Philadelphia (1993)
15. Fisher, W.M., Doddington, G.R., Goudie-Marshall, K.M.: The DARPA speech recognition research database: specifications and status. In: Proceedings of DARPA Workshop on Speech Recognition, pp. 93–99 (1986)
16. Fernandez, S., Graves, A., Schmidhuber, J.: Phoneme recognition in TIMIT with BLSTM-CTC (2008)
17. Lopes, C., Perdigao, F.: Phoneme Recognition on the TIMIT Database (2011)

Digital Filter Implementation in Hadoop Data Mining System

Dariusz Czerwinski[(✉)]

Lublin University of Technology, 38A Nadbystrzycka Street,
20-618 Lublin, Poland
d.czerwinski@pollub.pl

Abstract. Paper presents Savitzky-Golay digital filter implementation using the R software environment in private Hadoop Data Mining System. The idea of the research was the implementation of the digital filter to handle big data came from measurements. The study focused on proper filter implementation and comparison of the results with National Instruments Diadem – enterprise class system for data management and mining.

Keywords: Digital signal processing · Savitzky-Golay filter · Data mining system · Cloudera Hadoop · R programming

1 Introduction

Data mining systems play a very important role in the IT industry and science. The popularity of data mining is due the wide applicability of mining methods in different domains. Discovered knowledge is used in the optimization of business processes, detecting irregularities and peculiarities, prediction of future events and behaviours as also as in the science for analysing the huge amount of measured data [1–8].

Data Mining Applications in Digital Signal Processing can be located into following fields [9–12]:

- classification – which can distinguish fragments of the signal according to constructed model and defined classes,
- clustering – which can bind the discovered segments into a number of classes (taxones, clusterers) with the use of special algorithms (K-median, K-means, etc.),
- segmentation – it can split the signal into fragments of generally different sizes, which possess homogeneous properties,
- mining and sequential analysis – the logical regularities in signal's structure and relationship between different events in signals can be found, it includes also finding time patterns which characterize the nature of signal,
- multi-dimensional visualization – it is advantageous for plotting n-dimensional dataset with a specific end goal to watch their nearness or remoteness.

© Springer International Publishing Switzerland 2015
P. Gaj et al. (Eds.): CN 2015, CCIS 522, pp. 410–420, 2015.
DOI: 10.1007/978-3-319-19419-6_39

R, without any modifications is not very suitable for data mining processes. Data are limited to the available to the amount of user RAM, since R stores in memory [9,13]. Cloud computing allows to extend the capabilities of running the R on much better hardware, as also as using the parallelization. The following projects are focused on using R in clouds infrastructure [9]:

- RevoScaleR package and RHadoop by Revolution Analytics – the company offers their solutions by enabling their software as a rentable service on Amazon Ec2 and through RevoDeployR [14],
- Oracle R Enterprise – provides the set of tools allowing to handle big data in Oracle solutions [15],
- Renjin – is a JVM-based interpreter for the R language and enables developers to deploy their code to Platform-as-a-Service providers like Google App engine, Amazon Beanstalk, or Heroku [16],
- RAmazon – set of packages (RAmazonS3, RAmazonDBREST) to interact with Amazon cloud [17,18].

There are some other works related to Digital Signal Processing and Data Mining or Cloud Computing techniques [9–12,19–21]. The closest research, related to this article, is that one presented by Adam Laiacano in [10]. However those research was made in Scalding, which is an extension to Cascading a Java library abstracting away low-level Hadoop details.

In this article Savitzky-Golay digital filter implementation using R in Hadoop was presented. The filter implementation was made in the private Data Mining System, but it can be directly ported to the public clouds running on Hadoop. The filter was applied to acquisition data derived from second generation High Temperature Superconducting Tape measurement system. AC superconducting devices measurement systems often generates huge amount of data. Usually it is crucial to process the acquired data almost in real time, because of superconducting device stability and quench procedures. The results of implemented in R filter were compared with the results coming from commercial NI Diadem data mining software.

2   Test Platform Configuration

To implement Savitzky-Golay digital filter in private Data Mining System the Quick Start VM with Cloudera CDH 5.3 distribution was used. Cloudera system allows building structures of both computing clusters as well as private clouds. Thanks to this, users of the system use the resources in the same way as public cloud resources offered by Amazon. This allows for easy integration of this solution in a hybrid cloud projects [22–24]. Hadoop, which is part of the Cloudera distribution consists of the following elements [25,26]:

- HDFS – Hadoop Distributed File System it is a distributed file system, which is designed to run on common hardware, as also the library providing mechanisms for access the distributed resources,

- Hadoop YARN – application skeleton made for the management of distributed resources and calculations (load balancing, failure recover mechanisms, scheduling of computational processes),
- Hadoop MapReduce – implementation of the processing model using YARN skeleton,
- Hadoop Common – a set of tools and facilities to support the other modules.

The guest OS include an implementation of Cloudera's Distribution Including Apache Hadoop (CDH) version CDH 5.3 [27].

Testbed architecture is shown in Fig. 1. Hardware and software system components have been assigned as follows:

- Host machine (PC – CPU – AMD Athlon X2 240, 2 cores, 6 GB RAM, 250 GB HDD SATA), host system Windows Professional x64, VMware Player v.7.0 with VMware Tools installed,
- Guest – OS Cloudera CDH 5.3.0.0 (Centos 6.4 × 64) with 4 GB RAM, 2 cores, 40 GB.

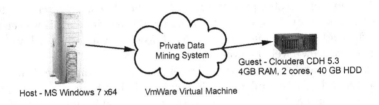

Fig. 1. The architecture of the testbed

Savitzky-Golay digital filter was implemented using R programming language, which is not included in Cloudera distribution therefore the additional configuration steps were needed. Revolution Analysis has created the set of connectors for Hadoop called RHadoop, that will allow to manipulate Hadoop data stores directly from HDFS and HBASE using R language. It gave R programmers the ability to write MapReduce jobs in R using Hadoop Streaming [28,29]. RHadoop consists of following R packages allowing the users to manage and analyze data with Hadoop in R language, including the creation of MapReduce jobs [30]:

- rhdfs – package for basic connectivity to the HDFS file system. This package allows browsing, reading, writing, and modifying files stored in HDFS from within R. Should be installed only on the node that will run the R client;
- rhbase – provides basic connectivity to HBASE, using the Thrift server. Tables stored in HBASE can be browsed, read, written, and modified from within R. Installed only on the node that will run the R client;
- plyrmr – enables to perform common data manipulation operations, as found in popular packages such as plyr and reshape2, on very large data sets stored on Hadoop. It relies on the MapReduce to perform the tasks. Should be installed on every node in the cluster;

– rmr2 – package that allows to perform statistical analysis in R via Hadoop MapReduce functionality on a Hadoop cluster. Should be installed on every node in the cluster;
– ravro – adds the ability to read and write avro files from local and HDFS file system and adds an avro input format for rmr2. Installed only on the node that will run the R client.

Designed by Revolution Analysis framework (called RHadoop) allows the specification of an operation to be applied to a huge data set. The problem and data can be divided and run in parallel (Fig. 2).

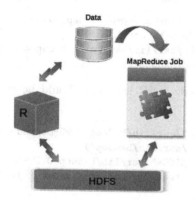

Fig. 2. Idea of the RHadoop main elements

Very large dataset can be reduced into a smaller subsets where calculations and analysis can be made. In a traditional solution in Hadoop, these kinds of operations are written as MapReduce jobs in Java or higher level languages like Hive and Pig. Executing R code in the context of a MapReduce job allows for additional functionality (i.e. digital filtering, visualization of data on charts and maps) [28].

To set up the RHadoop framework the R language and R development packages should be installed with command:

```
# yum install R R-devel
```

Next within the R environment run by command

```
# R
```

additional packages should be installed from Cran repository with:

```
> install.packages(c("rJava","Rcpp","RJSONIO","bitops",
    "digest","functional","stringr","plyr","reshape2",
    "caTools"))
```

Next step is to download from Revolution Analytics GitHub the packages rhdfs and rmr2, as also as from Cran project site the signal package. They need to be installed within R with commands:

```
> install.packages("/home/cloudera/Downloads/
rhdfs_1.0.8.tar.gz", repos = NULL, type="source")
> install.packages("/home/cloudera/Downloads/
rmr2_3.3.0.tar.gz", repos = NULL, type="source")
> install.packages("/home/cloudera/Downloads/
signal_0.7-4.tar.gz", repos = NULL, type="source")
```

Some additional system environmental variables should be set in ~/.bashrc:

```
export HADOOP_HOME=/usr/lib/hadoop-0.20-mapreduce
export HADOOP_CMD=/usr/bin/hadoop
export HADOOP_STREAMING=/usr/lib/hadoop-0.20-mapreduce/
        contrib/streaming/hadoop-streaming.jar
```

Configuration file .Rprofile for R language should be created and include the following lines:

```
Sys.setenv(HADOOP_HOME="/usr/lib/hadoop-0.20-mapreduce")
Sys.setenv(HADOOP_CMD="/usr/bin/hadoop")
Sys.setenv(HADOOP_STREAMING="/usr/lib/hadoop-0.20-
        mapreduce/contrib/streaming/hadoop-streaming.jar")
library(rJava)
library(rmr2)
library(rhdfs)
```

3 Implementation of Savitzky-Golay Filter in RHadoop

Developed test platform allows for quick and comprehensive implementation of Savitzky-Golay digital filter. Within the R system first the hdfs should be started with command:

```
> hdfs.init()
```

The data stored in csv file can be assigned to the my.data structure:

```
> my.data=read.csv("/home/cloudera/filter/sample.csv")
```

From the multicolumn data, there can be filtered one column and stored as

```
> I=my.data[,"IO"]
```

and in next step passed to HDFS with following command

```
> I.index=to.dfs(I)
```

Above command passes the data into the mapper record by record. After this digital filter in RHadoop can be applied with following command:

```
>sg= values(from.dfs(mapreduce(
                 input=I.index,
                 map=function(k,v) sgolayfilt(v))))
```

In this case input is the variable I.index which contains the big data object. The map function is applied to the data and in the case of this filter implementation the reduced function is not used. In the map process the sgolayfilt function is used.

The usage for this function is:

```
sgolayfilt(x, p = 3, n = p + 3 - p%%2, m = 0, ts = 1)
```

where arguments are:

x – signal to be filtered,
p – filter order,
n – filter length (must be odd),
m – return the m-th derivative of the filter coefficients,
ts – time scaling factor.

This function can be called with one parameter – signal to be filtered, with other parameters set as defaults (p=3, n=5, m=0, ts=1). The results of the mapreduce function are stored in sg variable, which is the effect of transfer big data object using from.dfs function.

4   Experimental Results

The experiment applying Savitzky-Golay filter implemented in RHadoop. Input data came from measurements and results has been compared with one obtained from applying the digital filter in National Instruments Diadem environment.

Measured data come from the system shown in Fig. 3, where the High Temperature Superconducting tape SF12050, produced by Super Power, placed in liquid nitrogen bath were overheated in chosen segments. The current I0 (frequency 50 Hz) flowing through the tape and voltages on segments 1 to 7 were recorded with NI DAQ 6251 PCI card with sampling rate 50 points per 1 ms. The measurements are further described by the authors in [31].

Eight channels recording, with sampling rate set as above, gave in 15 seconds the 6 million measured data points and during 12 min above 231 (2 billion) data points. This number of data points exceed the capabilities of NI Diadem system, which is specially designed for data mining obtained during data acquisition and/or generated during simulations [32]. Data finally written in CSV format produced over 7 GB output file.

Sample raw measured data of instantaneous current recorded during 15 seconds are presented in Fig. 4. Inset shows the zoomed top of chosen sine waveform, it can be noticed that curves are not smoothed. First test performed in the scenario was the applying Savitzky-Golay filter to the measured data in private Data Mining System.

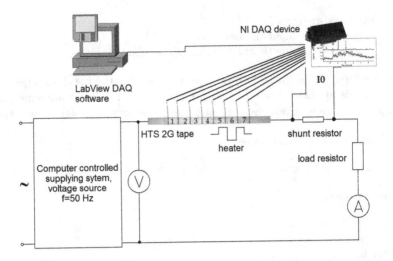

Fig. 3. Measurements of high temperature superconducting tape response on overheating

Filter was implemented in Hadoop environment according to the description in previous section. The results from sg variable are transferred to the CSV file with command:

```
> write.csv(sg, file="sg_result.csv")
```

Next performed test was the applying Savitzky-Golay filter with the same parameters to the measured data, but in the NI Diadem environment. After calculations obtained results were compared. Overall, both calculated waveforms look almost the same, however there are slight differences. Comparison of zoomed smoothed waveforms is shown in Fig. 5. The differences appear mostly on the top of sine waveforms.

Comparing the results of smoothed waveforms calculated in RHadoop and NI Diadem the value of the relative error versus the measurements time can be designated and shown in Fig. 6. Relative error is calculated according to the formula:

$$|\delta| = \frac{|x - x_0|}{|x_0|} \cdot 100\,\% \tag{1}$$

where:

δ – relative error,
x – value of the smoothed data coming from RHadoop,
x_0 – value of the smoothed data coming from NI Diadem.

Analysing the values of relative error it can be noticed that the minimum value is about 0.5 % and maximum one is about 1.5 %. These values are very low and much smaller comparing to the voltage and current measurements errors which were in that experiment estimated on 5 %.

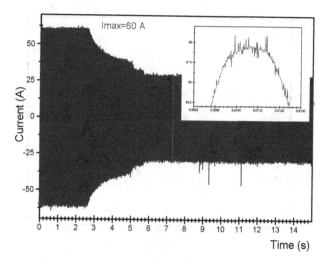

Fig. 4. Measured data of instantaneous current, inset – zoomed top of sample sine waveform

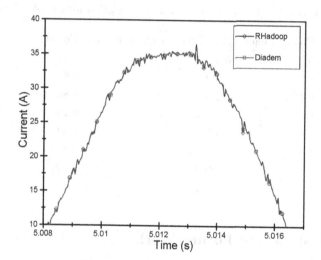

Fig. 5. Comparison of the measured data smoothed by savitzky-golay filter in RHadoop and NI diadem environments

To compare the similarity of the current I0 waveforms shapes the correlation coefficient was calculated. The value of this coefficient is equal 0.81, which means that there is very strong correlation between the compared smoothed curves.

Data from other recorded channels were also compared and summarized in Table 1. The relative error in other channels stays on the similar level, however the correlation coefficient varies from strong one (values below 0.8) to the very strong (values equal or above 0.8).

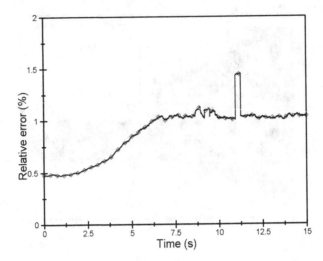

Fig. 6. Value of relative error versus time of measurements

Table 1. Summary of relative error and correlation for smoothed curves

Channel name	Relative error [%] min/max	Correlation
I0	0.5/1.5	0.81
U1	0.4/1.6	0.8
U2	0.2/1.2	0.85
U3	0.5/1.8	0.79
U4	0.3/1.1	0.84
U5	0.35/1.2	0.83
U6	0.2/1.1	0.82
U7	0.2/1.2	0.83

5   Conclusions and Future Work

The article presents Savitzky-Golay digital filter implementation in Cloudera
Hadoop distribution using the modified R programming environment. Obtained
results were compared with results acquired from NI Diadem system. Filter
implemented in test environment gave very good results and allows for big data
handling.

Visualizations of vast datasets can give vital experiences that help compre-
hend the information. Making a calculation in R that is executed as a MapReduce
job can deliver an important data, that can be used by an R customer to make
visualizations or other computations. The other difficult algorithms can be also
implemented in this environment. There can be included data mining algorithms
like logistic regression, K-Means clustering as also as digital filters algorithms.

Experimental results showed, that it is possible, with positive attempt, to build the digital filter in Data Mining System using R programming language and environment. It can be very suitable tool for measured data analysis, especially in systems which demand big data recording for example modern superconducting devices.

There is ongoing work to implement digital filter using reduce stage for introducing the filter equation and convolution coefficients and compare the results with earlier one. The other work which can be done in the future is the filter implementation in Java language and verification of obtained results.

References

1. Zyla, J., Finnon, P., Bulman, R., Bouffler, S., Badie, C., Polanska, J.: Investigation for genetic signature of radiosensitivity - data analysis. In: Gruca, A., Czachorski, T., Kozielski, S. (eds.) Man-Machine Interactions 3, vol. 242, pp. 219–227. Springer, Switzerland (2014)
2. Woźniak, M., Marszałek, Z., Gabryel, M., Nowicki, R.K.: Modified merge sort algorithm for large scale data sets. In: Rutkowski, L., Korytkowski, M., Scherer, R., Tadeusiewicz, R., Zadeh, L.A., Zurada, J.M. (eds.) ICAISC 2013, Part II. LNCS, vol. 7895, pp. 612–622. Springer, Heidelberg (2013)
3. Folkert, K., Bochenek, M., Huczała, L.: The concept of using data mining methods for creating efficiency and reliability model of middleware applications. In: Kwiecień, A., Gaj, P., Stera, P. (eds.) CN 2012. CCIS, vol. 291, pp. 55–62. Springer, Heidelberg (2012)
4. Czajkowski, K., Drabowski, M.: Semantic data selections and mining in decision tables. In: Czachórski, T., Kozielski, S., Stańczyk, U. (eds.) Man-Machine Interactions 2. AISC, vol. 103, pp. 279–286. Springer, Heidelberg (2011)
5. Jestratjew, A., Kwiecień, A.: Using cloud storage in production monitoring systems. In: Kwiecień, A., Gaj, P., Stera, P. (eds.) CN 2010. CCIS, vol. 79, pp. 226–235. Springer, Heidelberg (2010)
6. Borzemski, L.: Data mining in evaluation of internet path performance. In: Orchard, B., Yang, C., Ali, M. (eds.) IEA/AIE 2004. LNCS (LNAI), vol. 3029, pp. 643–652. Springer, Heidelberg (2004)
7. Yao, K.T., Lucas, R., Gottschalk, T., Wagenbreth, G., Ward, C.: Data analysis for massively distributed simulations. In: Interservice/Industry Training, Simulation, and Education Conference IITSEC 2009 Paper No. 9350 (2009)
8. Krauzowicz, L., Szostek, K., Dwornik, M., Oleksik, P., Piórkowski, A.: Numerical calculations for geophysics inversion problem using apache hadoop technology. In: Kwiecień, A., Gaj, P., Stera, P. (eds.) CN 2012. CCIS, vol. 291, pp. 440–447. Springer, Heidelberg (2012)
9. Ohri, A.: R for Cloud Computing, An Approach for Data Scientists. Springer, New York (2014)
10. Laicano, A.: Digital signal processing in Hadoop with Scalding (2013). https://github.com/alaiacano/dsp-scalding
11. Ferzli, R., Khalife, I.: Mobile cloud computing educational tool for image/video processing algorithms. In: Digital Signal Processing Workshop and IEEE Signal Processing Education Workshop (DSP/SPE), pp. 529–533 (2011)

12. Sheng, C., Zhao, J., Leung, H., Wang, W.: Extended kalman filter based echo state network for time series prediction using mapreduce framework. In: Ninth IEEE International Conference on Mobile Ad-hoc and Sensor Networks, pp. 175–180. IEEE (2013)
13. Torgo, L.: Data mining with R: learning with case studies. Chapman and Hall, CRC Data Mining and Knowledge Discovery Series (2010)
14. Revolution Analytics, Enterprise Deployment, January 2015. http://www.revolutionanalytics.com/enterprise-deployment
15. R Technologies from Oracle, January 2015. http://www.oracle.com/technetwork/topics/bigdata/r-offerings-1566363.html
16. Renjin: The R Programming Language on JVM, January 2015. http://www.rnejin.org
17. RAmazonS3, January 2015. http://www.omegahat.org/RAmazonS3/
18. Amazon's Simple DB API, January 2015. http://aws.amazon.com/simpledb/
19. Kampf, M., Kantelhardt, J.W.: Hadoop.TS: large-scale time-series processing. Int. J. Comput. Appl. **74**(17), 1–8 (2013)
20. Li, L., Ma, Z., Liu, L., Fan, Y.: Hadoop-based ARIMA algorithm and its application in weather forecast. Int. J. Database Theory Appl. **6**(5), 119–132 (2013)
21. Stokely, M., Rohani, F., Tassone, E.: Large-scale parallel statistical forecasting computations in R. In: JSM Proceedings (2011)
22. Nurmi, D., Wolski, R., Grzegorczyk, Ch., Obertelli, G., Soman, S., Youseff, L., Zagorodnov, D.: The eucalyptus open-source cloud-computing system. In: 9th IEEE/ACM International Symposium on Cluster Computing and the Grid (CCGRID), pp. 124–131 (2009)
23. Donnelly, P., Bui, P., Thain, D.: Attaching cloud storage to a campus grid using parrot, chirp, and hadoop. In: IEEE Second International Conference on Cloud Computing Technology and Science (CloudCom), pp. 488–495 (2010)
24. Kreps, D.: The time of our lives: understanding irreversible complex user experiences. In: Kimppa, K., Whitehouse, D., Kuusela, T., Phahlamohlaka, J. (eds.) HCC11 2014. IFIP AICT, vol. 431, pp. 47–56. Springer, Heidelberg (2014)
25. Strata 2012: Doug cutting, the apache hadoop ecosystem, June 2013. http://youtu.be/Ttu3ZQ58ovo
26. Cloudera, the platform for Big Data, January 2015. http://www.cloudera.com
27. CDH Version and Packaging Information - Cloudera Support, December 2014. https://ccp.cloudera.com/display/DOC/CDH+Version+and+Packaging+Information
28. Revolution analytics: advanced 'Big Data' analytics with R and Hadoop. Whitepaper (2011). http://www.revolutionanalytics.com/whitepaper/advanced-big-data-analytics-r-and-hadoop
29. Adler, J.: R in a Nutshell, 2nd edn. O'Reilly, Sebastopol (2012)
30. Revolution Analytics, Packages in RHadoop Toolkit, January 2015. http://projects.revolutionanalytics.com/documents/rhadoop/rhadooppkgs/
31. Czerwinski, D., Jaroszynski, L., Janowski, T., Kozak, J., Majka, M.: Analysis of alternating overcurrent response of 2G HTS tape for SFCL. IEEE Trans. Appl. Supercond. **24**(3), Article no. 5600104 (2014). doi:10.1109/TASC.2013.2281494
32. National Instruments: What Limitations Exist With Channel and File Sizes in DIAdem? Knowledge Base (2012, updated). http://digital.ni.com/public.nsf/allkb/B391603F3CD86AE486256FAC00780122

Decentralized Social Networking Using Named-Data

Leonid Zeynalvand$^{(\boxtimes)}$, Mohammed Gharib, and Ali Movaghar

Sharif University of Technology, Tehran, Iran
{zeynalvand,gharib}@ce.sharif.edu
movaghar@sharif.edu

Abstract. Online social networks (OSNs) can be considered as huge success. However, this success costs users their privacy and loosing ownership of their own data; Sometimes the operators of social networking sites, have some business incentives adverse to users' expectations of privacy. These sort of privacy breaches have inspired research toward privacy-preserving alternatives for social networking in a decentralized fashion. Yet almost all alternatives lack proper feasibility and efficiency, which is because of a huge mismatch between aforementioned goal and today's network's means of achieving it. Current Internet architecture is showing signs of age. Among a variety of proposed directions for a new Internet architecture is Named Data Networking (NDN), focused on retrieving content by name, which names packets rather than end-hosts. NDN characteristics greatly facilitate development of applications tailored for today's needs. In this paper a decentralized architecture for social networking is proposed that provides strong privacy guarantees while preserving the main functionalities of OSNs, in a content-based paradigm. The simulation results show that not only it is feasible to have decentralized social networking over content centric networks, but also it is significantly more efficient from a global network point of view.

Keywords: Privacy · Social networking · Decentralization · Named data · Information centric networks

1 Introduction

Today, a significant part of users' online activities involve social networking. Becoming the dominant mechanism for information sharing between users, Online social networks (OSNs) can be considered as huge success. However, this success costs users their privacy and loosing ownership of their own data; Sometimes the operators of social networking sites, have some business incentives adverse to users' expectations of privacy, which can result in selling user data for monetary benefits [1]. Dynamic and complex nature of privacy policies further magnifies this problem. Besides, centralized architectures form a single point of failure, motivating adversaries to exploit any vulnerability in these systems to gain access over users' data. Decentralized architectures can solve these

© Springer International Publishing Switzerland 2015
P. Gaj et al. (Eds.): CN 2015, CCIS 522, pp. 421–430, 2015.
DOI: 10.1007/978-3-319-19419-6_40

problems, but they also present new challenges, as ensuring certain levels of availability and efficiency that is necessary to support main OSN functionalities.

Yet almost all proposed decentralized architectures lack proper feasibility and efficiency, which is mainly because of a huge mismatch between aforementioned goal and today's network's means of achieving it.

The Internet's current host-to-host model, does not properly satisfy today's needs, since content is what a user wants. it is clear that data-oriented architectures can provide a good fit to the massive amounts of data exchanged via the World Wide Web [2,3]. Among a variety of proposed directions for a new Internet architecture is Named Data Networking (NDN), focused on retrieving content by name, which names packets rather than end-hosts. NDN characteristics greatly facilitate development of applications tailored for today's needs.

In this paper a decentralized architecture for social networking is proposed that provides strong privacy guarantees while preserving the main functionalities of OSNs, in a content-based paradigm. The simulation results show that not only it is feasible to have decentralized social networking over content centric networks, but also it is significantly more efficient from a global network point of view.

2   Background of OSNs

There has been a significant growth in the popularity of OSNs like Facebook. According to Nielson Online [4], OSN sites are visited by 67 % of global on-line population, most primarily for sharing contents, which often include photos and videos, with friends. This enormous growth in amount of personal data being shared rises the expectation that a considerable fraction of Internet traffic is used only for social networking.

2.1   Current Social Networks' Architecture

OSN users today share their personal contents by uploading them to the servers of OSN service providers. These servers which are centrally managed, host users' contents and enable other users to retrieve them, making completely separate data flows per each audience. In addition to enormous amount of traffic overhead, all caused by multiple flows carrying same data, a more important drawback is that users do not completely control their own data anymore [5,6]. Among examples of this control loss are specific constraints limiting type of the content being shared (Facebook and Flickr don't allow voice or music), and complex and continuously changing privacy policies, and potential user privacy violations.

2.2   Exploring Alternative Architectures

Different design decisions has been considered by previous works on alternative OSN architectures [5,7–11]: To what degree should it be decentralized (from home-based sharing to hosting data on user-chosen trusted servers), how should nodes be organized (in a structured distributed hash table (DHT) or with links between social contacts), what mechanism should be used for content dissemination (push or pull) and how access control should be applied. It is realized that to effectively preserve privacy while supporting the essential functionalities of an OSN, current Internet architecture lacks proper capabilities, thus a new Internet architecture is needed in which needs of today's users fit well. For example, in a decentralized architecture high availability can be achieved by storing redundant instances of same data at random DHT nodes; however, to support the newsfeed functionality huge amount of DHT lookups are needed which take a lot of time as observed in [10,12]. Furthermore, decentralization schemes that rely on the concept of trusted servers lack enough degree of decentralization therefore fail to provide strong privacy guarantees [9].

3   Architecture

One strong motivation for content-oriented networking is that additional translations are not needed. Alice should be able to share her photos, videos and all other social contents with Bob and other intended consumers, directly from her own home gateway with a feasible performance and high availability, as shown in Fig. 1.

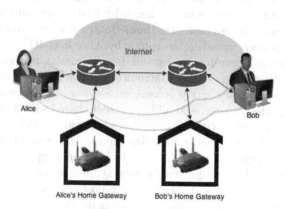

Fig. 1. Data flows in decentralized social networking over NDN

The architecture proposed in this paper for decentralized social networking using named data (so on referred to as DSNND) achieves this. There are a couple of points that must be considered in order to implement main functionalities of OSNs into a privacy preserving content-oriented model.

First. In this work the definition of social networking does not include public blogs, celebrity fan pages and other public pages like that. Since these are not private usages and have a public nature, they are best to remain on central architectures.

Second. DSNND must support decentralized identity since users should be able to use their account (e.g. see their newsfeed, comment on photos, etc.) from anywhere with any device without requiring any form of central authentication. From a cryptographic point of view the suggested mechanism is to use a Password-Based Key Derivation Function like PBKDF2 to derive the concatenation of domain name (user name) and password into a big enough (e.g. 128 bits) secret key, then the secret key can be used as seed for a Pseudo-random Number Generator (PRNG), since the PRNG is deterministic, same seed implies same output sequence and it can be used in the key pair generation of any asymmetric cryptography algorithm and for a given domain name and password, user can run it again every time she needs the private key. Using a key pair for authentication eliminates the need for central identity management and can be considered as a paradigm shift compared to today's web based models where a central service authenticates the user.

All main functionalities of OSNs can be categorized into four fundamental classes: contact management, content retrieval, content sharing, content feedback. But to do any of these, first the user have to sign in. We made sign in process feasible from anywhere with any device without requiring a central authentication, using the decentralized identity we explained. As illustrated in Fig. 2, when Alice wants to log into her account, she issues a signed interest named as `/alice/dsnnd/config/.../<Alice's_signature>`, her home gateway responds with a content object encrypted with Alice's public key and signed with Alice's private key containing all needed initial configurations like friends' domains, public keys, access lists, etc. (home router does not compute this content object, it has been stored in the router by Alice herself)

To share content securely and benefit from granular privacy in proposed DSNND architecture, a user can define multiple privacy groups and assign contacts to them. Each privacy group is assigned a symmetric key which should be renewed after a period of time, and all the content shared with a specific privacy group should be encrypted with the corresponding symmetric key. As illustrated in Fig. 3, to share a content with a specific group, Alice should encrypt it with group's symmetric key, sign it and send it to her home repository, and in case Alice has encrypted the content with a new symmetric key, she also needs to send the key to all intended audiences' home repositories using NDN Repo data insertion protocol (in Fig. 3 just one audience is shown).

In order for Bob to add Alice to a proper privacy group in his contact list (Fig. 4), Bob's social application which has implemented the proposed DSNND protocol should send the latest symmetric keys to Alice's home repository, depending on which access list he puts Alice on. Since any change in contact list results in a change in local config file, to make it persistent, the local config file should be stored in Bob's home repository using NDN Repo data insertion protocol, after being encrypted with Bob's public key and signed with Bob's private key. Although privacy is our main design goal as we described earlier, but achieving that with minimum changes to NDN home gateway is another important matter we should consider.

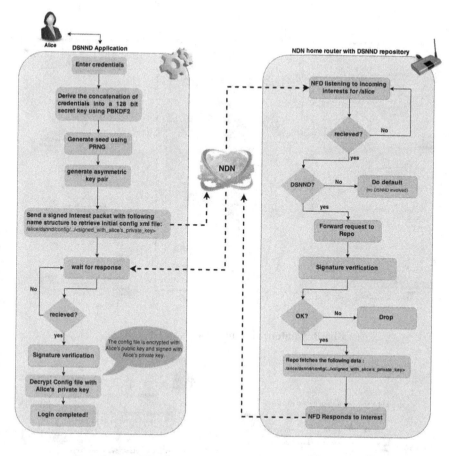

Fig. 2. Login process in DSNND

Content retrieval, forms the vast majority of social networking traffic and it can be simply achieved by issuing an interest for the intended content, which is the basic communication model in NDN, thus in proposed architecture a tremendous portion of social networking traffic can benefit significantly from content oriented paradigm.

And finally the last necessary functionality of a social network is the capability to put feedback on a content Fig. 5 (e.g. to put comment or like). To clarify this let's assume Bob wants to put a comment on a content named as /alice/dsnnd/<groupKeyName>/<contentName>. The comment itself is a content object, Bob's DSNND enabled social application encrypts Bob's comment with the same symmetric key used to decrypt the content (so that others in content's privacy group will be able to see the comment), signs it and sends it to Alice's home repository, Alice's home repository stores the received content object under the name /alice/dsnnd/<groupKeyName>/<contentName>/comments/<commentName>/.

426   L. Zeynalvand et al.

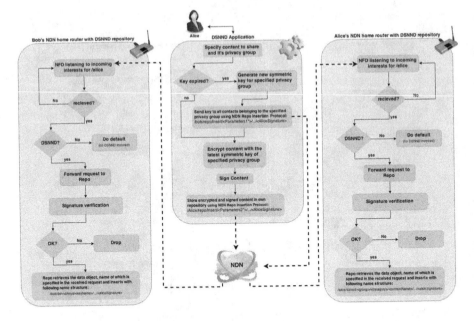

Fig. 3. Content sharing process in DSNND

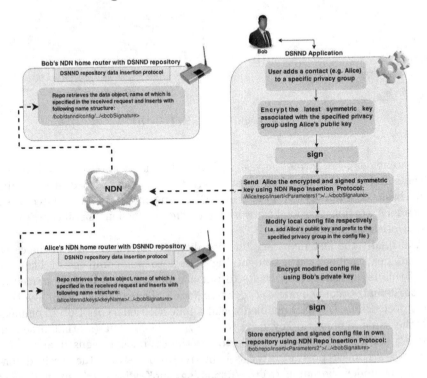

Fig. 4. Contact management process in DSNND

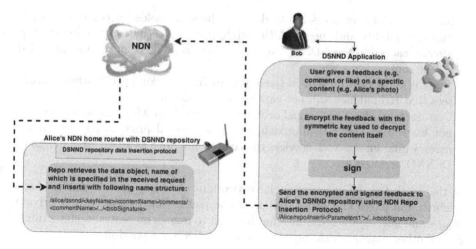

Fig. 5. Content feedback process in DSNND

4 Advantages and Drawbacks

DSNND approach adds appealing properties beyond simply supporting social networking in a content based paradigm:

- In all main functionalities explained above, secrecy is guaranteed using proper utilization of symmetric or asymmetric cryptography and integrity is guaranteed using signature.
- Decentralization gives users full control over their own data and provides strong privacy guarantees.
- Temporal locality of social contents benefits notably from NDN's built in caching mechanisms since interests for a social content rise around certain time slots (e.g. moments after content is shared).
- Spatial locality of social contents benefits significantly from decentralized nature of our architecture since the vast majority of users issuing interests for a content are geographically close to the content itself.

Generally decentralizing an online social network comes with known drawbacks such as accessibility, inconsistency, redundancy, content dissemination reliability and speed, however, many of these potential drawbacks are significantly reduced in the proposed DSNND approach:

- Since each user hosts her content from her own home-router device, the accessibility of data can be guaranteed by ensuring that the router is plugged in and connected to Internet. In other words in DSNND approach the accessibility degree is self-managed. Sure, the overall accessibility will be slightly less than what it is in today's centralized OSNs in which the provider's business incentives depend on it but it is still far better than approaches in which the user has no control over accessibility degree. The same holds for content dissemination reliability and speed. (Although in most cases the requested content is already cached in the NDN network and there is no need for it to be provided by the root).

– The freshness field present in the standard NDN packet specification guarantees that redundancy will not cause any inconsistency.

– Certain fancy features (e.g. Facebook's People You May Know service) do not fit well in DSNND approach, which is a small price to pay for strong privacy guarantees. Such services thus can be provided as an external plugin to DSNND architecture.

In this work the definition of social networking does not include public blogs, celebrity fan pages and other similar public pages. Since these are not private usages and have a public nature, they are best to remain on central architectures. For all other usages which require privacy, DSNND is a good fit.

5   Evaluation

In this section, the performance of proposed content oriented decentralized architecture is evaluated compared to existing centralized one. Comparison with other decentralized mechanisms [5, 7–11] is not necessary, since they do not outperform existing centralized architecture when it comes to performance, according to simulation results they have published. Formal privacy evaluation of DSNND is too big an issue to be included in this paper, and is best to be left for a future paper, however, since in DSNND no third party is involved in sharing social content with a group of trusted people, thus it provides far better privacy than today's central OSNs or decentralized ones which require user data to be hosted from a third party.

5.1   Implementation and Simulation Setup

The ndnSIM (an NS-3 based simulator which implements the basic components of an NDN network in a modular way [13]) is used on a standard machine with 2.40 GHz Intel Core 2 Duo, 4 GB memory, and running Ubuntu 12.04. To simulate the social graph, Facebook friendship graph from the New Orleans regional network [14] is utilized. This data set contains a list of all user-to-user links from the Facebook New Orleans network and consists of 63 732 nodes and 1.54 million edges. In addition to this, as the selected network topology often influences the outcome of the simulation, realistic topologies are needed to produce realistic simulation results, [15] had a good influence in forming a proper topology for simulation. Performance is measured using the following metrics:

– Mean Node Load: the average amount of social networking traffic, each node (user/router) transmits or receives in a given network topology. Let e be a single experiment performed on network topology t, containing n nodes, in which the total social networking traffic each node deals with (consume /produce) respectively for users and routers equals u_i and r_j bytes, then:

$$\text{MeanNodeLoad}(e) = \frac{\sum_{i \in Users} u_i + \sum_{j \in Routers} r_j}{n}. \qquad (1)$$

This metric gives an indication of how much stress is caused for a single node in the network as a consequence of social networking.

- Work Load: the total amount of social networking traffic, users transmit or receive in a given network topology. Let e be a single experiment performed on network topology t, containing n nodes, in which the total social networking traffic each user consumes or produces equals u_i bytes, then:

$$\text{WorkLoad}(e) = \sum_{i \in Users} u_i. \qquad (2)$$

5.2 Results

Figure 6, depicts Mean Node Load as a function of Work Load and the architecture used for social networking (decentralized over NDN or centralized over IP). As it is expected in proposed decentralized content centric architecture the average pure social networking stress each node tolerates is significantly less than what it is in a centralized IP architecture. Although the results may overestimate the out-performance of the proposed architecture since the social graph used in simulation is based on public profiles only and may over estimate the popularity of contents, yet it is well clarified that not only it is feasible to have decentralized social networking over content centric networks, but also it is significantly more efficient from a global network point of view.

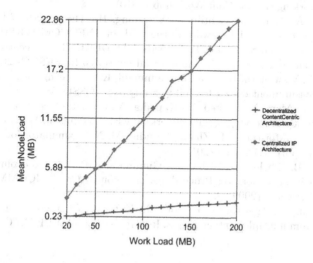

Fig. 6. Mean load of each node for different workloads

6   Conclusion and Future Work

In this paper a decentralized architecture for social networking over NDN has been presented that provides strong security and privacy guarantees while efficiently supporting the main functionalities of OSNs. The simulation results demonstrate that a decentralized approach to privacy-preserving social networking is practical in a content centric paradigm. In a future paper we will provide a formal privacy evaluation of the proposed DSNND architecture compared to existing methods then we will provide a library that implements the protocol which can be used to develop social applications over NDN.

References

1. Steel, E., Vascellaro, J.E.: Facebook, MySpace confront privacy loophole. Wall Street J. Online (2010). http://online.wsj.com/article/SB10001424052748 7045
2. Jacobson, V., Smetters, D.K., Thornton, J.D., Plass, M., Briggs, N., Braynard, R.: Networking named content. In: CoNext (2009)
3. Koponen, T., Chawla, M., Chun, B.G., Ermolinskiy, A., Shenker, S., Stoica, I.: A data-oriented (and beyond) network architecture. In: SIGCOMM (2007)
4. Nielsen Online Report. Social Networks Now 4th Most Popular Activity (2009)
5. Lucas, M., Borisov, N.: Flybynight: mitigating the privacy risks of social networking. In: ACM Workshop on Privacy in the Electronic Society (2008)
6. Shakimov, A., Varshavsky, A., Caseres, R.: Privacy, cost and availability tradeoffs in decentralized OSNs. In: ACM SIGCOMM WOSN (2009)
7. Cutillo, L.A., Molva, R., Sturfe, T.: Safebook: feasibility of transitive cooperation for privacy on a decentralized social network. In: WOWMOM (2009)
8. Mani, M., Nguyen, A.M., Crespi, N.: SCOPE: a prototype for spontaneous P2P social networking. In: PerCom Workshop (2010)
9. Bielenberg, A., Helm, L., Gentilucci, A., Zhang, H.: The growth of Diaspora - a decentralized online social network in the wild. In: INFOCOM WKSHPS (2012)
10. Jahid, S., Nilizadeh, S., Mittal, P., Kapadia, A.: DECENT: a decentralized architecture for enforcing privacy in online social networks. In: SESOC (2012)
11. Marcon, M., Viswanath, B., Cha, M., Gummadi, K.P.: Sharing social content from home: a measurement-driven feasibility study. In: NOSSDAV 2011 (2011)
12. Nilizadeh, S., Alam, N., Husted, N., Kapadia, A.: Cachet: a decentralized architecture for privacy preserving social networking with caching. In: WPES 2011 (2011)
13. Afanasyev, A., Moiseenko, I., Zhang, L.: ndnSIM: NDN simulator for NS-3. NDN, Technical Report NDN-0005 (2012)
14. Viswanath, B., Mislove, A., Cha, M., Gummadi, K.P.: On the evolution of user interaction in Facebook. In: Proceedings of Second ACM SIGCOMM Workshop on Social Networks (2009)
15. Subramanian, L., Agarwal, S., Rexford, J., Katz, R.: Characterizing the internet hierarchy from multiple vantage points. In: Proceedings of IEEE INFOCOM (2002)

USB Data Capture and Analysis in Windows Using USBPcap and Wireshark

Wojciech Mielczarek[1][✉] and Tomasz Moń[2]

[1] Institute of Informatics, Silesian University of Technology,
Akademicka 16, 44-100 Gliwice, Poland
wmielczarek@polsl.pl
[2] Espotel Poland Sp. z o.o., Wrocław Technology Park,
Muchoborska 18, 54-424 Wrocław, Poland
tomasz.mon@espotel.com

Abstract. The USB device designers, as well as the advanced users often need a software utility to capture and analyse USB data exchanged between a device and an application in host. The software USB analysers are designed for this purpose.

Software USB analyser solution designed for Windows XP, Vista, 7 and 8 is presented in the paper. The solution consists of USBPcap responsible for USB data capture and Wireshark responsible for data analysis. The solution was used to record and analyse messages exchanged between DigiTech RP250 guitar multi-effect processor and host-side software package X-Edit supplied by vendor (DigiTech). For this purpose Wireshark was extended with two dissectors: USB Audio and X MIDI SysEx.

Keywords: USB · USB analyzer · USBPcap · Wireshark · USB Audio · X MIDI SysEx · DigiTech RP250

1 Introduction

The paper presents an Open Source software USB analyser for Windows XP, Vista, 7 and 8. The analyser consists of two parts:

- USBPcap – the USB data capture engine,
- Wireshark – the protocol analyser.

USBPcap is a unique software written for this project. Wireshark is a multi-platform versatile packet analyser. In order to use Wireshark to analyse USBPcap's capture files, USBPcap's capture file format support was added to Wireshark's dissection engine.

The operation of the USBPcap/Wireshark analyser was tested in different USB system configurations, in particular, to analyse data exchanged between a USB host and the DigiTech RP 250 Guitar Effect Processor. For this purpose, two dissectors were designed and added to Wireshark: MIDI System Exclusive dissector and USB Audio dissector.

P. Gaj et al. (Eds.): CN 2015, CCIS 522, pp. 431–443, 2015.
DOI: 10.1007/978-3-319-19419-6_41

The DigiTech RP250 guitar processor was an inspiration for our project, because we wanted to design an application for Linux working in similar way as the vendor X-Edit for Windows. We knew nothing how X-Edit interacts with RP250, so we decided to use reverse engineering and sniff messages exchanged between vendor application and RP250. We couldn't use hardware analyzer, although we have one (the Ellisys USB Explorer). USB Explorer is useful to test the USB protocol and three USB class protocols only (hub, HID and mass storage). Moreover, it is not able to dissect any application protocols. That's why we have designed USBPcap and extended Wireshark to analyse USB Audio (USB class protocol) and X MIDI SysEx (application protocol).

In a such design there are plenty of problems. In this paper we focused on explanation how the row USB data can be captured in Windows and what should be done to input them to Wireshark for further dissection.

2   Software USB Analysers for Windows

There are different software USB analysers for Windows available, most of them are commercial and some are Open Source. The commercial, for instance, are: HHD Software USB Monitor, USBlyzer, busTrace. These analysers are designed for testing Windows USB device drivers and provide information of the USB data, I/O Request Packets and USB system mechanisms. The analysers are usually able to decode the USB class commands. Unfortunately, in almost every case it is only one level decoding and most of the analysers do not perform protocol-in-protocol analysis.

Before the development of USBPcap was started, there were few Open Source USB analysers for Windows available. Unfortunately, these tools were rather outdated, unsuitable to use in modern systems or problematic due to wrong design choices. Usbsnoopy (last update in 2001) and its successors SnoopyPro (last update in 2002) and SniffUSB (last update in 2007) work with Windows versions up to Windows XP only. They do not have friendly user interface for data analysis (data is written to a text file). Usbsnoopy and its successors perform monitoring on the individual device instance level (hardware key). In 2009 the busdog project (last update in 2010) was released. Busdog uses KMDF (Kernel-Mode Driver Framework) platform which is unsuitable for bus filter drivers (required for effective capture of all devices connected to USB root hub).

In February 2012 Deka Prikarna described how the IRP (I/O Request Packet) requests monitoring in Windows can be performed [1]. His example project (DkURBMon) does not reveal any information about USB bus operation, but only about URB (USB Request Block) requests. DkURBMon is able to analyse only the communication of the USB device that was last connected to user selected USB hub. DkURBMon does not have any legacy or design issues and therefore it was established as starting point for USBPcap.

In order to use Wireshark for USB analysis in Windows we had to:

- write USB data capture driver,
- design USB data capture file format convenient for processing in Wireshark,
- write Wireshark dissectors for the capture file format and USB classes under interest.

3 USB Data Capture in USBPcap

USBPcap is responsible for the USB data acquisition (capture) only. The data is later analysed in Wireshark. USBPcap consists of a filter driver (USBPcap-Driver) and a user application (USBPcapCMD). The filter driver works in Windows kernel mode and captures data transferred to a USB device. The data capture is based on monitoring the IRP requests, which are transferred between the FDO (Functional Device Object) and PDO (Physical Device Object). Captured data is written inside a cyclic buffer. The user application reads the USB data stream from the cyclic buffer and writes it to a pcap file with the link-layer header type of USBPcap. It is the input file for Wireshark. Wireshark analyses the USB data and presents it to a user in dissected form.

3.1 USBPcapDriver

The USBPcapDriver has 3 major functions:

- USB device filter driver (USBPcapDevice),
- USB root hub filter driver (USBPcapRootHub),
- user-space control device (USBPcapController).

The Fig. 1 shows the place in the USB stack, where the USBPcapDevice instance performs USB data stream acquisition. The lightly shaded blocks are described in two ways: the normal type represents USB specification terms, and the italic represents the Microsoft Windows Driver Model terms. "The bus drivers" mean all the drivers that process an IRP request from PDO, before the USB transfer is executed on the bus (Root hub PDO and Host Controller Driver). There is one USBPcapDevice instance for each Functional Device Object (FDO), one USBPcapRootHub instance for each USB Root Hub present in system and one USBPcapController instance for each USB Root Hub.

The Fig. 2 shows the USB data stream flow between the USBPcap driver and the USBPcapCMD application.

The tasks of drivers shown in the Figs. 1 and 2 can be explained by analogy to a transport company. The FDO is a customer. The USB device is a place, where the customer's goods must be delivered (write operation), or taken from (read operation). The PDO is a customer service department that collects requests from customers. The PDO perfectly knows how the company operates, but does not know anything about the application of transported goods. The bus drivers correspond to dispatchers responsible for delivery scheduling. The host controller performs delivery by entering data packets (consecutive data units) into the USB. USBPcap is an "industrial spy network" that captures information on customer's requests. USBPcapRootHub manages the central information point (the cyclic buffer) to which filter drivers (USBPcapDevice; individual spies) write USB data and the control device (USBPcapController) transfers it to user space (spy network client).

The root hub filter driver task is to create a control device and to monitor devices connected to USB. More than one root hub can be present in a single

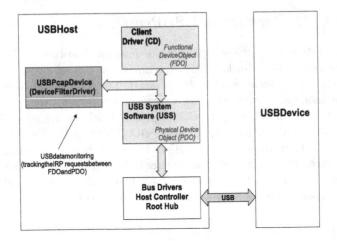

Fig. 1. The USB data acquisition point

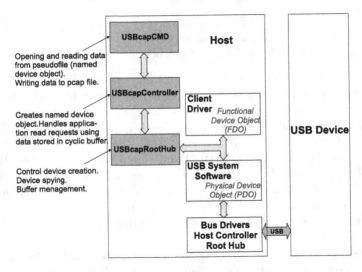

Fig. 2. The USB data flow between USBPcap driver and USBPcapCMD application

USB host (for example computer). A separate root hub filter driver instance is created for each root hub. For every FDO present on filtered root hub, separate device filter instance is created.

The control device (USBPcapController) is responsible for communication with user application (USBPcapCMD). The communication is based on a named device object. An application working in user space gains access to the named device object by opening the pseudofile (the path to pseudo-file starts from "\\.\"). Each of the named device objects created by the USBPcap has its individual name \\.\USBPcapX, where X is a consecutive integer (starting at 1) assigned to the individual USB root hub.

After the pseudo-file is successfully opened, the returned handle is used by USBPcapCMD for reading data and sending the IOCTL (input/output control) messages. Writing data is forbidden and returns an error. Any reading from the control device returns pcap data stream from the cyclic buffer. Data, that was once read, can't be retrieved again. After each read operation, USBPcapCMD application writes data to output pcap file. The control device processes the IOCTL messages responsible for:

– setting up cyclic buffer size,
– data acquisition (capture) start,
– data acquisition (capture) stop.

USBPcapCMD application discovers all USBPcap's named device objects and determines which USB hub corresponds to a given named device object. After that, the complete USB system topology is displayed using device names that appear in Windows Device Manager.

USBPcapDevice monitors data transferred to a device. This process is based on inspecting all IRPs that are exchanged between the FDO and PDO.

Windows USB IRP does not directly include any information about USB device address and endpoint number. The IRP structure contains pipe handle (the virtual communication channel handle). The only one documented method to associate pipe handle with USB device address and endpoint number is reading them during enumeration [2]. Windows IOCTLs for SET CONFIGURATION and SET INTERFACE USB commands contains information that makes it possible to associate the handle with an endpoint number.

The USB filter driver inspects all IRPs. SET CONFIGURATION and SET INTERFACE IRPs are analysed regardless of data acquisition state (stopped or started). The information essential for handles is taken from the IRPs and recorded in the special table associating a handle with the endpoint number. That table is later used (during data acquisition) to determine the endpoint the transfer request is directed to.

3.2   USB Requests Intercepting

In order to explain USB IRPs capture let's use specific example of controlling the DigiTech RP250 (the USB device) with X-Edit application. DigiTech RP250 is a modelling guitar processor. The device is used by guitarists to model their tone using plenty of different effect types. The device is equipped with USB port that can be used to record the (processed) sound or add backing track. DigiTech RP250 comes with X-Edit application for Windows and Mac OS X. X-Edit enables users to change the device settings, save presets, create configuration backups. X-Edit communicates with a device using the X MIDI SysEx protocol. As a USB device, the DigiTech RP250 has two input and two output endpoints: one input/output pair for isochronous transfers (playing and recording sound), and the other for bulk transfers (reading and changing settings).The Fig. 3 shows, how the MIDI packet is encapsulated before it is transmitted via USB. When a user changes a parameter in X-Edit, the application creates MIDI System Exclusive message corresponding to that change and transfers it to DigiTech RP250 driver's FDO. The FDO splits the message into USB MIDI packets that can be transferred to USB device (USB Audio in the Fig. 3). The USB MIDI maximum packet size is 3 bytes only [3]. The FDO driver generates IRP requests containing USB transfer requests (the USB URB in the Fig. 3), and transfers them to a physical driver (PDO). One request can contain several USB MIDI packets (no more than 4 in practice).

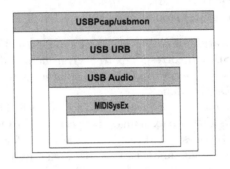

Fig. 3. The USB MIDI packets

Each request transferred between FDO and PDO is intercepted by USBPcap. USBPcap records it in cyclic buffer with an addition of pseudo-header containing IRP information. The pseudo-header in Fig. 3 is indicated as "USBPcap/usbmon" (in Linux, usbmon can be used instead of USBPcap).

After the information is recorded, the IRP is passed down to the PDO driver (and later to bus drivers). At last, the USB transaction is executed on the USB bus and the IRP is completed. The IRP is passed back to the PDO, and then it reaches USBPcap filter driver. USBPcap doesn't write any additional information this time (because it was write operation) and passes IRP to FDO.

In the case of reading data from a USB device, the IRP round trip is almost the same. The only difference is that the IRP created by the driver has the read flag set. This flag is used by USBPcap to distinguish if the USB data is to be written to the cyclic buffer when passed from FDO to PDO (in the case of writing data to a device), or when passed from PDO to FDO (in the case of reading data from a device).

3.3   USBPcap Capture File Format

The Fig. 4 shows the pcap file format. At the beginning of the pcap file there is the global header (structure pcap_hdr_s as shown in Listing 1).

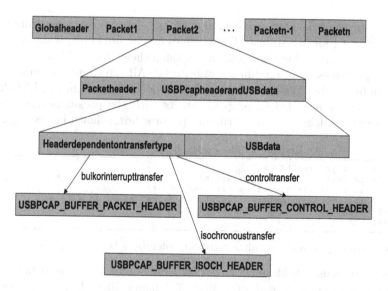

Fig. 4. The pcap file format. This is the format of output file created by USBPcap.

Listing 1. The pcap file global header

```
typedef struct pcap_hdr_s {
  guint32 magic_number; /* magic number */
  guint16 version_major; /* major version number */
  guint16 version_minor; /* minor version number */
  gint32  thiszone; /* GMT to local correction */
  guint32 sigfigs; /* accuracy of timestamps */
  guint32 snaplen; /* max length of captured packets, in octets */
  guint32 network; /* data link type */
} pcap_hdr_t;
```

The pcap_hdr_s structure is written in a pcap file in binary form using the native order of bytes (either Little or Big Endian) of a device that created the file. Individual members of pcap_hdr_s structure are described below:

- magic_number field is used to determine byte ordering in the file. Writing application writes 0xA1B2C3D4 in its native byte order. If the reader application reads this as 0xA1B2C3D4, then it uses the same endianness as writing application. If reader application reads this as 0xD4C3B2A1, it knows that the writing application was run on host with different endianness and there is a need to swap all multi-byte values in the file.
- version_major and version_minor fields determine main number and additional number of format version respectively (USBPcap uses version 2.4).
- thiszone field stores the offset from the universal time (UTC) and local time in seconds.
- sigfigs field informs about the time stamps accuracy (in practice, all tools set it to 0).
- snaplen field determines maximum packet size.
- network field determines link-layer header type.

Link-layer header type (network field in pcap_hdr_s structure) values from 147 to 162 are reserved for private use. These values can be used only, if the pcap files will never be sent outside the organization using them. The full list of link-layer header type values is on tcpdump's website [4]. All the captures generated by USBPcap have value 249 in this field (249 is the identifier assigned to USBPcap).

Every packet recorded in pcap file starts with the packet header. Packet header format is defined by the structure pcaprec_hdr_s shown in Listing 2.

Listing 2. The packet header structure

```
typedef struct pcaprec_hdr_s {
  guint32 ts_sec;   /* timestamp seconds */
  guint32 ts_usec;  /* timestamp microseconds */
  guint32 incl_len; /* number of octets of packet saved in file */
  guint32 orig_len; /* actual length of packet */
} pcaprec_hdr_t;
```

The meaning of pcaprec_hdr_s structure members is following:

- ts_sec and ts_usec fields include packet time stamp. ts_sec field determines number of seconds elapsed since first of January 1970 (Unix time). ts_usec field determines the number of microseconds, as an offset to ts_sec. This field can't reach the values greater or equal of one second (1 000 000).
- incl_len and orig_len fields store the number of packet bytes recorded in the file and the full packet size, respectively. The values in these fields are different, when a packet is cut as a result of snaplen limit.

Directly behind the packet header, there is incl_len bytes including the USBPcap header and data. The USBPcap header is shown in Listing 3. Following data types defined in the Windows Driver Development Kit were used in Listing 3:

- UCHAR – 8 bit value without sign.
- USHORT – 16 bit value without sign.
- UINT32 – 32 bit value without sign.
- UINT64, ULONG – 64 bit value without sign.
- USBD_STATUS – 32 bit value defining status values for USB requests [5].

Listing 3. USBPcap packet header

```
#pragma pack(1)
typedef struct
{
  USHORT       headerLen; /* This header length */
  UINT64       irpId;     /* I/O Request packet ID */
  USBD_STATUS  status;    /* USB status code
                            (on return from host controller) */
  USHORT       function;  /* URB Function */
  UCHAR        info;      /* I/O Request info */

  USHORT       bus;       /* bus (RootHub) number */
  USHORT       device;    /* device address */
  UCHAR        endpoint;  /* endpoint number and transfer direction */
  UCHAR        transfer;  /* transfer type */

  UINT32       dataLength;/* Data length */
} USBPCAP_BUFFER_PACKET_HEADER, *PUSBPCAP_BUFFER_PACKET_HEADER;
```

The #pragma pack(1) preprocessor directive enables structure packing. This means that the fields are stored in memory (therefore also in the capture file) directly one after another without any padding.

Directly after USBPCAP_BUFFER_PACKET_HEADER, additional data dependent on transfer type value can appear. For control transfers, it is one byte informing of the transfer stage (0 – setup, 1 – data, 2 – status). For isochronous transfers USBPCAP_BUFFER_ISOCH_HEADER structure defined in Listing 4 is used. Please note, that this structure starts with USBPCAP_BUFFER_PACKET_HEADER.

Listing 4. The isochronous USBPcap packet header

```
#pragma pack(1)
typedef struct
{
    ULONG           offset;
    ULONG           length;
    USBD_STATUS     status;
} USBPCAP_BUFFER_ISO_PACKET, *PUSBPCAP_BUFFER_ISO_PACKET;

#pragma pack(1)
typedef struct
{
    USBPCAP_BUFFER_PACKET_HEADER    header;
    ULONG                           startFrame;
    ULONG                           numberOfPackets;
    ULONG                           errorCount;
    USBPCAP_BUFFER_ISO_PACKET       packet[1];
} USBPCAP_BUFFER_ISOCH_HEADER, *PUSBPCAP_BUFFER_ISOCH_HEADER;
```

For USB bulk transfers and USB interrupt transfers there isn't any additional header data besides USBPCAP_BUFFER_PACKET_HEADER. Immediately after USBPcap packet header, there is data that was transmitted on the USB bus.

3.4   USBPcap's Cyclic Buffer

USBPcap data acquisition code can be executed on different IRQL (Interrupt Request Level) depending on transfer type. For example, isochronous data acquisition is executed on the DISPATCH_LEVEL. All the applications working in user space are executed on the PASSIVE_LEVEL (the lowest priority level). When page fault occurs in a code executing on level lower than DISPATCH_LEVEL, page manager handles this error and loads the page into RAM. The code executing on DISPATCH_LEVEL or higher cannot cause page fault (if it does, the Blue Screen of Death occurs).

To prevent page faults, the driver must (before rising IRQL) lock in memory all pagable data it is going to use when IRQL is equal or higher than DISPATCH_LEVEL. In the case of USBPcap filter driver, this solution is not possible because the data acquisition IRQL is not changed by the filter driver itself (it depends on the FDO). Therefore, the capture buffer must be allocated from nonpaged memory pool. Data allocated from nonpaged pool is always present in RAM (is never paged to disk). In Windows XP, nonpaged pool size is in the range of [128,256] MiB, the exact value depends on the amount of computer's RAM. Nonpaged pool is common for all kernel mode code (drivers and kernel). If drivers allocate all memory from this pool, the operating system will stop working correctly.

USBPcap's data packets size ranges from several bytes to several kibibytes. Because the capture buffer must be allocated from limited resource (non-paged pool), cyclic buffer with data overwrite protection was chosen. The driver keeps writing pcap file data into the cyclic buffer and the user-space application keeps reading from it. Data overwrite protection is crucial to keep pcap file contents valid, because pcap format does not contain any synchronisation flags. If there is not enough space left in the buffer, whole packet is lost. The cyclic buffer with overwrite protection implementation uses the following variables:

- KSPIN_LOCK bufferLock is used to synchronize access to the buffer. When spinlock is acquired, the priority level gets risen to DISPATCH_LEVEL (if the code is already executed at DISPATCH_LEVEL or higher, the priority level is not changed). Due to that, the code executed within critical section protected in this way should be as short as possible [6].
- PVOID buffer is the pointer to the allocated buffer.
- UINT32 bufferSize is the size of the allocated buffer.
- UINT32 readOffset is the index of first unread byte in the buffer.
- UINT32 writeOffseti is the index of first not occupied location in the buffer.

Figure 5 shows a hypothetical cyclic buffer operation cycle. The variables readOffset and writeOffset are represented by R and W respectively. The values of R and W indexes are in range of $[0, N-1]$ where N is bufferSize. The buffer is empty if $R = W$. Then $N - 1$ bytes can be written to buffer. When $R > W$, there are $N - R + W$ unread bytes in the buffer and $R - W - 1$ more bytes can be written. When $R < W$, there are $W - R$ unread bytes in buffer and $N - W - R - 1$ more bytes can be written.

The cycle in Fig. 5 consists of six steps. The first step is buffer reset: the N bytes buffer is allocated, R and W are set to 0. In the second step 2 bytes were written to the buffer ($R = 0$; $W = 2$). In the third step 2 bytes were read from the buffer. The result is an empty buffer ($R = W = 2$). In the fourth step $(N - 2) - 2$ bytes were written to the buffer ($R = 2$; $W = N - 2$). There is free space for only 3 bytes in the buffer. Attempts to write more bytes to the buffer will fail. In the fifth step 3 bytes were written. The result is full buffer ($R = 2$; $W = 1$) and no more bytes can be written. In the sixth step $(N - 2) - 1$ bytes were read from buffer ($R = N - 1$; $W = 1$). Maximum of $N - 3$ bytes can be written to the buffer.

4   USB Data Analysis in Wireshark

Wireshark (GPLv2 licence) is used to analyse USB data captured by USBPcap. Wireshark's input is the pcap file created by USBPcap. This file contains USB packets sent/received to/from all USB devices connected to captured USB Root

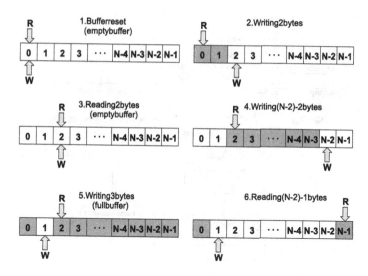

Fig. 5. Capture buffer hypothetical operation cycle. Grey boxes represent unread data

Hub (user-selected prior capture start). USBPcapCMD application reads the data from the cyclic buffer and writes it to the pcap file.

Wireshark is an Open Source software licensed under GPLv2. The source code and the complete history (in the form of git repository) are available online free of charge. Wireshark is commonly used for protocol analysis in computer networks. At the time of writing over 800 developers from all around the world contributed to Wireshark. In 2006 USB analysis support (back then capture files could only be created in Linux) was added to Wireshark. Wireshark was then able to dissect following USB classes: Audio, CCID, HID, Hub, MassStorage, Video.

Wireshark's dissectors are analysers of packet data. Wireshark is capable to analysing many different communication protocols. The Fig. 6 shows Wireshark dissector architecture overview.

Wireshark's input is stream of packets. Packets are analysed (dissected) and described in detail. Moreover, some dissectors are capable of modifying the data (for example: reassembling higher level protocol data or decrypting the communication) and passing such modified data to other dissectors. For example, the USB data packet 80 06 02 03 09 04 1 A 00 (hex) representing the USB Setup transaction will be presented and described as follows:

```
bmRequestType:  0x80
    1... .... = Direction: Device-to-host
    .00. .... = Type: Standard (0x00)
    ...0 0000 = Recipient: Device (0x00)
bRequest: GET DESCRIPTOR (6)
Descriptor Index: 0x02
bDescriptorType: STRING (3)
Language Id: English (United States) (0x0409)
wLength: 26
```

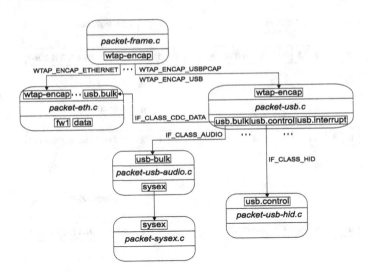

Fig. 6. Wireshark dissector architecture overview

Such accurate analysis is possible by dissectors cooperation. Each dissector is responsible for analysis of data transferred at a given protocol level. Each rectangle in Fig. 5 represents one dissector. In the middle part of dissector, there is a dissector's source code file name. In the upper part, there are possible sources of input data for this dissector and in the lower part, there are the names of lower layer dissectors that can perform further analysis. Depending on lower layer characteristics an additional dissector can be selected. For example, USB bulk transfers depending on USB device class can be audio data (IF_CLASS_AUDIO) or network data (IF_CLASS_CDC_DATA).

5   Summary

Our main goal was to explain how the raw USB data can be captured in Windows (we designed the USBPcap for this purpose) and what should be done to input them to Wireshark for analysis (we designed data capture file format convenient for processing in Wireshark). This part is universal and independent of a USB device class.

USB data acquisition was the first, rather difficult step, because it had to be done in Windows kernel. The second step was to discover the captured data meaning (this part is performed in user space). To analyse data in Wireshark, the proper dissectors were written (dissectors for the capture file format, USB classes and application protocol).

We checked our solution by observing messages exchanged between DigiTech RP250 guitar processor and host-side software package X-Edit. To interpret them we wrote USB Audio and X MIDI SysEx dissectors for Wireshark. In our paper it was only an example. Someone else could connect any other USB class device, use our USBPcap to capture data, use our dissector for the capture file format and at last use proper Wireshark USB class and application dissectors (if they are available or write them if not) to analyse messages exchanged between host and that device. Such a tool can be useful for those who write (or sniff existing) USB drivers to check if a driver is in accordance with a USB class specification, as well as for those who use reverse engineering to examine message exchange between a host application and USB device.

USBPcap is Open Source software. The filter driver is GPLv2 licensed. User application (USBPcapCMD) is licensed under BSD simplified license. The BSD license (unlike GPLv2) is not copyleft license. Therefore, USBPcapCMD code can be used in applications released under other licenses (including commercial licenses). The purpose of adopting different licenses for the individual components in this project was:

- to protect USBPcapDriver code from use in (modified) commercial data acquisition modules without publishing the changes (commercial use is welcome as long as the modifications are made public),
- to enable anybody to use the USBPcap in any tool for data analysis.

USBPcap project homepage address is: http://desowin.org/usbpcap. The project was announced on the 20th march 2013 at Wireshark discussion group [6] and presented at Sharkfest conference at University of California in Berkely in June 2013. Official USBPcap releases are digitally signed by Tomasz Moń.

This is an open and living project that can be still improved, for instance by adding statistics of lost packets (possible in pcap-ng file format). We believe that our USBPcap for Windows together with Wireshark form a perspective tool for USB data capture and analysis that can be further expanded with another USB device classes and applications dissectors.

References

1. Deka Prikarna, A.: Simple URB (USB Request Block) Monitor. http://www.code project.com/Articles/335364/Simplde-URB-USB-Request-Block-Monitor (2009). Accessed 6 Apr 2013
2. Oney, W.: Programming the Microsoft Windows Driver Model. Microsoft Press, Redmond (2003)
3. USB Implementers Forum, Universal Serial Bus Device Class Definition for MIDI Devices (1999)
4. Link-Layer Header Types. http://www.tcpdump.org/linktypes.html. Accessed 7 Apr 2013
5. USB Reference, USB Constants and Enumerations – USBD_STATUS. http://msdn. microsoft.com/en-us/library/windows/hardware/ff539136(v=vs.85).aspx. Accessed 7 Apr 2013
6. Mon, T.: RFC: USBPcap. http://www.wireshark.org/lists/wireshark-dev/201303/ msg00189.html. Accessed 21 May 2013

Reliability of Bluetooth Smart Technology for Indoor Localization System

Andrzej Kwiecień, Michał Maćkowski[(✉)], Marek Kojder,
and Maciej Manczyk

Institute of Computer Science, Silesian University of Technology,
Gliwice, Poland
{akwiecien,michal.mackowski}@polsl.pl,
{marekkojder,maciej.manczyk}@gmail.com

Abstract. The main objective of the paper was to test whether the devices compatible with Bluetooth Low Energy are reliable for indoor localization system. To determine the reliability of this technology several tests were performed to check if measured distance between Bluetooth transmitter and mobile device is close to the real value. Distance measurement focused on Bluetooth technology based mainly on received signal strength indicator (RSSI), which is used to calculate the distance between a transmitter and a receiver. As the research results show, the Bluetooth LE signal power cannot be the only reliable source of information for precise indoor localization.

Keywords: Bluetooth Low Energy · Bluetooth smart · Navigation system · Localization system · Accuracy of localization · Positioning techniques · RSSI

1 Introduction

In recent years the dynamic development of mobile devices can be observed. Mobile devices have been accepted by consumers and represent a strong branch of technology market. They give the users the opportunities similar to those offered by personal computers. Such devices can be used anywhere for e-mail, browsing the web, online banking, etc. Moreover, smartphones give also the possibility for individual navigation so that a user can be navigated to and from a particular address. The most widespread type of navigation is applied for outdoor purposes, however in the last few years there has been a need for navigation inside a building.

The outdoor navigation standards usually rely on GPS (*Global Positioning System*) satellites to determine the user position [1,2]. Despite the fact that this solution is very popular and precise for outdoor navigation, it is generally not well suited for indoor use. It is because:

- the GPS signal is strongly attenuated and disturbed inside the building,
- the GPS accuracy is not sufficient for such solution.

© Springer International Publishing Switzerland 2015
P. Gaj et al. (Eds.): CN 2015, CCIS 522, pp. 444–454, 2015.
DOI: 10.1007/978-3-319-19419-6_42

The idea of indoor positioning is not completely a new approach, there are several solutions which base on different kinds of technologies [3–6]. Most of them use the waypoints calling also anchors whose position is static and determined. The example of such anchors inside the building can be Wi-Fi Access Points or other network devices [7–9]. Despite of growing popularity of these solutions none of them has become so far a universal standard for indoor localization. Although some indoor localization systems exist, but in most cases they are limited to the particular areas or used for highly specific purposes [10, 11]. Wi-Fi has high power demands and is difficult to set-up and maintain than Bluetooth technology.

This paper focuses on Bluetooth wireless technology standard for exchanging data over short distances from fixed and mobile devices. Such technology has existed on the market nearly 20 years, however a new edition of this standard Bluetooth smart or Bluetooth Low Energy (BLE) from 2010 [12] open a new widely available possibilities of applying it for indoor positioning. The new versions 4.1 and 4.2 released in 2013–2014 introduced only some improvements and updates in software and hardware.

Compared to classic Bluetooth, BLE is intended to provide considerably reduced power consumption and cost while maintaining a similar communication range. The new standard defines also several profiles – specifications how a device should work in a particular application [13]. One of this applications is proximity sensing which should be designed to operate for a long period of time (months or even years) powered by a single coin cell battery.

Apart from the already mentioned works there are also other related works on using BLE technology. For example, the authors of [14] propose several empirical propagation models for BLE in different conditions: indoor/outdoor, line-of-sight (LOS). They also compare the propagation characteristics between BLE and Wi-Fi which indicates that BLE can be more accurate (around 27 percent) when is used in localization scenarios.

Paper [15] proposes a localization method which uses calculated values of a defined error function to estimate the positions of unknown transmitters. In this case the error function is based on a modified Root-Mean-Square-Error (RMSE) metric. The error function is calculated by the authors by using Received Signal Strength Indicator (RSSI) measurements at each point under consideration. Adding such error function is caused by unsufficient and inaccurate correlation between the received signal strength and the distance.

This paper focuses on reliability of using BLE technology for indoor localization, but there are also several solutions which use the older version of this standard [16, 17]. These papers present the set of algorithms to transform Bluetooth data in order to estimate or improve the location process (fingerprinting-based positioning algorithms).

The main objective of the paper is to test whether the devices compatible with Bluetooth LE are reliable for indoor localization. To determine the reliability of this technology several tests will be performed to check if measured distance between BLE transmitter and mobile device is close to the real value. This measured distance is based on received signal strength indicator (RSSI), which represents the relationship between a transmission and a received power. On the receiver side RSSI is represented by an integer value used to calculate the distance between a transmitter and a receiver [18,19]. RSSI is also accessible in BLE by simply receiving a broadcasted message. In the present paper the RSSI is the only input to determine the location of a mobile device in an indoor environment.

The research results are carefully analyzed and explained if BLE technology is suitable enough for precisely indoor localization.

2 Bluetooth 4.0 Low Energy Standard

Bluetooth Low Energy operates on band 2.4 GHz with 40 channels located every 2 MHz (Fig. 1). The transmission speed is about 1 Mb/s and Gaussian Frequency Shift Keying (GFSK) modulation is used to select which data channels are to be utilized. Three of all available channels number 37, 38 and 39 are used to detect the devices [13]. Their frequency is not random and was selected to minimize the collision with channels 1, 6 and 11 in Wi-Fi standard. Selection of only the three channels with constant frequency (advertising channels) simplify and speed up the process of detecting other devices – the entire frequency spectrum scanning is no longer required.

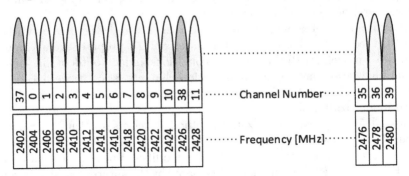

Fig. 1. Overview of Bluetooth LE channel spectrum

After detecting and connecting with the device, the remaining 37 channels are used for data transmission. BLE has four basic operate modes: master, slave, advertising and scanning. Advertising mode is used for cyclic sending advertising information from a particular device, needed to make a proper connection with this device. This mode can be also used to response to additional questions send by other devices. Scanning mode is used to receive advertising information that are sent by devices operating in advertising mode. Slave and master modes are used when two devices are already connected together. The principal function is to provide the possibility to read, write and poll. After connection the devices operating in advertising mode and scanning mode switch respectively to slave and master mode.

The Bluetooth LE stack is completely new solution and it is not compatible with the traditional Bluetooth stack (previous Bluetooth version). The stack protocol is divided into controller and host. The controller includes lower stack layers responsible for receiving the physical packets and radio operation. The host operates in upper stack layers and includes applications, and protocol of attributes. Moreover, the host runs the following services: L2CAP (*Logical Link Control and Adaptation Layer*), GAP (*Generic Access Profile*), SM (*Security Manager*), ATT (*Attribute protocol*), GATT (*Generic Attribute Profile*).

Thanks to several improvements, the energy consumption in BLE is much lower comparing to previous generations. First of all, it has the shorter work cycle which means that the device is more often in a sleeping mode. Additionally, using GATT profiles the module can send smaller segments of data in small packets which significantly increase the energy saving [20]. Another improvement for saving energy by the device is controlling the communication time. More precisely, when the communication between two devices is over the device switches to sleeping mode. In case of another data exchange the connection is quickly restored.

3 Test Bench and Research Procedure

The main point of the test bench is one of the commercially available Bluetooth LE transmitter (beacon node) which, in fact is also compatible with iBeacon[1]. The research results can be extended also to beacon nodes of other vendors because of very similar construction. For the research requirements monitoring application for iOS[1] system was developed (Fig. 2). This application helps to collect the measurements of distance from beacon node and then saves the data that can be later analyzed.

The application allows also to change particular parameters of beacon nodes, such as: advertising interval, broadcasting power. Application enables to choose one of three power options: -30, -12, 4 dBm and three interval values: 50, 200, 1000 ms.

After preparing application for distance measurements collection, tests that check the accuracy of the results were conducted. The received data were then verified with the real distance which was determined with measurement tape. The research was divided into several scenarios. First type includes data collected in open area so that the fewest number of obstacles influenced the results. Therefore this phase was conducted outside the building away from disturbances [21]. The measurements were for distances 1 and 3 meters. The devices responsible for receiving data (smartphone or tablet) were located in three positions (Fig. 3), lying (in this position a device is parallel to the ground, screen to the top, upper side directed to BLE transmitter), vertical and horizontal positions.

The next scenario is a room simulating a natural environment with different obstacles (walls, other electronic devices, etc.) where beacon nodes can be

---

[1] iBeacon and iOS are trademarks of Apple Inc., registered in the U.S. and other countries.

Fig. 2. Application for iOS system to configure and measure the distance from the Bluetooth LE transmitter

normally used. All the combinations of broadcasting powers of −12, 4 dBm and advertising intervals 200 and 50 ms were tested. The measured distance between the BLE transmitter and mobile device is the same as in scenario for open area.

The last test were conducted indoor for combinations of three broadcasting powers: −30, −12, 4 dBm and three values of advertising intervals: 1000, 200 and 50 ms. The mobile devices were set in the distance of 10 and 20 cm next to and then above the beacon node. All tests were conducted on Apple iPad mini and iPad mini Retina Wi-Fi.

4   The Research Results

In the first phase BLE transmitters were tested in open area to eliminate the influence of disturbances on collected measurements. For each set of parameters 30 samples were collected, which were used for preparing charts and statistics as follows:

- Broadcasting power: −12 dBm, advertising interval: 200 ms, distance: 1 m (Fig. 4 and Table 1),
- Broadcasting power: 4 dBm, advertising interval: 200 ms, distance: 1 m (Fig. 5 and Table 2),
- Broadcasting power: −12 dBm, advertising interval: 200 ms, distance: 3 m (Fig. 6 and Table 3),
- Broadcasting power: 4 dBm, advertising interval: 200 ms, distance: 3 m (Fig. 7 and Table 4).

During the research it was noticed that for parameters of beacon node: broadcasting power: −12 dBm, advertising interval: 200 ms, distance: 3 m and horizontal

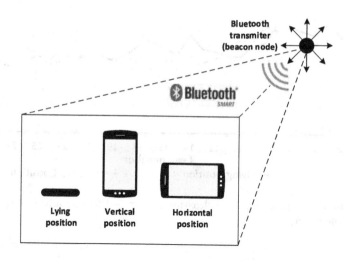

Fig. 3. Mobile device orientation in relation to beacon node

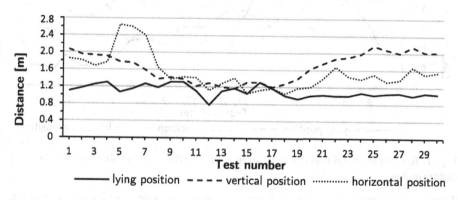

Fig. 4. Results for open area, broadcasting power: −12 dBm, advertising interval: 200 ms, distance: 1 m

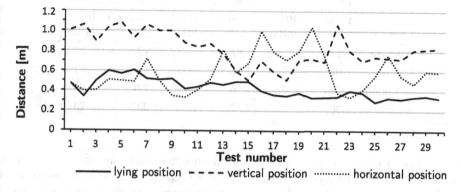

Fig. 5. Results for open area, broadcasting power: 4 dBm, advertising interval: 200 ms, distance: 1 m

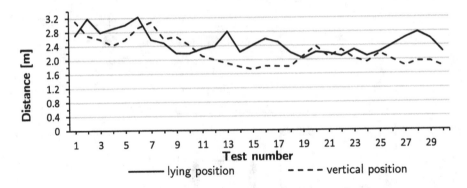

Fig. 6. Results for open area, broadcasting power: $-12\,$dBm, advertising interval: $200\,$ms, distance: $3\,$m

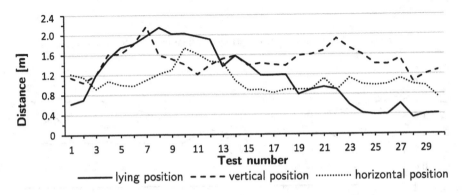

Fig. 7. Results for open area, broadcasting power: $4\,$dBm, advertising interval: $200\,$ms, distance: $3\,$m

Table 1. Measured values for open area, broadcasting power: $-12\,$dBm, advertising interval: $200\,$ms, distance: $1\,$m

	Lying		Vertical		Horizontal	
	Distance [m]	RSSI [dBm]	Distance [m]	RSSI [dBm]	Distance [m]	RSSI [dBm]
Average	1.10	−74.8	1.67	−78.03	1.53	−77.1
Median	1.07	−74.5	1.75	−78	1.44	−77
Minimum	0.78	−78	1.16	−83	1.05	−83
Maximum	1.30	−70	2.17	−74	2.65	−73

orientation, the Core Location framework from iOS system supplying the information about the distance between the devices (beacon – smartphone) returned in this case the value −1, which according to framework specification means that the signal power was too low to determine the right distance. That is why Fig. 6

Table 2. Measured values for open area, broadcasting power: 4 dBm, advertising interval: 200 ms, distance: 1 m

	Lying		Vertical		Horizontal	
	Distance [m]	RSSI [dBm]	Distance [m]	RSSI [dBm]	Distance [m]	RSSI [dBm]
Average	0.42	−67.13	0.82	−72.53	0.58	−70.23
Median	0.40	−66.5	0.81	−72.5	0.54	−70.5
Minimum	0.29	−76	0.50	−79	0.33	−79
Maximum	0.60	−62	1.08	−68	1.04	−62

Table 3. Measured values for open area, broadcasting power: −12 dBm, advertising interval: 200 ms, distance: 3 m

	Lying		Vertical		Horizontal	
	Distance [m]	RSSI [dBm]	Distance [m]	RSSI [dBm]	Distance [m]	RSSI [dBm]
Average	2.49	−80.93	2.22	−79.93	—	—
Median	2.44	−81	2.09	−80	—	—
Minimum	2.04	−84	1.73	−83	—	—
Maximum	3.23	−78	3.11	−77	—	—

Table 4. Measured values for open area, broadcasting power: 4 dBm, advertising interval: 200 ms, distance: 3 m

	Lying		Vertical		Horizontal	
	Distance [m]	RSSI [dBm]	Distance [m]	RSSI [dBm]	Distance [m]	RSSI [dBm]
Average	1.15	−73.93	1.48	−76.97	1.08	−74.8
Median	1.16	−74	1.49	−77	1.00	−74
Minimum	0.33	−81	1.05	−81	0.74	−83
Maximum	2.15	−64	2.16	−72	1.73	−71

and Table 3 do not contain the measurement for these parameters and horizontal orientation.

The next series of tests were conducted indoor to simulate more real conditions where beacon nodes operate. As in the previous case, 30 samples for each configurations were collected. Table 5 presents the results of the conducted research.

Having established that for greater distances the measurements are not accurate, tests for smaller distances 10 and 20 cm were performed (Table 6). Such distance is an alternative for NFC (*Near Field Communication*) technology.

Table 5. Measurement results for indoor environment for different orientations (Broadcasting Power – BP, Advertising Interval – AI, Distance – D)

	Measured distance [m] for orientations:		
	Lying	Vertical	Horizontal
BP: –12 dBm, AI: 200 ms, D: 1 m	1.09	1.93	1.42
BP: –12 dBm, AI: 200 ms, D: 3 m	1.62	2.31	2.1
BP: –12 dBm, AI: 50 ms, D: 1 m	5.13	9.83	9.46
BP: –12 dBm, AI: 50 ms, D: 3 m	8.43	14.46	2.49
BP: 4 dBm, AI: 200 ms, D: 1 m	0.14	0.97	0.26
BP: 4 dBm, AI: 200 ms, D: 3 m	0.33	0.66	0.72
BP: 4 dBm, AI: 50 ms, D: 1 m	0.67	2.25	1.27
BP: 4 dBm, AI: 50 ms, D: 3 m	1.59	3.81	3.39

Table 6. Measurement results for a short distance and different advertising intervals (Broadcasting Power – BP, Distance – D, Location – L)

	Measured distance [cm] for advertising intervals:		
	1 s	200 ms	50 ms
BP: –30 dBm, D: 10 cm, L: next to	42.9	22.5	19.9
BP: –30 dBm, D: 10 cm, L: above	35.1	46.3	44.7
BP: –12 dBm, D: 10 cm, L: next to	7.1	7.6	13.5
BP: –12 dBm, D: 10 cm, L: above	22.2	8.5	8.1
BP: 4 dBm, D: 10 cm, L: next to	7.9	5.3	3.8
BP: 4 dBm, D: 10 cm, L: above	5.1	6.7	9.5
BP: –30 dBm, D: 20 cm, L: next to	47.6	60.1	59.9
BP: –30 dBm, D: 20 cm, L: above	43.1	42.8	38.4
BP: –12 dBm, D: 20 cm, L: next to	54.5	70.4	13.6
BP: –12 dBm, D: 20 cm, L: above	19.8	8.3	15.9
BP: 4 dBm, D: 20 cm, L: next to	15.1	5.5	8.7
BP: 4 dBm, D: 20 cm, L: above	7.2	12.3	20.2

Additionally two the most possible relative positions were taken into consideration, which means when the mobile device is above and next to BLE transmitter.

The conducted research shows that increasing the number of sent packets from Bluetooth transmitter within 1 second resulted in reducing the differences between following readings, and decreasing the signal strength resulted in large drop of the measured distances. The most appropriate results were achieved for average signal power and for the highest frequency of transmission.

5   Conclusion

Comparing the received results for longer distances it can be stated that especially for indoor environment the results are not very precise, and in consequence they cannot be applied for exact positioning. To determine indoor position the distance of at least several Beacons is required that are next used for searching the user position. The differentiation in measurement values however is so great, that it makes in some cases it impossible to determine such point.

The authors intended also to check if the device orientation has influence on the distance readings. It has been proved that an additional obstacle (e.g. a user hand), can decrease significantly the accuracy of measurements. It is because a user holds a device in various positions and at the same time Bluetooth antenna can be covered and the readings are then inaccurate.

Distance measurement focused on Bluetooth technology based mainly on received signal strength indicator RSSI, which is used to calculate the distance between a transmitter and a receiver. As the measurements results of RSSI shows, using this parameter for distance calculations can be problematic. A good correlation between distance and RSSI is observable only in case when a distance between a beacon node and mobile device is very small, especially in open area. Above the value of 1 m, huge fluctuations of measured distance were noticed.

According to the observations, Bluetooth LE signal power cannot be the only reliable source of information for precise indoor localization. The authors in their future work intend to focus on issue referring to the use of additional signals and information form the indoor environment in order to increase the accuracy of indoor positioning.

Acknowledgments. This work was supported by the European Union from the European Social Fund (grant agreement number: UDA-POKL.04.01.01-00-106/09).

References

1. Bulusu, N., Heidemann, J., Estrin, D.: GPS-less low-cost outdoor localization for very small devices. IEEE Pers. Commun. **7**(5), 28–34 (2000)
2. Magnusson, C., Rassmus-Gröhn, K., Szymczak, D.: Navigation by pointing to GPS locations. Pers. Ubiquit. Comput. **16**(8), 959–971 (2012)
3. Woo-Yong, L., Kyeong, H., Doo-Seop, E.: Navigation of mobile node in wireless sensor networks without localization. In: IEEE International Conference on Multisensor Fusion and Integration for Intelligent Systems, pp. 1–7 (2008)
4. Seovv, C., Seah, W., Liu, Z.: Hybrid mobile wireless sensor network cooperative localization. In: IEEE 22nd International Symposium on Intelligent Control, pp. 29–34 (2007)
5. Ng, M.L., Leong, K.S., Hall, D.M., Cole, P.H.: A small passive UHF RFID tag for livestock identification. In: IEEE International Symposium on Microwave, Antenna, Propagation and EMC Technologies for Wireless Communications, MAPE 2005, vol. 1, pp. 67–70 (2005)

6. Dong, G.-F., Chang, L., Fei, G., et al.: Performance enhancement of localization in wireless sensor network by self-adaptive algorithm based on difference. In: Mobile Congress (GMC), pp. 1–5 (2010)
7. Bhargava, P., Krishnamoorthy, S., Nakshathri, A.K., Mah, M., Agrawala, A.: Locus: an indoor localization, tracking and navigation system for multi-story buildings using Heuristics derived from Wi-Fi signal strength. In: Zheng, K., Li, M., Jiang, H. (eds.) MobiQuitous 2012. LNICST, vol. 120, pp. 212–223. Springer, Heidelberg (2013)
8. Canalda, P., Cypriani, M., Spies, F.: Open Source OwlPS 1.3: Towards a Reactive Wi-Fi Positioning System Sensitive to Dynamic Changes. In: Chessa, S., Knauth, S. (eds.) EvAAL 2012. CCIS, vol. 362, pp. 95–107. Springer, Heidelberg (2013)
9. Yucel, H., Yazici, A., Edizkan, R.: A survey of indoor localization systems. In: Signal Processing and Communications Applications Conference (SIU) IEEE Conference, pp. 1267–1270 (2014)
10. Dagtas, S., Natchetoi, D., Wu, H.: An integrated wireless sensing and mobile processing architecture for assisted living and healthcare applications. In: Proceedings of the 1st ACM SIGMOBILE, pp. 70–72 (2007)
11. Xu, X., Zheng, P., Li, L., Chen, H., Ye, J., Wang, J.: Design of underground miner positioning system based on ZigBee technology. In: Wang, F.L., Lei, J., Gong, Z., Luo, X. (eds.) WISM 2012. LNCS, vol. 7529, pp. 342–349. Springer, Heidelberg (2012)
12. Specification of the bluetooth system. Technical report, Bluetooth special interest group, ver. 4.0 (2010)
13. Georgakakis, E., Nikolidakis, S.A., Vergados, D.D., Douligeris, C.: An analysis of bluetooth, zigbee and bluetooth low energy and their use in WBANs. In: Lin, J. (ed.) MobiHealth 2010. LNICST, vol. 55, pp. 168–175. Springer, Heidelberg (2011)
14. Zhao, X., Xiao, Z. et al.: Does BTLE measure up against WiFi? A comparison of indoor location performance. In: 20th European Wireless Conference on European Wireless 2014, pp. 1–6. IEEE VDE (2014)
15. Oksar, I.: A Bluetooth signal strength based indoor localization method. Systems, Signals and Image Processing (IWSSIP), pp. 251–254. IEEE (2014)
16. Mair, N., Mahmoud, Q.H.: A collaborative bluetooth-based approach to localization of mobile devices. In: 8th International Conference on Collaborative Computing: Networking, Applications and Worksharing (CollaborateCom), pp. 363–371. IEEE (2012)
17. Perez Iglesias, H.J., Barral, V., Escudero, C.J.: Indoor person localization system through RSSI bluetooth fingerprinting. In: Systems, Signals and Image Processing (IWSSIP), pp. 40–43. IEEE (2012)
18. Dong, Q., Dargie, W.: Evaluation of the reliability of RSSI for indoor localization. In: 2012 International Conference, ICWCUCA, pp. 28–30. IEEE (2012)
19. Zemek, R., Anzai, D., Hara, S., et al.: RSSI-based localization without a prior knowledge of channel model parameters. Int. J. Wireless Inf. Networks 15(3–4), 128–136 (2008)
20. Siekkinen, M., Hiienkari, M., Nurminen, J. et al.: How low energy is bluetooth low energy? Comparative measurements with ZigBee/802.15.4. In: Wireless Communications and Networking Conference Workshops (WCNCW), pp. 232–237. IEEE (2012)
21. Maćkowski, M.: The influence of electromagnetic disturbances on data transmission in USB standard. In: Kwiecień, A., Gaj, P., Stera, P. (eds.) CN 2009. CCIS, vol. 39, pp. 95–102. Springer, Heidelberg (2009)

Author Index

Printed in the United States
By Bookmasters